Computational Physics

Computational Physics

Problem Solving with Computers

RUBIN H. LANDAU

Professor of Physics

Oregon State University

MANUEL JOSÉ PÁEZ MEJÍ A

Professor of Physics

University of Antioquia

Contributors

Hans Kowallik and Henri Jansen

A Wiley-Interscience Publication

JOHN WILEY & SONS, INC.

New York / Chichester / Weinheim / Brisbane / Singapore / Toronto

Library of Congress Cataloging in Publication Data:

Landau, Rubin H.
Computation physics : problem solving with computers / by Rubin
Landau, Manuél J. Páez Mejiá, with contributions by Hans Kowallik
and Henri Jansen.
p. cm.
Includes bibliographical references and index.
ISBN 0-471-11590-8 (cloth : alk. paper)
1. Physics—Problems, exercises, etc.—Data processing.
2. Physics—Computer simulation. 3. Mathematical physics. I. Páez
Mejiá, Manuél J. II. Title
QC20.82.L36 1997
530′.078′5—DC21 96-51776

Printed in the United States of America

10 9 8 7 6 5 4 3 2

With memory of **Bertha Israel Landau**
and "how's your book coming along?"

Contents

Part III PARTIAL DIFFERENTIAL EQUATIONS

Part IV NONLINEAR PARTIAL DIFFERENTIAL EQUATIONS

Preface

Applying computer technology is simply finding the right wrench to pound in the correct screw.

—Anonymous

This is not the book I thought I'd be writing. When, about a decade ago, I initiated the discussions that led to our Computational Physics course, I thought we would teach mainly physics in it. The Computer Science Department, I thought, would teach the students what they needed to know about computers, the Mathematics Department would teach them what they needed to know about numerical methods and statistics, and I would teach them what I knew about applying that knowledge to solve physics problems using computers. That's how I thought it would be. But, by and large, I have found that the students taking our Computational Physics course do not carry the subject matter from these other disciplines with them, and so a lot of what I have put into this book is material that, in a more perfect world, would be taught and written by experts in other fields.

While that is why I feel this is not the book I thought I would be writing, I believe it's probably for the better. On the one hand, having a physicist with extensive basic research experience tell students they need to know "this" in computer science and "that" in mathematics, gets the message across that "this stuff really matters." And on the other hand, it's useful to have the physics, mathematics, and computer science concepts conveyed in the language of a natural scientist and within the context of solving a problem scientifically.

The official and detailed aims of this book are given in Chapter 1, and the subjects covered are listed in the Table of Contents. This book differs from many others in its underlying philosophy:

> *I hear and I wonder,*
> *I see and I follow,*
> *I do and I understand.*
>
> —as recalled from a fortune cookie

By this I mean that this book is designed to help students learn by having them solve a very wide class of computational physics problems. When I teach from it I continually emphasize that it is the student's job to solve each problem, which includes understanding the results the computer gives. In the process, the students are excited by applying scientific, mathematical, and computational techniques that are often new to them (at least in combination).

As a consequence of having to interact with the materials from a number of viewpoints, and often on their own, the materials become part of the personal experiences of the students. For example, I have heard students comment that "I never understood what was dynamic in thermodynamics until after this simulation," and "I would never have imagined that there could be such a difference between upward and downward recursion," as well as "Is that what a random walk *really* looks like?" or "Why is the pendulum jumping around like that?" In my experience, a teacher just doesn't hear students express such insight and interest in course material in lecture courses. As a consequence, the students are often stimulated to learn more about these subjects, or to understand them at greater depth when encountered elsewhere.

There is a price to pay for my unusually broad approach: the students must work hard and cannot master material in great depth. The workload is lightened somewhat by providing "bare bones" programs, and the level is deepened by having references readily available. There are appendixes listing the C and Fortran programs, as well as source code on the diskette and the World Wide Web (the "Web"). The names, but not the listings, of these programs are also included in the titles of the Implementation sections. By eliminating time-consuming theoretical background (more properly taught in other places), and by deemphasizing timidity-inducing discussions of error in every single procedure, the students get many challenging projects "to work" for them and enjoy the stimulation of experiencing the material. In the process, the students gain pride and self-confidence immediately, and this makes for a fun course for everyone.

A sample of the problems actually solved during each of two 10-week quarters is given in an appendix. I require the students to write up a mini-lab report for each problem solved containing

Equations solved	Numerical method	Code listing
Visualization	Discussion	Critique

The emphasis is to make the report an executive summary of the type given to a boss or manager; just tell enough to get across that you know what you are talking about, and be certain to convey what you did and your evaluation of it. Recently, some students have written their reports as hypertext documents for the Web. This medium appears ideal for computational projects; the projects are always in a centralized place for the students and faculty to observe, the original code is there to run or modify, and the visualizations are striking in three-dimensional (3-D) color or animation.

An unusual aspect of this book is that, in one form or another, and to an increasing degree, parts of it are available on the Web. This is an experimental research project aimed at better incorporating high-performance computing techniques into science. In §1.3 we give the addresses that should lead you to this material. Even though we cannot recommend the Web as a pleasant way to read a book, we do recommend that the students browse through some of the multimedia tutorials and obtain corrections or updates to the programs from the Web. Your comments and corrections regarding the book and its Web tutorials are welcomed.

There is more than enough material to use this book for a full year's course (I used it for a two-quarter course and had to pick and choose). It is possible to teach a 10-week class, in which case I would advise moving rapidly through the basic materials so that the students can experience some applications. Chapters 2 and 3, *Computing Software Basics* and *Errors and Uncertainties in Computation*, in Part I are essential background material. Although placed in the middle of the book, students who are not familiar with computer hardware may benefit from reading Chapter 18, *Computing Hardware Basics: Memory and CPU*, before progressing too far into the materials. Those chapters and sections marked with the symbol ⊙ I view as optional.

Regardless of how the course is taught, most of the learning occurs when the students sit down and experience the physics with their computers, referring to the book or Web tutorials for assistance. Nevertheless, I have found that one or two lectures a week in which the instructor emphasizes the big points and makes it clear what the students "have" to do (and when it is due), appears necessary to keep the course moving at a good pace.

RUBIN H. LANDAU

rubin@physics.orst.edu
http://www.physics.orst.edu

Acknowledgments

This book was compiled for and tested on a decade's worth of students in the Computational Physics course at Oregon State University. I am deeply indebted to them for their good-willed cooperation and enthusiasm, and codes. Some materials developed by Henri Jansen have ingrained themselves into this book, and his contributions are gratefully acknowledged. Hans Kowallik has contributed as a student in the course, as a researcher preparing Web tutorials based on these materials, and as a teacher developing materials, figures and codes; his contributions kept the project on time and improved its quality.

I have also received helpful discussions, valuable materials, and invaluable friendship and encouragement from Paul Fink, Melanie Johnson, Al Stetz, Jon Maestri, Tom Marchioro II, Cherri Pancake, Pat Canan, Shashi Phatak, Paul Hillard, and Al Wasserman. The support of the UCES project and their award for the Web implementations of our projects, is sincerely acknowledged. In many ways, this book was made possible by the U.S. National Science Foundation (through laboratory development grants and through the NACSE Metacenter Regional Alliance), the U.S. Department of Energy, and Oregon State University. Thanks also go to the people at Wiley-Interscience, in particular Greg Franklin, John Falcone, Lisa Van Horn, Rosalyn Farkas, and Amy Hendrickson. My final formatting of this book was done at the San Diego Supercomputer Center and the Institute for Nonlinear Science, both at UCSD. Finally, I am especially grateful to my wife Jan, whose reliable support, encouragement (and proofreading) is lovingly accepted.

RHL

Acronyms

cc	Complex conjugate
CISC	Complex instruction set computer
CPU	Central processing unit
DFT	Discrete Fourier transform
DRAM	Dynamic RAM
FFT	Fast Fourier transform
FPU	Floating point (arithmetic) unit
HPC	High-performance computing
HTML	Hypertext markup language
KdeV	Korteweg–deVries
LAN	Local area network
LAPACK	Linear algebra package
LHS	Left-hand side
MIMD	Multiple instruction, multiple data
NACSE	Northwest Alliance for Computational Science and Engineering
NAN	Not a number
ODE	Ordinary differential equation

1-D	One-dimensional
OOP	Object-oriented programming
PDE	Partial differential equation
PC	Personal computer (usually IBM)
ps	PostScript
PVM	Parallel virtual machine
RAM	Random access memory
RHS	Right-hand side
RISC	Reduced instruction set computer
RK4	Fourth-order Runge–Kutta
SIMD	Single instruction, multiple data
SISD	Single instruction, single data
SLATEC	Sandia, Los Alamos, Air Force Weapons Laboratory Technical Exchange Committee
SRAM	Static RAM
TCP/IP	Transmission control protocol/internet protocol
WWW	World wide web
UCES	Undergraduate Computational Engineering and Science Project
URL	Universal resource locator

Computational Physics

Part I

GENERALITIES

1

Introduction

1.1 THE NATURE OF COMPUTATIONAL SCIENCE

Computational science explores models of the natural and artificial world with the aim of understanding them at depths greater than otherwise possible. This is a modern field in which computers are used to solve problems whose difficulty or complexity places them beyond analytic solution or human endurance. Sometimes the computer serves as a super–calculating machine, sometimes as a laboratory for the numerical simulation of complex systems, sometimes as a lever for our intellectual abilities, and optimally as all of the above.

The focus of a computational scientist is science. The aim of this book is to teach how to do science with computers, and, in the process, to teach some physics with computers. This is computational science but not "computer science." Computer sciences studies computing for its own intrinsic interest and develops the hardware and software tools computational scientists use. This difference is not just semantic or academic. Computational scientists are interested in computer applications in science and engineering, and their values, prejudices, tools, organizations, goals, and measures of success reflect that interest. For example, a computational scientist may view a particular approach as reliable, self-explanatory, and easy to port to sites throughout the world, while a computer scientist may view it as lengthy and inelegant; both are right, because both are viewing it from their different disciplines.

Computational science is a team sport. It draws together people from many disciplines via a commonality of technique, approach, and philosophy. A computational scientist must know a lot about many things to be successful.

But because the same tools are used for many problems in different fields, he or she is not limited to one specialty area. A study of computational science helps broaden horizons, which is a welcome exception to the stifling subspecialization found in so much of science.

Traditionally, physics divides into experimental and theoretical approaches; computational physics requires the skills of both and contributes to both. Transforming a theory into an algorithm requires significant theoretical insight, detailed physical and mathematical understanding, and mastery of the art of programming. (The sections in this book are labeled to reflect these steps.) The actual debugging, testing, and organization of scientific programs is like an experiment. The simulations of nature with programs are virtual experiments. Throughout the entire process, the synthesis of numbers into generalizations, predictions, and conclusions requires the insight and intuition common to both experimental and theoretical science. And as visualization techniques advance,[1] computational science enters into and uses psychology and art; this, too, makes good science because it reveals the beauty contained within a theoretical picture of nature and permits scientists to use extensive visual processing capabilities of their brains to "see" better their discipline.

1.1.1 How Computational Scientists Do It

A computational scientist uses computers in a number of distinct ways, with new ways not necessarily eliminating old ones.

- In the classic approach, a scientist formulated a problem and solved as much as possible analytically. Only then was the computer used to determine numerical solutions to some equations or to evaluate some hideously complicated functions. In many cases, computing was considered a minor part of the project with little, if any, discussion of technique or error.

- A computational scientist formulates and plans the solution of a problem with the computer and program libraries as active collaborators. Use is made of past analytic and numerical advances during all stages of work. And, as the need arises, new analytic and numerical studies are undertaken.

- In a different, but by now also classic scientific approach, computers play a key role from the start by *simulating* the laws of nature. In these simulations, the computer responds to input data as a natural system might to different initial conditions. Examples are the computer tracing of rays through an optical system and the numerical generation of random numbers to simulate the radioactive decay of nuclei.

[1]There are Web tutorials describing visualization and animation techniques.

- Another modern use of computers is to create *problem-solving environments*, such as *Maple, Mathematica, Macsyma,* and *Matlab*, which hide most of the details from the user and which often include symbolic manipulations as might be done analytically.

- One of the most rewarding uses of computers is *visualizing* the results of calculations with 2-D and 3-D plots or pictures, and sometimes with color shading and animation. This assists the debugging process, the development of physical and mathematical intuition, and the enjoyment of the work. Visualization is incorporated into as many of our projects as possible and especially in the Web tutorials associated with this book.

- Finally, many personal computer applications also have value in computational science. For example, a numerical *spreadsheet* is a helpful way to analyze data as well as the results of calculations, and *hypertext* and World Wide Web documents are true advances in storing various types of information that supplement, even if they do not replace, the lab notebook and research paper.

1.2 AIMS OF THIS BOOK

To emphasize our general purpose of teaching how to do science with computers, the paradigm suggested by the Undergraduate Computational Science and Engineering Project [UCES] will be followed:

Problem	**Model**	**Method**	**Implementation**
(Physics)	(Discrete)	(Symbolic)	(C/Fortran)
(Life science)	(Continuous)	(Numeric)	(High-performance)

↕

Assessment
(Visualization)
(Experimentation)

This is not easy to do when developing basic skills, but it will work well once projects deal with physical problems.

When the students are relieved of the burden of extensive programming, they should be able to "pass lightly" through the background material and have a personal experience with many projects. This personalization of the material acts as stimulation for further study, discussion, and exploration.

The specific aims of the projects are

- To teach the use of scientific computers in thinking creatively and solving problems in the physical sciences through direct experience.

- To advance the development and organization of thinking about physical systems in a manner compatible with advanced computational analysis.
- To use the graphic capabilities of scientific computers to study and teach the visualization of numerical solutions into highly interpretable forms.
- To instill attitudes of independence, personal communication, and organization, all of which are essential for mastery of complex systems.
- To understand physical systems at a level often encountered only in a research environment, and to use programming to deepen that understanding.
- To understand why hard work and properly functioning and powerful software and hardware do not guarantee meaningful results. As with experimental physics, there are accuracy and applicability limits that often determine when viable results are generated.
- To instill an *objected-oriented* view of problem solving.

1.3 USING THIS BOOK WITH THE DISK AND WEB

There are references throughout this book to programs and tutorials available on the floppy diskette accompanying the text and through the World Wide Web (the "Web"). These are meant as a supplement to the text, to be used at the discretion of the student and instructor.

Programming is a valuable skill for all scientists, but it is also incredibly demanding and time consuming. For this reason the diskette and two appendixes provide both C and Fortran programs as the basic implementation part for most of the **Problems**. It is suggested that the student read through the given programs and modify or rewrite them for the project at hand. Not only will this save time, but it is a valuable lesson in learning how to read someone else's code (real-world scientists seldom have the luxury of writing their own). Note, the C and Fortran programs are not direct translations of each other, and for some problems a program in only one language is provided. We provide an appendix that tabulates analogous elements in the C and Fortran languages. This should help those readers having to struggle with a foreign language.

Most of the **problems** we examine can also be worked in a problem solving environment such as *Maple, Mathematica, Matlab*, or *Mathcad.* If you use those packaged systems, you may not learn the same programming skills, your program may be less flexible, and they may be much slower; but then again, you may end up being able to spend more time understanding the science and mathematics.

Referring to as rapidly changing a resource as the Web in a textbook is somewhat risky, yet it also adds a new dimension that is just too good to pass up. References in this book to the Web are primarily to the resources maintained by the Northwest Alliance for Computational Science and Engineering and the Undergraduate Computational Science and Engineering Project:

Computational Science Web Sites	
NACSE	http://www.nacse.org
UCES	http://uces.ameslab.gov/uces

As a research project aimed at better incorporating the techniques of high-performance computing into science, these two groups have supported the conversion of some of the computational physics projects in this book into interactive Web tutorials. On the Web you will find running codes, figures, animations, sonifications, corrected code listings, and control-panel interfaces. While these are not meant to be a substitute for studying the text or for your running you own codes, they provide some stimulating examples of what can de done and of how the physics can be "seen" in differing ways. For example, not only can you see coordinate- and phase-space plots of a chaotic pendulum, but you can actually see the pendulum swing and hear the oscillations!

Additional Web resources of interest are given by a Computational Physics Resource Letter [DeV 95] and by the list of URLs (universal resource locators) on Landau's home page. Particularly recommended are the Web sites of the U.S. National Science Foundation Supercomputer Centers [NSF].

2

Computing Software Basics

In this chapter we explore basics of computing languages, number representation, and programming. Related topics dealing with hardware basics are found in Chapter 18, *Computing Hardware Basics: Memory and CPU.* We recommend that you glance through Chapter 18 now. If you find that you really have no idea what it's about, then you would benefit by studying it as soon as possible, and especially before getting involved in heavy-duty computing.

2.1 PROBLEM 1: MAKING COMPUTERS OBEY

You write your own program, wanting to have the computer work something out for you. Your **problem** is that you are beginning to get annoyed because the computer repeatedly refuses to give you the correct answers.

2.2 THEORY: COMPUTER LANGUAGES

As anthropomorphic as your view of your computer may be, it is good to keep in mind that computers always do exactly as told. This means that you must tell them exactly and everything they have to do. Of course, the programs you write may be so complicated and have so many logical paths that you may not have the endurance to figure it out in detail, but it is always possible in principle. So the real **problem** addressed in this chapter is how to give you enough understanding so that you feel well enough in control, no matter how illusionary, to figure out what the computer is doing.

Before you tell the computer to obey your orders, you need to understand that life is not simple for computers. The instructions they understand are in a *basic machine language*[1] that tells the hardware to do things like move a number stored in one memory location to another location, or to do some simple, binary arithmetic. Hardly any computational scientist really talks to a computer in a language it can understand. When writing and running programs, we usually talk to the computer through *shells* or in *high-level languages*. Eventually these commands or programs all get translated to the basic machine language.

A *shell* (*command-line interpreter*) is a set of medium level commands or small programs, run by a computer. As illustrated in Fig. 2.1, it is helpful to think of these shells as the outer layers of the computer's *operating system*. While every general-purpose computer has some type of shell, usually each computer has its own set of commands that constitute its shell. It is the job of the shell to run various programs, compilers, linkage editors, and utilities, as well as the programs of the users. There can be different types of shells on a single computer, or multiple copies of the same shell running at the same time for different users. The nucleus of the operating system is called, appropriately, the *kernel*. The user seldom interacts directly with the kernel.

The *operating system* is a group of instructions used by the computer to communicate with users and devices, to store and read data, and to execute programs. The operating system itself is a group of programs that tells the computer what to do in an elementary way. It views you, other devices, and programs as input data for it to process; in many ways, it is the indispensable office manager. While all this may seem unnecessarily complicated, its purpose is to make life easier for you by letting the computer do much of the nitty-gritty work to enable you to think higher-level thoughts and communicate with the computer in something closer to your normal, everyday language. Operating systems have names such as *Unix*, *VMS*, *MVS*, *DOS*, and *COS*.

We will assume you are using a *compiled* high-level language like *Fortran* or *C*, in contrast to an *interpreted* one like *BASIC* or *Maple*. In a compiled language the computer translates an entire subprogram into basic machine instructions all at one time. In an interpretive language the translation is done one statement at a time. Compiled languages usually lead to more efficient programs, permit the use of vast libraries of subprograms, and tend to be portable.

When you submit a program to your computer in a high-level language, the computer uses a *compiler* to process it. The compiler is another program that

[1]The "BASIC" (Beginner's All-purpose Symbolic Instruction Code) programming language should not be confused with basic machine language.

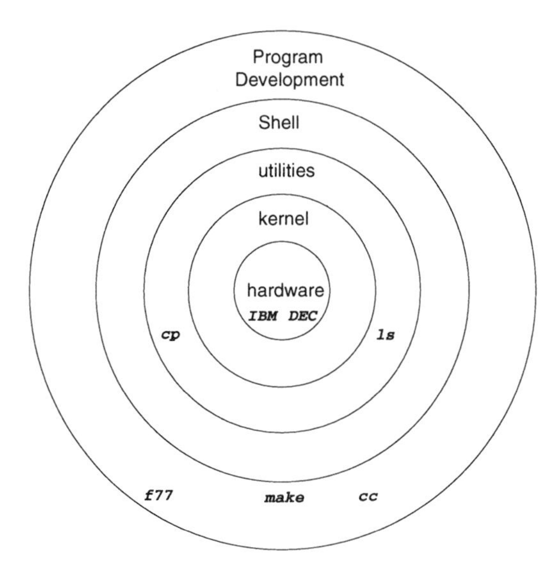

Fig. 2.1 A schematic view of a computer's kernel and shells.

treats your program as a foreign language and uses a built-in dictionary and set of rules to translate it into basic machine language. As you can imagine, the final set of instructions are quite detailed and long, and the compiler may make several passes through your program to decipher your convoluted logic and to translate it into a fast code. The translated statements form an *object* code, and when *linked* together with other needed subprograms, form a *load module.* A load module is a complete set of machine language instructions that can be *loaded* into the computer's memory and read, understood, and followed by the computer.

2.3 IMPLEMENTATION: PROGRAMMING CONCEPTS

Before we discuss general programming techniques, we need to be sure that you can talk to your computer. Here is a tutorial to get you communicating. Begin by assuming that calculators have not been invented and you need a program to calculate the area of a circle. Rather than using any specific language, we will discuss how to write that program in *pseudocode* that can be converted to your favorite language later. The first program tells the computer:[2]

```
calculate area of circle                         Do this, computer!
```

[2]Comments placed in the field to the right are for you and *not* for the computer to view.

This program cannot really work because it does not tell the computer which circle to consider and what to do with the area. A better program would be

```
read radius                                    Input
calculate area of circle                       Numerics
print area                                     Output
```

The instruction *calculate area of circle* has no meaning in most computer languages, so we need to specify an *algorithm*[3] for the computer to follow:

```
read radius                                    Input
calculate area of circle                       Comment
π = 3.141593                                   Set constant
area = π× radius²                              The algorithm
print area                                     Output
```

This is a better program. When we cannot think of any more embellishments, we convert this pseudocode to a language the computer can understand.

2.4 IMPLEMENTATION: FORTRAN, AREA.F

A Fortran version of our *area* code is found on the disk and in Appendix D under the name *area.f* (we usually indicate the appropriate program name in the title of Implementation sections, as you may note here). Because beginnings are so hard, we will be nice to you this time and list[4] the program here:

```
      Program area                     Tell compiler it's a main program
c                                                  Space helps readability
c     area of circle, input r                        Say what's happening
      Double Precision pi, r, A                     Uppercase for clarity
c     calculate pi                                                Comment
       pi = 3.141593                                      Set value of π
c     Read r from standard input (terminal)
       Write(*,*) 'specify radius'                     Appears on terminal
       Read (*, *) r                                   Input from terminal
c     calculate area
       A = pi * r**2
c     Write area onto terminal screen
      Write (*,10) 'radius r =', r, ' A =', A                 * for terminal
   10 Format (a20, f10.5, a15, f12.7)
       Stop 'area'                           Stop program and write 'area'
       End
```

[3]An *algorithm* is a set of rules for doing mathematics.

[4]Beware, our typeset spaces may not be perfect. In Fortran, comments usually have a c or C in column 1, statement numbers must be in columns 2–5, continuation characters must be in column 6, and executable statements begin in column 7 (or higher).

Notice that the variable and program names are meaningful and similar to standard nomenclature (even with an uppercase A), there are plenty of comments, and the input and output are self-explanatory.

2.5 IMPLEMENTATION: C, AREA.C

A C version of our area program is found on the disk and in Appendix C under the name *area.c* (we usually indicate the appropriate program name in the title of Implementation sections, as you may note here). Because beginnings are so hard, we will be nice to you this time and list the program here:

```
/* Calculate area of a circle */                    A comment, for reader only
                                                                  A blank line
#include  <stdio.h>                                Need standard I/O routines
#define pi 3.14159265369                                      Define constant
main()                                        Tell compiler it's a main program
{                                                               Begin program
double r, A;                                       Double-precision variables
printf("Enter the radius of a circle \n");                            Request
input  scanf("%lf", &r);                              Read from standard input
A = r * r * pi;                                                Calculate area
printf("radius r= %f, area A = %f\n", r, A);                   Print results
}                                                                 End program
```

2.6 IMPLEMENTATION: SHELLS, EDITORS, AND PROGRAMS

1. To gain some experience with your computer system, enter one of the preceding programs into a file. Then

   (a) Compile and execute it (in one command).

   (b) Check that the results are correct. Good input datum for testing is $r = 1$, because then $A = \pi$.

   (c) Try $r = 2$ and see if the area increases by a factor of 4. Then experiment (e.g., see what happens if you leave off decimal points, if you feed in blanks, if you feed in a letter, ...).

2. The programs given here take input from and place output on the terminal screen. Revise one of these programs so that the input and output come from and are placed into two separate files.

3. Revise this program so that it uses a main program (which does the input and output) and a subroutine (which does the calculation). Check that it still runs properly.

2.7 THEORY: PROGRAM DESIGN

Now that you have warmed up on the computer, let's get back to the theory that should be behind your actions. Even with a perfect set of physical laws, a perfect algorithm, and a perfect computer, there still remains the challenge of *programming*. Programming is viewed as a written art that blends elements of science, mathematics, and computer science into a set of instructions so that the computer can accomplish a scientific goal (for example; generating the cross section for the scattering of an electron from a krypton atom). Sooner or later, a scientist who wants to do something new or different has to write his or her own programs. Computational scientists who place a high value on collaboration with other people, as well as making contributions to the development of science, write programs that

- Are simple and easy to read, making the action of each part clear and easy to analyze. (Just because it was hard for you to write that program, doesn't mean that you should make it hard for others to read.)
- Document themselves so that the programmer and others understand what the programs are doing.
- Are easy to use.
- Are easy and safe to modify for different computers or systems.
- Can be passed on to others to use and further develop.
- Give the correct answers.

The lack of program readability leads to credibility problems and the stifling of creativity. It is in the interests of the science to write clear programs even for the complicated problems encountered in modern science and engineering. Keep in mind, the program is the ultimate documentation of a computational science project, and the human and economic savings in being able to reuse someone else's work is often tremendous.

True creative artists follow their own rules. Nonetheless, here are some suggested ideas for *modular* and *top–down* programming that may help you on the road to becoming a creative programmer:

1. A *modular* approach breaks up the tasks of a program into subprograms. In general, your programs will be clearer and simpler, and easier to write, if you make them modular. While you may be able to view small programs in a single glance, the complexity of hundreds or thousands of uninterrupted lines of code boggles the mind and makes a single-glance understanding impossible.[5]

[5] This may not be good for vectorization on a supercomputer, but you can always recombine the subprograms after they are debugged and running.

(a) Write many small subprograms, each of which accomplishes limited tasks.
(b) Give each subunit well-defined input and output that gets passed as arguments.
(c) Make each subprogram reasonably independent of the others. You can then test them independently and use them again and again in other programs.
(d) Do not become overzealous about writing subroutines. If a subroutine is very small and is often called, the overhead time for the calls may be relatively expensive. In that case, the compiler will optimize better if you combine often-called and related program units into one.

2. Put off as long as possible the actual writing of your program. Concentrate instead on clarifying, understanding, and defining the problem to be solved and the logic to be used.

3. Try to choose the most reliable and simple algorithm. Speed matters, but not if you get the wrong answers.

4. Be aware that an algorithm that is best for scalar architecture may not be best for parallel architecture.

5. A program that is clear and simple will usually end up being less buggy. While the clear program may take more time to write and run, this usually saves you time in the long run. More importantly, it may help a project reach a successful completion rather than being abandoned in frustration.

6. The planning of your program should be from *top down to bottom*. This means you first outline the major tasks of the algorithm, always keeping the big picture visible.

   (a) Arrange the major tasks in the order in which they need to be accomplished. This is the most basic outline.
   (b) Plan the details of each major task, making sure to break these tasks into subtasks (which may turn out to be subprograms or groups of subprograms). This will be the next level of complexity in your outline.
   (c) Continue breaking up your tasks into smaller ones until you are at the subroutine level.

7. Keep the flow through the program *linear*, as indicated in Fig.2.2, with a minimal amount of jumping around.[6] Avoid *go to*'s and especially computed *go to*'s.

[6]This principle is modified for a parallel computer where multiple, central processors work simultaneously on one problem.

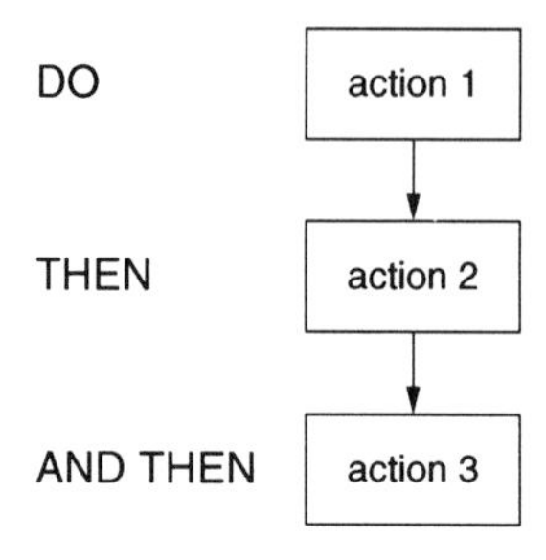

Fig. 2.2 A simple (linear) sequence of actions.

2.8 METHOD: STRUCTURED PROGRAMMING

You should always be striving to build *structures* into your program that clearly reveal the content and logic of the program. The physical structuring of your code can be built by using successive indentations for different sections (ignored by the compiler); by the frequent and judicious use of comments and spacings; and by using upper- and lowercase letters to improve clarity (the Fortran compiler is case-insensitive). You may, nonetheless, wish to avoid non-commented blank lines because they will not always be accepted by compilers.

On a more conceptual level, modern high-level languages contain building blocks to provide structure to programs; they have been proven to be logically sufficient for all programming needs. The elements that are used to construct the logical building blocks are given in Fig.2.3. Some common structures are illustrated in Figs.2.2–2.7. These logical building blocks start with the linear sequence, shown in Fig. 2.2, and include:

Repeat N times, Fig.2.4: *Do* all instructions up to the end of loop indicator *Endo* N times.

If-then-else, Fig.2.5: *If* a certain condition is met, *Then* execute some instructions, or *Else*, do something else. When one of these possibilities is finished, this sequence ends.

Repeat unknown number of times, Fig.2.6: *While* some condition is met, keep repeating these instructions. *If* the condition is no longer met, *go to* statements beyond *Endwhile.*

You will note that these structures have well-defined beginnings, endings, and conditions for their actions, all of which help clarify the program flow. Of

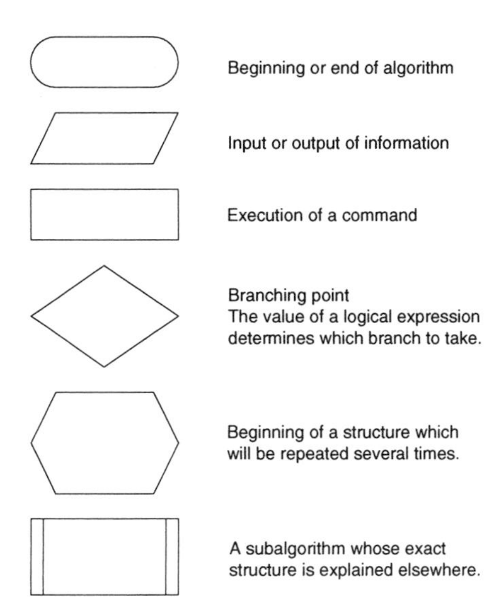
Beginning or end of algorithm
Input or output of information
Execution of a command
Branching point
The value of a logical expression
determines which branch to take.
Beginning of a structure which
will be repeated several times.
A subalgorithm whose exact
structure is explained elsewhere.

Fig. 2.3 The logical elements used to construct the requisite logical structures of programs.

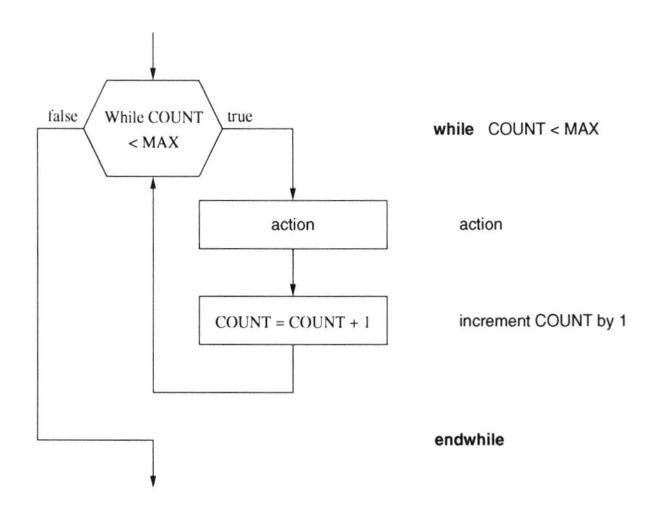
false
While COUNT
< MAX
true
action
COUNT = COUNT + 1
while COUNT < MAX
action
increment COUNT by 1
endwhile

Fig. 2.4 The *Repeat Loop MAX times* structure and pseudocode. This is a special case of the *While* loop.

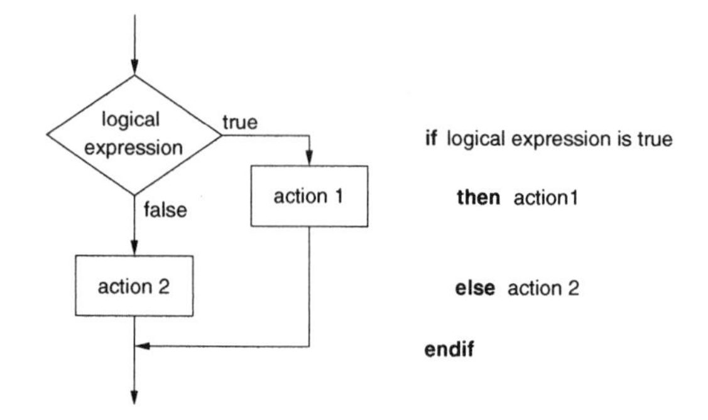

Fig. 2.5 The *If-then-else* structure and pseudocode.

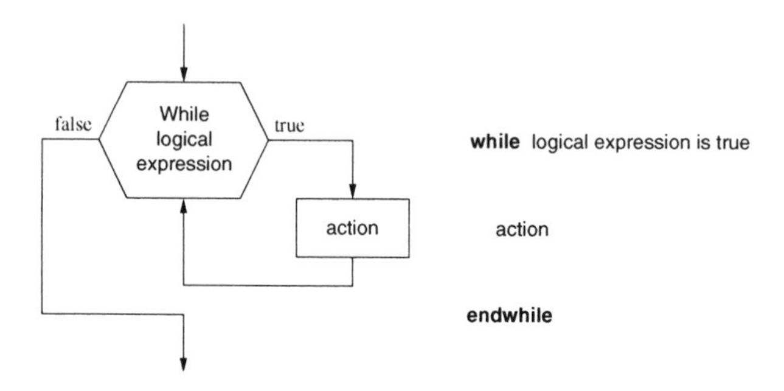

Fig. 2.6 The *While* structure and pseudocode.

course, actual programs often contain more complicated logic, yet as we see in Fig.2.7, these basic structures may be combined to provide a richer structure.

2.9 METHOD: PROGRAMMING HINTS

Some specific programming hints that may help you implement the preceding general rules for writing programs are

- Always keep an updated, working version of your program; make modifications on a copy.
- Use the standard version of the programming language if you want to *port* your code to another computer or immediately run it when future systems become available. (Avoid local language extensions.)
- Add plenty of comments and documentation as you write the code, with at least a short description about each subprogram. This will help keep

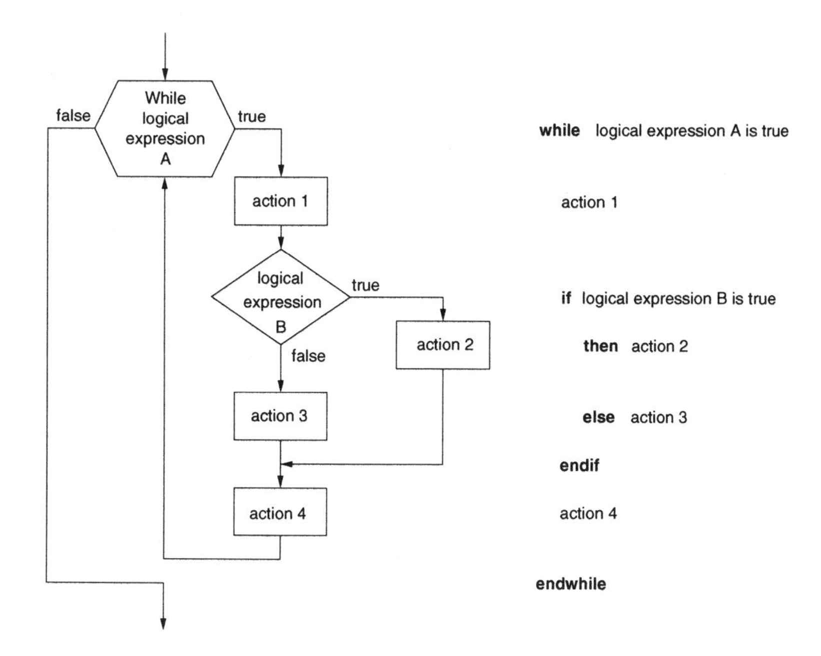

Fig. 2.7 A combination of *While* and *If-then-else* structures.

your mind on track and be useful later (a natural action for those who talk to themselves or keep a diary).

- Use descriptive names for variables and subroutines, like *mass* and *temp*, and keep them similar to the variables employed in standard texts and papers. Describe your variables in comment statements.

- Declare all variables (do not use the *Implicit* statement in Fortran unless you make it an *Implicit none*). Combined with cross-referenced maps, this helps you pick up spelling, typographical, and forgetfulness errors.

- Avoid global variables.

- In Fortran, use statement labels only for *Continue* and *Format* statements.

- Avoid *Equivalence* statements. They make the logic too hard to follow.

- Remember that compilers make errors, too. You'll want to be particularly careful if your programming is subtle or clever or highly convoluted. Comparing results derived with different levels of *optimization*, different flags, or reorganized program parts may help reveal compiler bugs (the least-optimized answer is usually your best bet). Using program checkers like *lint* is also recommended. And if you are fortunate enough to have different machines around, you can try different compilers.

2.10 PROBLEM 2: LIMITED RANGE OF NUMBERS

Computers may be powerful, but they are finite. A **problem** in computer design is how to represent a general number in a finite amount of space, and then how to deal with the approximate representation that results.

2.11 THEORY: NUMBER REPRESENTATION

If we are given the digits 0 and 1 as the microscopic units of memory (*bits*), it should be no great surprise that all numbers are ultimately represented in *binary* form. Correspondingly, there are only 2^N integers that can be represented with N bits. Because the sign of the integer is represented by the first bit (a zero bit for positive numbers), this leaves the remaining $N-1$ bits to represent the value of the integer. Therefore N-bit integers will be in the range $[0, 2^{N-1}]$. Already we begin to see the limitations.

Long strings of zeros and ones are fine for computers, but are awkward for people. Consequently, binary strings are converted to *octal*, *decimal*, or *hexadecimal* numbers before results are communicated to people. Octal and hexadecimal numbers are nice because the conversion loses no precision, but not so nice because our decimal rules of arithmetic do not work for them. Converting to decimal numbers make the numbers easier for us to work with, but we often lose precision.

A description of a particular computer system will normally state the number of bits used to store a number (also called *word length*). This word length is often expressed in *bytes*, where

$$1\text{byte} \equiv 1\text{B} \stackrel{\text{def}}{=} 8\text{bits}. \tag{2.1}$$

Conventionally, storage size is measured in bytes or kilobytes. Be careful, not everyone means the same thing by a thousand:

$$1\text{K} \stackrel{\text{def}}{=} 1\text{KB} = 2^{10}\text{bytes} = 1024\text{bytes}. \tag{2.2}$$

This is often (and confusingly) compensated for when memory size is stated, for example

$$512\text{K} = 2^{9}\text{bytes} = 524,288\text{bytes} \times \frac{1\text{K}}{1024\text{bytes}}. \tag{2.3}$$

Conveniently, 1 byte is also the amount of memory needed to store a single character, like the letter "a" or "b." This adds up to a typical typed page requiring ~3 KB.

The memory chips in some of the older personal computers used 8-bit words. This means the maximum integer was $2^7 = 128$ (7 because one bit is used for the sign). Trying to store a number larger than possible (*overflow*) was common on these machines, sometimes accompanied by an informative error message and sometimes not. At present, most workstation-class computers use 32 bits for an integer, which means that the maximum integer is $2^{31} \simeq 2 \times 10^9$. While at first this may seem a large range for numbers, it really isn't compared to the range of sizes encountered in the physical world. For example, the ratio of the size of the universe to the size of a proton is, for example, 10^{24}.

2.12 METHOD: FIXED AND FLOATING

Real numbers are represented on computers in either *fixed-point* or *floating-point* notation. In fixed-point notation, the number x is represented as

$$x_{\text{fix}} = \text{sign} \times (\alpha_n 2^n + \alpha_{n-1} 2^{n-1} + \cdots + \alpha_0 2^0 + \cdots + \alpha_{-m} 2^{-m}). \tag{2.4}$$

That is, one bit is used to store the sign and the remaining $N-1$ bits are used to store the α_i values ($n + m = N - 2$). The particular values for N, m, and n are machine-dependent. For a 32-bit machine, the integers are typically 4 bytes in length and in the range

$$-2147483648 \leq \text{integer} * 4 \leq 2147483647\ . \tag{2.5}$$

The advantage of the representation (2.4) is that you can count on all fixed-point numbers to have the same absolute error of 2^{-m-1} [the term left off the right-hand end of (2.4)]. The corresponding disadvantage is that *small* numbers (those for which the first string of α values are zeros) have large *relative* errors. Because in the real world relative errors tend to be more important than absolute ones, fixed-point numbers are used mainly in special applications (like business).

Your scientific work will mainly use floating-point numbers. In floating-point notation, the number x is stored as a sign, a mantissa, and an exponential field *expfld*. The number is reconstituted as

$$x_{\text{float}} = (-1)^s \times \text{mantissa} \times 2^{\text{expfld - bias}}. \tag{2.6}$$

Here the mantissa contains the significant figures of the number, s is the sign bit, and the actual exponent of the number has the *bias* added to it and is then stored as the exponential field *expfld*.

Just as introducing a sign bit guarantees that the mantissa is always positive, so introducing the bias guarantees that the number stored as the exponent field in (2.6) is always positive (the actual exponent of the number can, of course, be negative). The use of bias is rather indirect. For example, a single-precision 32-bit word may use 8 bits for the exponent in (2.6) and represent it as an integer. This 8-bit integer "exponent" has a range $[0, 255]$. Numbers with actual negative exponents are represented by a bias equal to 127, a fixed number for a given machine. Consequently, the exponent has the range $[-127, 128]$ even though the value stored for the exponent in (2.6) is a positive number. Of the remaining bits, one is used for the sign and 23 for the mantissa.

It is important to remember that single-precision (4-byte) numbers have 6–7 decimal places of precision (1 part in 2^{23}) and magnitudes typically in the range

$$10^{-44} \leq \text{single precision} \leq 10^{38}. \tag{2.7}$$

The mantissa of a floating number is represented in memory in the form

$$\text{mantissa} = m_1 \times 2^{-1} + m_2 \times 2^{-2} + \cdots + m_{23} \times 2^{-23}, \tag{2.8}$$

with just the m_i stored, similar to (2.4). As an example, the number 0.5 is stored as

```
0    0111 1111    1000 0000 0000 0000 0000 000,
```

where the bias is $0111\ 1111_2 = 127_{10}$.

In order to have the same relative precision for all floating-point numbers, it is standard to normalize the number so that the leftmost bit is unity, $m_1 = 1$. Once this convention is adopted, the m_1 does not even have to be stored and the computer only needs to recall that there is a *phantom bit*. During the processing of numbers in a calculation, the first bit of an intermediate result may become zero, but this will be corrected before the final number is stored.

Typically, the largest possible floating-point number for a 32-bit machine

```
0    1111 1111    1111 1111 1111 1111 1111 111
```

has the value 1 for all its bits (except sign) and adds up to $2^{128} = 3.4 \times 10^{38}$. Typically, the smallest possible floating-point number,

```
0    0000 0000    1000 0000 0000 0000 0000 000
```

has the value 0 for almost all its bits and adds up to $2^{-128} = 2.9 \times 10^{-39}$. As built in by the use of bias, the smallest number possible to store is the inverse of the largest.

If you write a program requesting *double precision*, then 64-bit (8-byte) words will be used in place of the 32-bit (4-byte) words. With 11 bits used for the exponent and 52 for the mantissa, double-precision numbers have about 16 decimal places of precision (1 part in 2^{52}) and typically have magnitudes in the range

$$10^{-322} \leq \text{double precision} \leq 10^{308} . \tag{2.9}$$

2.13 IMPLEMENTATION: OVER- AND UNDERFLOWS, OVER.F (.C)

Write a program to test for the underflow and overflow limits (within a factor of 2 at least) of your computer system and of your favorite computer language. A sample pseudocode is

```
under = 1.
over = 1.
  begin do N times
    under = under/2.
    over = over * 2.
    write out:  loop number, under, over
  end do
```

You may need to increase N if your initial choice does not lead to underflow and overflow. Be careful to notice whether your computer's implementation of your programming language converts overflows, as well as underflows, to zero. (Converting underflows to zero is usually a good thing to do; converting overflows to zero is usually a good way to cause a disaster.) Notice that if you want to be more precise regarding the limits of your computer, you may want to multiply and divide by a number smaller than 2.

1. Check where under- and overflow occur for single-precision floating-point numbers.

2. Check where under- and overflow occur for double-precision floating-point numbers.

3. Check where under- and overflow occur for integers (you need to multiply and subtract 1 to see the effect).

2.14 MODEL: MACHINE PRECISION

One consequence of a computer's memory scheme for numbers is that the numbers can be recalled with only a limited precision. While the exact precision depends on the computer, single precision is usually 6–7 decimal places for a 32-bit word machine, and double precision is usually 15–16 places. (Some symbolic manipulation programs can store numbers with infinite precision; that is, the word size increases as the requisite precision increases.) To see how *machine precision* affects calculations, consider the simple addition of two 32-bit words:

$$7 \ + \ 1.0 \times 10^{-7} = ? \tag{2.10}$$

The computer fetches these numbers from memory and stores the bit patterns

$$\begin{aligned} 7 &= 0 \ 10000010 \ 1110\ 0000\ 0000\ 0000\ 0000\ 000, & (2.11) \\ 10^{-7} &= 0 \ 01100000 \ 1101\ 0110\ 1011\ 1111\ 1001\ 010, & (2.12) \end{aligned}$$

in *working registers* (pieces of fast-responding memory). Because the exponents are different, it would be incorrect to add the mantissas. So the exponent of the smaller number is made larger while progressively decreasing the mantissa by *shifting bits* to the right (inserting zeros) until both numbers have the same exponent:

$$\begin{aligned} 10^{-7} &= 0 \ 01100001 \ 0110\ 1011\ 0101\ 1111\ 1100101\ (0) \\ &= 0 \ 01100010 \ 0011\ 0101\ 1010\ 1111\ 1110010\ (10) & (2.13) \\ \cdots &, \\ &= 0 \ 10000010 \ 0000\ 0000\ 0000\ 0000\ 0000\ 000\ (0001101\cdots) \\ \Rightarrow &\ 7 \ + \ 1.0 \times 10^{-7} = 7 & (2.14) \end{aligned}$$

Because there is no more room left to store the last digits, they are lost. After all this hard work, the addition gives 7. This means a 32-bit computer only stores 6–7 decimal places and in effect ignores the 10^{-7}.

The preceding loss of precision is categorized by defining the *machine precision* ϵ_m as the maximum positive number that, on the computer, can be added to the number stored as 1 without changing the number stored as 1:

$$1_c + \epsilon_m = 1_c, \tag{2.15}$$

where the subscript c is a reminder that this is the number stored in the computer's memory. Likewise, x_c, the computer's representation of x, and the actual number x, are related by

$$x_c = x(1 + \epsilon), \quad |\epsilon| \leq \epsilon_m \ .$$

In other words, $\epsilon \simeq 10^{-7}$ for single precision and 10^{-16} for double precision.

If a single-precision number x is larger than 2^{128}, an *overflow* occurs. If x is smaller than 2^{-128}, an *underflow* occurs. The resulting number x_c may end up being a machine-dependent pattern, or *NAN* (not a number), or unpredictable. Because the only difference between the representations of positive and negative numbers on the computer is the sign bit of one for negative numbers, the same considerations hold for negative numbers.

In our experience, serious scientific calculations almost always require double precision, especially on 32-bit machines. And if you need double precision in one part of your calculation, you probably need it all over, and that also means double-precision library routines.

2.15 IMPLEMENTATION: DETERMINING YOUR PRECISION, LIMIT.F (.C)

Write a program to determine the machine precision ϵ of your computer system (within a factor of 2 at least). A sample pseudocode is

```
eps = 1.
  begin do N times
    eps = eps/2.                                    Make smaller
    one = 1.  + eps
    write out:  loop number, one, eps
  end do
```

1. Check the precision for single-precision floating-point numbers.

2. Check the precision for double-precision floating-point numbers.

To print out a decimal number, the computer must make a conversion from its internal format. Not only does this take time, but if the internal number is close to being garbage, it's not clear what will get printed out. So if you want a truly precise indication of the stored numbers, you want to avoid conversion to decimals and, instead, print them out in octal or hexadecimal format.

2.16 PROBLEM 3: COMPLEX NUMBERS AND INVERSE FUNCTIONS

The language of physics is mathematics. Therefore using a computer to do physics ultimately means using it to do mathematics. But as we have seen, mathematics is not the native language of computers. The **problem** for you to investigate is the way your computer handles complex numbers and inverse trigonometric functions.

2.17 THEORY: COMPLEX NUMBERS

A complex number z is defined in terms of its real and imaginary parts as

$$z = x + iy. \tag{2.16}$$

It is also defined in terms of its magnitude r and phase ϕ as

$$z = re^{i\phi}, \tag{2.17}$$

$$\text{where} \quad r = \sqrt{x^2 + y^2}, \quad \phi = \tan^{-1}\left(\frac{y}{x}\right). \tag{2.18}$$

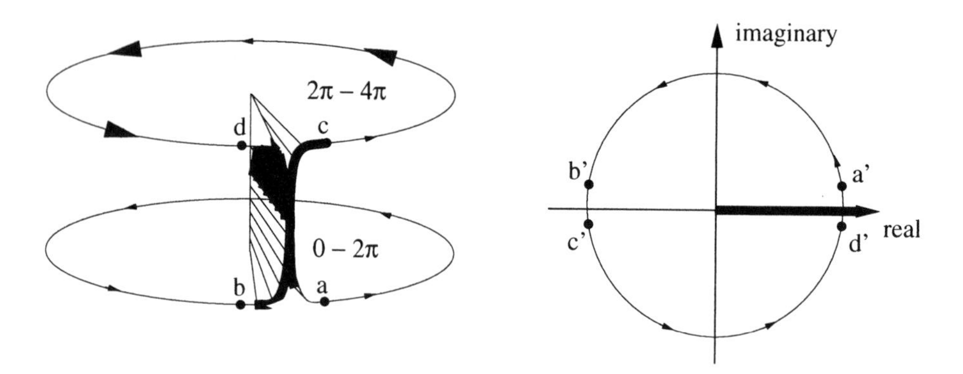

Fig. 2.8 (*Left*) The complex plane represented as two Riemann sheets attached through a cut appropriate to the $\sqrt{z}$ function. (*Right*) The cut complex plane. The cut prevents us from getting to a' from d' with a small rotation.

With this second definition we see that the square root and logarithms of z must be

$$\sqrt{z} = \sqrt{r}e^{i\phi/2}, \tag{2.19}$$

$$\ln z = \ln r + i\phi, \tag{2.20}$$

where $\ln r$ is the standard natural logarithm of a real number. That being so, while the phase ϕ can have an arbitrary multiple of 2π added to it *without* changing the location of z in the complex plane, this addition *does* change the values of $\sqrt{z}$ and $\ln z$. (Although adding 4π to the phase ϕ returns both z and $\sqrt{z}$ to the same complex plane location, the ln function never returns.)

A way of avoiding this apparent multivalueness of functions is to agree not to encircle the origin $z = 0$. To build this agreement into the mathematics, a *branch cut* or is drawn along the intersection of sheets, in this case from the *branch point* $z = 0$ to $z = \infty$, as shown in Fig. 2.8. One then agrees not to pass through this branch cut. Typically the cut lies along the real x axis, or the imaginary y axis, although any line will do.

On the left in Fig. 2.8 we show a complex plane made up of two *Riemann sheets* attached through the cut in the shaded region. A rotation of 2π takes us from a to b but not back to a. After point b, the rotation moves to point c

on a second Riemann sheet, then to d, and, after a total rotation of 4π, back down to the first sheet at a. On the right in Fig. 2.8 we show the conventional complex plane. Because there is a cut along the positive x axis (as occurs for the $\sqrt{z}$), the points a' and d' are not as close to each other as are the points c' and b'.

Once the complex plane is cut, we imagine crossing the cut, but without getting into multivalueness troubles, by passing onto another Riemann sheet that is joined to the original sheet along the cut. For a doubled-valued function like $\sqrt{z}$, passing through the cut twice, once from each sheet, returns us to our starting point. While this ambiguity may not be much of a problem for clever mathematicians, it is for operationally minded computers who only know to do what they are told to do. For you to uncover if this is a problem for your computer, or how some programmer has decided to resolve the ambiguity, you need to work out these exercises.

2.18 IMPLEMENTATION: COMPLEX NUMBERS, COMPLEX.C (.F)

Fortran is nice enough to do complex arithmetic and evaluate inverse functions for you. Some C compilers, like those from Borland and Linux, have extensions to handle complex numbers, but a standard C compiler does not. Problem-solving environments, like Maple and Mathematica, are usually also good with complex numbers.

Write a program that gets the computer to print a table of the form

ϕ	x	y	e^z	$\sqrt{z}$	$\ln z$	atan(y/x)	atan2(y,x)
-4π	*	*	*	*	*	*	*
⋮	⋮	⋮	⋮	⋮	⋮	⋮	⋮
$15\pi/4$	*	*	*	*	*	*	*
4π	*	*	*	*	*	*	*

Here ϕ will increase with uniform steps, and the columns marked $\sqrt{z}$, log, atan, and atan2 are to be the computer's output. (If your compiler cannot handle complex numbers, forget about the z terms.)

1. Make a plot of the output phases obtained with the arctangent functions versus the input phase ϕ. [Notice that if $r = 1$ in (2.20), then the ln function is purely imaginary.]

2. If your plotting program appears to be making some strange jumps, you may need to use more points near a multiple of $\pi/2$ and avoid being precisely "at" a multiple of $\pi/2$.

3. If your compiler is not bright enough to automatically use a complex library routine when you feed it a complex number, you may have to look up the particular function name required to evaluate a complex function.

4. State clearly where the computer has placed the cuts for sqrt, ln, atan, and atan2 functions.

2.19 EXPLORATION: COMPLEX ENERGIES IN QUANTUM MECHANICS

A particle *not* under the influence of a potential is called "free." In quantum mechanics, a free particle moving in one dimension is described by the *plane wave*

$$\phi(x,t) = e^{i(kx-\omega t)}. \tag{2.21}$$

We obtain the probability density ρ for finding this particle at position x at time t by computing the squared modulus of the wave function $\rho = |\phi(x,t)|^2$. A trick used to describe an *unstable* system that *decays exponentially* in time is to say it can still be described by (2.21), only now with the complex ω (energy)

$$\hbar\omega = E_r - i\Gamma/2. \tag{2.22}$$

(Of course, if it decays, it cannot truly be "free," but that is why this is a model and not a theory.)

1. Show analytically that for positive or negative values of Γ, the probability ρ decays or grows (respectively) with increasing time t.

2. Show analytically that if energy and momentum are related in the usual way

$$\hbar\omega = \frac{k^2}{2m}, \tag{2.23}$$

then the momentum k also becomes a complex number, and the probability ρ then decays or grows with increasing distance x.

3. Write a program to calculate k for arbitrary values of E_r and Γ. Make the program *interactive*; that is, have it read E_r and Γ from the terminal and print *all* possible k values on the screen of the terminal.

4. Check the momentum values predicted by your program as both E_r and Γ change sign. Describe the physical reasonableness of its predictions.

2.20 PROBLEM 4: SUMMING SERIES

A classic numerical problem is the summation of a series to evaluate a function. For this exercise we examine the power series for the exponential function:

$$e^{-x} = 1 - x + \frac{x^2}{2!} - \frac{x^3}{3!} + \cdots \quad (x^2 < \infty). \tag{2.24}$$

We want to use the series (2.24) to calculate e^{-x} for $x = 0.1$, 1, 10, 100, and 1000, with an absolute error in each case of less than one part in 10^8. But how do we know when to stop summing? (Do not dare say or even think of saying, "When the answer agrees with the table or with the built-in library function.")

2.21 METHOD: NUMERIC

While we really want to ensure a definite accuracy for e^{-x}, that is not so easy to do. What is easy is to assume that the error in the summation is approximately the last term summed (this also assumes no roundoff error, which we will discuss soon). To obtain an absolute error of one part in 10^8, we can stop the calculation when

$$\left|\frac{\text{term}}{\text{sum}}\right| < 10^{-8}, \tag{2.25}$$

where *term* is the last term in the series (2.24), and *sum* is the accumulated sum.

2.22 IMPLEMENTATION: PSEUDOCODE

A pseudocode for performing the summation is

```
term = 1, sum = 1, eps = 10**(-8)                 Initialize.
do
  term = -term * x/i                     New term in terms of old.
  sum = sum + term                                   Add in term.
  while abs(term/sum) > eps           Break iteration if accurate.
end do
```

You can see that this technique saves time by avoiding raising x to an ever-increasing power, and by never calculating the factorial.

2.23 IMPLEMENTATION: GOOD ALGORITHM, EXP-GOOD.F (.C)

Write a program that implements this pseudocode for the indicated x values. Present your results as a table of the form

x	imax	sum	$\frac{\lvert\texttt{sum - exp(-x)}\rvert}{\texttt{sum}}$

where `exp(-x)` is calculated with the built-in exponential function.

2.24 IMPLEMENTATION: BAD ALGORITHM, EXP-BAD.F (.C)

Modify your code that sums the series in a "good way" (no factorials) to one that calculates the sum in a "bad way" (explicit factorials). A sample is given by `exp-bad.f (.c)`.

2.25 ASSESSMENT

1. Observe how the good and bad series summations fail for large x. In particular, notice whether there are underflows or overflows.
2. Produce a table as above.
3. Use a built-in timing function on the computer to compare the time for each method.

3

Errors and Uncertainties in Computations

Whether you like it or not, errors and uncertainties are a part of computation. Some errors are the ones humans inevitably keep on making, but many are introduced by the computer. Computer errors arise either because of the limited *precision* with which a computer stores numbers or because sometimes it really does make mistakes (particularly in sophisticated chores like compilation with optimization). Although it stifles creativity to keep thinking *error* when approaching a computation, it certainly is a waste of time to generate meaningless results because of errors. In this chapter we examine some of the errors and *uncertainties* introduced by the computer.

3.1 PROBLEM: LIVING WITH ERRORS

Let's say you have a program of significant complexity. To gauge why errors are such a concern, let us assume that your program has the logical flow

$$\text{start} \rightarrow U_1 \rightarrow U_2 \rightarrow \ldots \rightarrow U_n \rightarrow \text{end}, \tag{3.1}$$

where each unit U might be a step. If each unit has probability p of being correct, then the probability P of the whole program being correct is $P = p^n$. Let's say we have a large program, say, with $n = 1000$ steps, and that the probability of each step being correct is $p = 0.9993$. This means that you end up with $P = \frac{1}{2}$; that is, a final answer that is as likely wrong as right (not what you want to bring to your boss). The **problem** is that, as a scientist, you want a result that is correct—or at least in which the uncertainty is small.

3.2 THEORY: TYPES OF ERRORS

Four general types of errors exist to plague your computations:

Blunders: typographical errors entered with your program or data, running the wrong program, using the wrong data file, and so on. (If your blunder count starts increasing, it is time to go home or take a break.)

Random errors: those caused by events such as fluctuation in electronics due to power surges, cosmic rays, or someone pulling a plug. These may be rare but you have no control over them and their likelihood increases with running time; while you may have confidence in a 20-second calculation, a week-long calculation may have to be run several times to check reproducibility.

Approximation errors: those arising from simplifying the mathematics so that a problem can be solved or approximated on the computer. They include the replacement of: infinite series by finite sums, infinitesimal intervals by finite ones, and variable functions by constants. For example

$$e^x = \sum_{n=0}^{\infty} \frac{x^n}{n!} \tag{3.2}$$

$$\simeq \sum_{n=0}^{N} \frac{x^n}{n!} = e^x + \mathcal{E}(x, N), \tag{3.3}$$

where $\mathcal{E}(x, N)$ is the total absolute error. Because approximation error arises from the application of the mathematics, it is also called *algorithmic error*, the *remainder*, or *truncation error*.[1] The approximation error clearly decreases as N increases, and vanishes in the $N \to \infty$ limit. Specifically for (3.3), because the scale for N is set by the value of x, small approximation error requires $N \gg x$. So if x and N are close in value, the approximation error will be large.

Roundoff errors: those arising very much like the uncertainty in the measurement of a physical quantity encountered in an elementary physics laboratory. Because any stored number is represented by a finite number of bits (and consequently digits), the set of numbers that the computer can store exactly, *machine numbers*, is much smaller than the set of real numbers. In particular, there is a maximum and minimum to machine numbers. The overall error arising from using a finite number of digits

[1] The use of the term "truncation" may be somewhat confusing here because it is sometimes also used to describe the truncation of digits in the representation of a number, an effect more usually referred to as *roundoff error*.

to represent numbers accumulates as the computer handles more numbers; that is, as the number of steps in a computation increases. In fact, roundoff error causes some algorithms to become *unstable* with a rapid increase in error for certain parameters. In some cases, roundoff error may exceed the number itself, leaving what computer experts call *garbage*. For example, as computed (and you may try this at home)

$$2\left(\tfrac{1}{3}\right) - \tfrac{2}{3} = 0.6666666 - 0.6666667 = -0.0000001 \neq 0. \tag{3.4}$$

When dealing with roundoff error, you may be sensitive as to whether this error arises from "subtractive cancellation" or "multiplicative cancellation." And when considering these cancellations, it is good to recall those discussions of *significant figures* and scientific notation given in your early physics or engineering classes. For computational purposes let us consider how the computer may store the floating-point number

$$a = 11223344556677889900 = 1.12233445566778899 \times 10^{19}. \tag{3.5}$$

Because the exponent is stored separately and is a small number, we can assume that it will be stored in full precision. The mantissa may not be stored completely, depending on the word length of the computer and whether we declare the word to be stored in single or double precision. In double precision (or *REAL*8* on a 32-bit machine or *doubles*), the mantissa of a will be stored as two words, the *most significant part* representing the decimal 1.12233, and the *least significant part* 44556677. The digits beyond 7 may be lost. As we see below, when we perform calculations with words of fixed length, it is inevitable that errors get introduced into the least significant parts of the words.

3.3 MODEL: SUBTRACTIVE CANCELLATION

An operation performed on a computer usually only approximates the analytic answer. The approximation arises because computers are finite. Let us use the notation in which the number x is represented on the computer as x_c. The representation of a simple subtraction is then

$$a = b - c \quad \Rightarrow \quad a_c = b_c - c_c, \tag{3.6}$$

$$a_c = b(1+\epsilon_b) - c(1+\epsilon_c), \tag{3.7}$$

$$\Rightarrow \quad \frac{a_c}{a} = 1 + \epsilon_b \frac{b}{a} - \frac{c}{a}\epsilon_c. \tag{3.8}$$

We see from (3.8) that, in the crudest sense, the average error in a is a weighted average of the errors in b and c. Yet there can also be exceptional cases. The error in a increases when $b \approx c$ because we subtract off (and thereby lose) the most significant parts of both numbers. This leaves the least significant parts. This is a general rule:

> *If you subtract two large numbers and end up with a small one, there will be less significance in the small one.*

In other words, if a is small it must mean $b \simeq c$ and so

$$\frac{a_c}{a} = 1 + \epsilon_a, \quad (3.9)$$

$$\epsilon_a \simeq \frac{b}{a}(\epsilon_b - \epsilon_c). \quad (3.10)$$

This shows that even if the relative errors in b and c cancel somewhat, they are multiplied by the large number b/a, which, in turn, can make a differ significantly from a_c, even for small ϵ. If the signs of the numbers are such that the magnitude of a turns out to be larger than those of either b or c, this means that the numbers have been added together, in which case there is no subtractive cancellation and we can expect an accurate representation.

A good example of subtractive cancellation occurs in the power series summation for e^{-x} studied in Chapter 2, *Computing Software Basics*. For very large x, the early terms in the series can be quite large, but because the final answer must be very small, most of the large terms must be cancelled out. Consequently, one approach is to calculate e^x for very large x, and then take its inverse to obtain e^{-x}. This eliminates the subtractive cancellation occurring between successive terms because all terms in e^x just add.

3.4 ASSESSMENT: SUBTRACTIVE CANCELLATION EXPERIMENT

1. Remember back in high school when you learned that the quadratic equation

$$ax^2 + bx + c = 0, \quad (3.11)$$

has the analytic solution

$$x_{1,2} = \frac{-b \pm \sqrt{b^2 - 4ac}}{2a}, \quad (3.12)$$

or alternatively

$$x'_{1,2} = \frac{-2c}{b \pm \sqrt{b^2 - 4ac}}. \quad (3.13)$$

Inspection of (3.12)–(3.13) indicates that subtractive cancellation (and consequently an increase in relative error) arises when $b^2 \gg 4ac$ because then the square root and its preceding term nearly cancel. If $b > 0$, this subtractive cancellation occurs in x_1 and x'_2, while for $b < 0$ it occurs in x'_1 and x_2.

(a) Write a program that calculates all four solutions for arbitrary values of a, b, and c.

(b) Investigate how errors in your computed answers become large as the subtractive cancellation increases, and relate this to the known machine precision. (*Hint*: A good test case employs $a = 1$, $b = 1$, $c = 10^{-n}$, $n = 1, 2, 3, \ldots$.)

(c) Extend your program so that it will always tell you which are the most precise solutions.

2. You also have to be careful to avoid subtractive cancellation when summing a series. For example, consider the finite sum with alternating signs:

$$S_N^{(1)} = \sum_{n=1}^{2N} (-1)^n \frac{n}{n+1}. \tag{3.14}$$

If you sum the even and odd values of n separately, you get two sums

$$S_N^{(2)} = -\sum_{n=1}^{N} \frac{2n-1}{2n} + \sum_{n=1}^{N} \frac{2n}{2n+1}. \tag{3.15}$$

All terms are positive in this form with just a single subtraction at the end of the calculation. Even this one subtraction and its resulting cancellation can be avoided by combining the series analytically:

$$S_N^{(3)} = \sum_{n=1}^{N} \frac{1}{2n(2n+1)}. \tag{3.16}$$

While all three summations are mathematically equal, this may not be true numerically.

(a) Write a single-precision program that calculates $S^{(1)}, S^{(2)}$, and $S^{(3)}$.

(b) Assume $S^{(3)}$ to be the exact answer. Make a log–log plot of the relative error versus number of terms, that is, of $\log_{10} |(S_N^{(1)} - S_N^{(3)})/S_N^{(3)}|$, versus $\log_{10}(N)$. Start with $N = 1$ and work up to $N = 1{,}000{,}000$.

(c) See whether straight-line behavior occurs in some region of your plot.

3. In spite of the power of your trusty computer, calculating the sum of even a simple series may require some thought and care. Consider the series

$$S^{(\mathrm{up})} = \sum_{n=1}^{N} \frac{1}{n}, \tag{3.17}$$

which is finite as long as N is finite. When summed analytically, it does not matter if you sum the series upward from $n = 1$ or downward from $n = N$,

$$S^{(\text{down})} = \sum_{n=N}^{1} \frac{1}{n}. \tag{3.18}$$

Nonetheless, because of roundoff error, when summed numerically, $S^{(\text{up})} \neq S^{(\text{down})}$.

(a) Write a program to calculate $S^{(\text{up})}$ and $S^{(\text{down})}$ as functions of N.

(b) Make a log–log plot of the relative difference divided by the relative sum versus N.

(c) Observe the linear regime on your graph and explain why the downward sum is more precise.

3.5 MODEL: MULTIPLICATIVE ERROR

Error in computer multiplication arises in the following way:

$$a = b \times c \quad \Rightarrow \quad a_c = b_c \times c_c, \tag{3.19}$$

$$\Rightarrow \quad \frac{a_c}{a} = \frac{(1+\epsilon_b)(1+\epsilon_c)}{(1+\epsilon_a)} \simeq 1 + \epsilon_b + \epsilon_c. \tag{3.20}$$

Since ϵ_b and ϵ_c can have opposite signs, the error in a_c is sometimes larger and sometimes smaller than the individual errors in b_c and c_c.

It often turns out that we can estimate an average roundoff error for a series of multiplications by assuming that the computer's representation of a number differs *randomly* from the actual number. In these situations we have the analog of a random walk (which is discussed in Chapter 6, *Deterministic Randomness*). If the direction of each step in the walk is random, then R, the average distance covered in N steps each of length r, is

$$R \approx \sqrt{N} r. \tag{3.21}$$

Equation (3.20) indicates that each step of a multiplication has a roundoff error of length ϵ_m, the machine precision. Imagine making physical steps of length ϵ_m. By analogy to a random walk, the average relative error ϵ_{ro} arising after a large number N steps is

$$\epsilon_{\text{ro}} \approx \sqrt{N} \epsilon_m. \tag{3.22}$$

We will find (3.22) useful when we examine the error in algorithms.

For those situations in which the roundoff errors do not occur in a random manner, a careful analysis is needed to predict the dependence of the error on the number of steps N. In some cases there may be no cancellation of error and the relative error may well increase like $N\epsilon_m$. Even worse, in some recursive algorithms where the production of errors is coherent (e.g., upward recursion for Bessel functions), the error increases like $N!\epsilon_m$.

Our discussion of errors has an important implication for a student to keep in mind before being impressed by a calculation requiring hours of supercomputer time. A fast computer may complete 10^{10} floating-point operations per second. This means a program running for 3 hours performs about 10^{14} operations. Therefore, even in the best case of random errors, after 3 hours we expect roundoff errors to have accumulated to a relative importance of $10^7\epsilon_m$. For the error to be smaller than the answer, this demands $\epsilon_m < 10^{-7}$. As a result, we can make the generalization that the results of a several-hours-long calculation with 32-bit arithmetic (which inherently possesses only six to seven places of precision) probably contains much noise. This fact is seldom appreciated by users of large amounts of computer time.

3.6 PROBLEM 1: ERRORS IN SPHERICAL BESSEL FUNCTIONS

Accumulating roundoff errors often limits the ability of a program to perform accurate calculations. Your **problem** is the computation of the spherical Bessel and Neumann functions j_l and n_l.

Spherical Bessel functions occur in many physical problems, for example; the j_l's are part of the partial wave expansion of a plane wave into spherical waves,

$$e^{i\mathbf{k}\cdot\mathbf{r}} = \sum_{l=0}^{\infty} i^l (2l+1) j_l(kr) P_l(\cos\theta), \tag{3.23}$$

where θ is the angle between $\mathbf{k}$ and $\mathbf{r}$. Fig. 3.1 shows what a number of j_l's look like, and Table 3.1 gives some explicit values. The spherical Bessel function $j_l(x)$ is the solution of the differential equation

$$x^2 f''(x) + 2xf'(x) + \left[x^2 - l(l+1)\right] f(x) = 0, \tag{3.24}$$

which is regular (nonsingular) at the origin. The spherical Neumann function $n_l(x)$ is a second, independent solution of (3.24). It is irregular (diverges at $x = 0$), and is chosen to contain just the right amount of j_l needed for proper asymptotic behavior. Specifically

$$\begin{array}{llll} j_l(x) & \rightarrow & x^l/(2l+1)!! & \text{for } x \ll l, \\ n_l(x) & \rightarrow & -(2l-1)!!/x^{l+1} & \text{for } x \ll l, \\ j_l(x) & \sim & \sin(x - l\pi/2)/x & \text{for } x \gg l, \\ n_l(x) & \sim & -\cos(x - l\pi/2)/x & \text{for } x \gg l, \\ \text{where} \quad (2l+1)!! & \equiv & 1\cdot 3\cdot 5\cdot\cdot(2l+1). & \end{array} \tag{3.25}$$

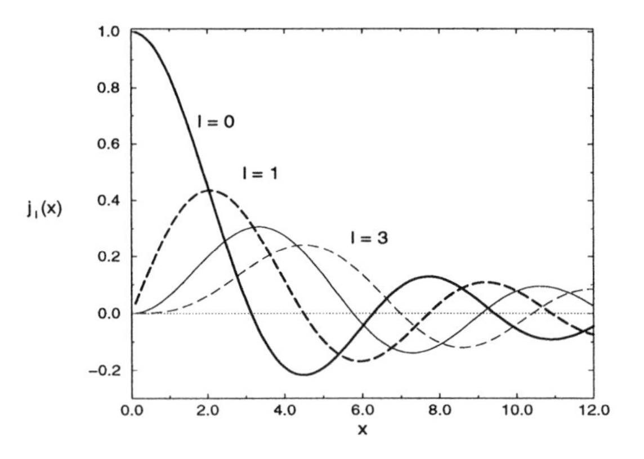

Fig. 3.1 The first four spherical Bessel functions, $j_l(x)$, as functions of x.

3.7 METHOD: NUMERIC RECURSION RELATIONS

One way to write a computer program to calculate $j_l(x)$ is to deduce its power series and asymptotic expansion. You then use these to evaluate $j_l(x)$ for small and large x/l, respectively [possibly augmented by a direct integration of the differential equation (3.24) for values in between]. The needed equations can be found in [Jack 75] and [A&S 64].

The approach we investigate here is often quicker than the use of series and has the advantage of generating the spherical Bessel functions for *all* l values at one time (for fixed x). It is based on the *recursion relation*:

$$j_{l+1}(x) = \frac{2l+1}{x} j_l(x) - j_{l-1}(x), \quad \text{(up)}, \tag{3.26}$$

$$j_{l-1}(x) = \frac{2l+1}{x} j_l(x) - j_{l+1}(x), \quad \text{(down)}. \tag{3.27}$$

Equations (3.26) and (3.27) both express the same relation, one written for recurring upward and the other for recurring downward. With just a few additions and multiplications, this recurrence relation permits a rapid and simple computation of the entire set of j_l's for fixed x and all l.

To recur upward we start with the fixed value of x and the known forms for j_0 and j_1:

$$j_0(x) = \frac{\sin(x)}{x}, \quad j_1(x) = \frac{\sin(x) - x\cos(x)}{x^2}. \tag{3.28}$$

We then use (3.26) to calculate $j_l(x)$ for all higher l values.

As you yourself will see, this upward recurrence usually starts working

Table 3.1 Approximate values for spherical Bessel functions of orders 3, 5, and 8 at $x = 0.1$, 1.0, and 10

x	$j_3(x)$	$j_5(x)$	$j_8(x)$
0.10	$+9.518510^{-6}$	$+9.616310^{-10}$	$+2.901210^{-16}$
1.00	$+9.006610^{-3}$	$+9.256110^{-05}$	$+2.826510^{-08}$
10.0	-3.949610^{-2}	$+5.553510^{-02}$	$+1.255810^{-01}$

pretty well but then fails. The reason for the failure can be seen from the plots of $j_l(x)$ and $n_l(x)$ versus x. If we start at $x \approx 2$ and $l = 0$, then we see from the graph that as we recur j_l up to larger l values with the relation (3.26), we are essentially taking the difference of two "large" numbers to produce a "small" one. This always reduces the precision. As we continue recurring, we are taking the difference of two "small" numbers to produce a smaller number yet, and this increases the relative error. After a while, the ever-increasing subtractive cancellations mean we are left with only roundoff error (garbage).

In contrast, if we use the upward recurrence relation (3.26) to produce the spherical Neumann function n_l, there is no problem. In that case, the graph makes clear that we are combining small numbers to produce larger ones, and in this way do not have any subtractive cancellation. In that case we are always working with the most significant parts of the numbers.

To be more specific, let us call $j_l^{(c)}$ the numerical value we compute as an approximation for $j_l(x)$. Even if we start with pure j_l, after a short while the computer's lack of precision effectively mixes in a bit of $n_l(x)$:

$$j_l^{(c)} = j_l(x) + \epsilon n_l(x). \tag{3.29}$$

This is inevitable because both j_l and n_l satisfy the same differential equation, and on that account, the same recurrence relation.

The admixture of n_l becomes a problem if the numerical value of n_l is much larger than that of j_l, because then even a miniscule amount of a very large number may be large. We can see from the limits (3.25) that if $l \gg x$, then the Neumann function is larger, $n_l > j_l$. This means that the error behaves like the spherical Neumann function, and consequently grows without bounds at the origin for upward recurrence.

The simple solution to this problem is *Miller's device*: use (3.27) for downward recursion starting at a large value of l. This essentially takes two small j values and produces a larger one by addition and in this way avoids subtractive cancellation. While the error may still behave like a Neumann function, the actual magnitude of the error will *decrease* quickly as we move downward to smaller l values. In fact, we start iterating downward with arbitrary values (garbage) for $j_{N+1}^{(c)}$ and $j_N^{(c)}$, and after a short while we arrive at very good answers. While the numerical value of $j_0^{(c)}$ so obtained will not be correct because it depends on the explicit value assumed for "garbage," the ratio of the $j_l^{(c)}$ values will be accurate. Therefore, after you have finished the downward recurrence, you use the analytic expression for $j_0^{(c)}$ (3.28) to normalize $j_0^{(c)}$ and all higher $j_l^{(c)}$ values.

3.8 IMPLEMENTATION: RECURSION RELATIONS, BESSEL.F (.C)

1. Write a program to calculate $j_l(x)$ that will give "good" values for the first 25 l values for $x = 0.1, 1.0, 10.0$. (For single precision, "good" means $\epsilon \simeq 10^{-6}$, and for double precision, "good" means $\epsilon \simeq 10^{-14}$.) See Table 3.1 for some sample values.

2. Try it with both upward and downward recursion, but don't try too hard for upward recursion. (Try using single precision in order to see error effects more quickly.)

3.9 ASSESSMENT

1. Give results of the downward recursion for different, large values of the starting l, showing the convergence and stability of your results.

2. Compare the upward and downward recursion methods, printing out l, $j_l^{(\mathrm{up})}$, $j_l^{(\mathrm{down})}$, and the relative difference

$$\frac{|j_l^{(\mathrm{up})} - j_l^{(\mathrm{down})}|}{|j_l^{(\mathrm{up})}| + |j_l^{(\mathrm{down})}|}. \tag{3.30}$$

3. The errors in the upward recursion depend on x, and for certain values of x, both up and down recursions give similar answers. Explain the reason for this and what it tells you about your program.

3.10 PROBLEM 2: ERROR IN ALGORITHMS

Numerical algorithms play a vital role in computational physics. You start with a physical theory or mathematical model, you use algorithms to convert the mathematics into a calculational scheme, and, finally, you convert your scheme into a computer program. Your **problem** is to take a general algorithm, and decide

1. Does it converge?

2. How precise are the results when it does converge?

3. How expensive (time consuming) is it to run ?

3.11 MODEL: ERRORS IN ALGORITHMS

An algorithm is often characterized by its step size h or by the number of steps N it takes to reach its goal. If the algorithm is "good," it should give the exact answer in the limit $h \to 0$ or $N \to \infty$. Every algorithm contains an *approximation error*; that is, there is a difference between the exact result and the result of the algorithm. If you know the approximation error as a function of the number of terms N used in the approximation, you may be able to judge "when enough is enough already." Yet do not be misled into believing the "error" has vanished because you made N so ridiculously large that the approximation error must be small. The *total* error in your calculation also includes roundoff errors, systematic errors, and possibly bad input data, all of which tend to increase when you make the computer work harder.

In general, as you continue to decrease the step size h or increase the number of steps N, you will reach a point where the roundoff error has grown large enough to exceed your approximation error. Clearly, the optimum choice of your parameters are those that minimize the total error. Unfortunately, in many cases there is no simple expression to minimize. Yet by using some of the methods described here, you may be able to determine the behavior of your error and so gain some control over it.

3.11.1 Total Error

Let us assume that an algorithm takes a large number N steps to get a good answer and that the approximation error approaches zero like

$$\epsilon_{\text{aprx}} \simeq \frac{\alpha}{N^{\beta}}. \tag{3.31}$$

Here α and β are empirical constants that would change for different algorithms, and may be "constant" only for $N \to \infty$. As indicated at the beginning of this chapter, the roundoff error keeps accumulating as you take more steps; that is, it increases with N. If the roundoff errors in the individual steps of your algorithm are not correlated, then we know from our previous discussion that

$$\epsilon_{\text{ro}} \simeq \sqrt{N}\epsilon_m, \tag{3.32}$$

where ϵ_m is the machine precision. The total error would be the sum of the two:

$$\begin{aligned}
\epsilon_{\text{tot}} &= \epsilon_{\text{aprx}} + \epsilon_{\text{ro}}, && (3.33)\\
&\simeq \frac{\alpha}{N^\beta} + \sqrt{N}\epsilon_m. && (3.34)
\end{aligned}$$

Although no discussion of errors is exciting (except maybe for masochists), it is useful. We assume we have a test case for which a good answer is known either analytically or from some other source. By comparing the test case answers to those computed, we deduce the total error ϵ_{tot} in the calculation. If we then plot $\log(\epsilon_{\text{tot}})$ against $\log(N)$, we can use the slope of this graph [the power of N in the expansion of the error (3.34)] to deduce which error term is dominant for differing N values. Alternatively, by starting at very large N values where we expect there to be essentially no approximation error, we can move in to smaller values of N and thereby deduce the N behavior of the approximation error.

If you run your test case with N much smaller than α/N^β, then the approximation error term in (3.34) should dominate and the slope should be $-\beta$. If N is much larger than $\sqrt{N}\epsilon_m$, then the roundoff error term should dominate and the slope should be $\frac{1}{2}$. If your test case does not have this behavior, there may be a problem in your program, or the model may be too simple.

3.12 METHOD: OPTIMIZING WITH KNOWN ERROR BEHAVIOR

In order to see more clearly how different kinds of errors balance off each other, let us now turn to the *relative* size of errors. We will assume the approximation error (3.31) has $\alpha = 1, \beta = 2$:

$$\epsilon_{\text{aprx}} \simeq \frac{1}{N^2}. \qquad (3.35)$$

If the total error is given by (3.34), then it will have an extremum when

$$\frac{d\epsilon_{tot}}{dN} = 0 \quad \Rightarrow \quad N^{\frac{5}{2}} = \frac{4}{\epsilon_m}. \qquad (3.36)$$

Because a maximum total error occurs for $N = \infty$, the extremum should be a minimum. For a computer with 32-bit words and single precision, $\epsilon_m \simeq 10^{-7}$, so the minimum total error (3.36) occurs when

$$\begin{aligned}
N^{\frac{5}{2}} \simeq \frac{4}{10^{-7}} \quad &\Rightarrow \quad N \simeq 1099, && (3.37)\\
\epsilon_{\text{tot}} &\simeq \frac{1}{N^2} + \sqrt{N}\epsilon_m && (3.38)
\end{aligned}$$

$$= \quad 8 \times 10^{-7} + 33 \times 10^{-7} \simeq 4 \times 10^{-6}. \tag{3.39}$$

This shows that for a typical algorithm, most of the error is due to roundoff. Observe, too, that even though this is the minimum error, the best we can do is to get some 40 times machine precision (the double-precision results are better).

Seeing that total error is mainly roundoff error, an obvious way to decrease the total error is to decrease roundoff error by using a smaller number of steps N. Let us assume we do this by finding another algorithm that converges more rapidly with N, for example, one for which the approximation error behaves like

$$\epsilon_{\rm aprx} \simeq \frac{2}{N^4}. \tag{3.40}$$

The total error is now

$$\epsilon_{\rm tot} = \frac{2}{N^4} + \sqrt{N}\epsilon_m. \tag{3.41}$$

The number of points for minimum error is found as before

$$\frac{d\epsilon_{tot}}{dN} = 0 \quad \Rightarrow \quad N^{\frac{9}{2}} = \frac{16}{\epsilon_m}, \tag{3.42}$$

$$\epsilon_m \simeq 10^{-7} \quad \Rightarrow \quad N \simeq 67, \tag{3.43}$$

$$\epsilon_{\rm tot} \quad = \quad \frac{2}{N^4} + \sqrt{N}\epsilon_m \tag{3.44}$$

$$= \quad 1 \times 10^{-7} + 8 \times 10^{-7} \simeq 9 \times 10^{-7}. \tag{3.45}$$

The error is now smaller by a factor of 4 with only $\frac{1}{16}$ as many steps needed. Subtle are the ways of the computer. In this case, it is not as important that the better algorithm is more elegant or more simple, but rather that its fewer steps produce less roundoff error.

3.13 METHOD: EMPIRICAL ERROR ANALYSIS

Let us say you have a program you want to optimize for minimum total error, yet you do not know (or do not want to go to the trouble of deriving) what the approximation error is. As just discussed, you know that in some approximate and general sense, the roundoff error $\epsilon_{\rm ro}$ is related to the machine precision ϵ_m and the number of calculational steps N by

$$\epsilon_{\rm ro} \simeq \sqrt{N}\epsilon_m. \tag{3.46}$$

Because the approximation error should get smaller with larger N, the roundoff error $\epsilon_{\rm ro}$ should dominate the total error for very large N.

Let us assume that the exact answer to your problem is $\mathcal{A}$, while that obtained by your algorithm after N steps is $A(N)$. The trick is to examine the behavior of $A(N)$ for values of N large enough for the approximation error to have its asymptotic value (the term with the smallest inverse power of N dominates), but not too large to dominate roundoff error. In this case we can write

$$A(N) \simeq \mathcal{A} + \frac{\alpha}{N^\beta}, \tag{3.47}$$

where α and β are unknown constants. We now run our computer program with a large number N of steps, and again with twice that number of steps. If roundoff error is not yet dominating, then

$$A(N) - A(2N) \approx \frac{\alpha}{N^\beta}. \tag{3.48}$$

To actually see if these assumptions are correct, and see graphically the number of decimal places to which the solution has converged, you plot $\log_{10}|A(N) - A(2N)|$ versus $\log_{10} N$. That part of your plot that is a straight line indicates the region in which the assumptions are valid, and the slope gives the value for $-\beta$.

If N is too small, we would not be in the asymptotic region for the approximation error, and the graph will not be a straight line. As N gets much larger, roundoff error begins to enter and the graph departs from the previous straight line (it should change slope to something like $+\frac{1}{2}$).

All this means that you can figure out what is happening with your algorithm by experimenting: start off with small N and increase it until you get a reasonably straight-line graph. Then increase N and watch as the graph changes slope to a positive one appropriate to roundoff error. Because the ordinate is the logarithm of the relative error to the base 10, it immediately tells you the number of decimal places of precision obtained.

3.14 ASSESSMENT: EXPERIMENT

Consider the series for the exponential function

$$e^{-x} = 1 - x + \frac{x^2}{2!} - \frac{x^3}{3!} + \cdots \quad (x^2 < \infty), \tag{3.49}$$

$$\simeq \sum_{n=0}^{N} \frac{(-x)^n}{n!}. \tag{3.50}$$

To most readily see the effects of error accumulation in this algorithm, use single precision for your programming.

1. Write a program that calculates e^{-x} as the finite sum (3.50).

2. Try $x = 1$, 10, and 100.

3. Examine the terms in the series for $x \simeq 10$ and observe the significant subtractive cancellations that occur when large terms add together to give small answers. See if better precision is obtained by using $\exp(-x) = 1/\exp(x)$ for large x values.

4. By progressively increasing N, use your program to experimentally determine whether there is a range of N for which the approximation error is asymptotic and yet larger than roundoff error. (You may assume that the built-in exponential function is exact.)

5. Determine whether (3.47) is valid and, if so, determine the values for β.

Because this series summation is such a simple and correlated process, the roundoff error does not accumulate randomly as it might for a more complicated computation, and we do not obtain the error behavior (3.47). To really see this error behavior, try this test with the integration rules discussed in Chapter 4, *Integration*.

4 Integration

In this chapter we show you how to carry out numerical integrations. We will derive the Simpson and trapezoid rules because they are so easy and give a good idea of how integration algorithms work, but we will just sketch the other algorithms and quote some error estimates.

4.1 PROBLEM: INTEGRATING A SPECTRUM

An experiment has measured $dN(t)/dt$, the number of particles entering a counter, per unit time, as a function of time. Your **problem** is to integrate this spectrum to obtain the number of particles $N(1)$ that entered the counter in the first second:

$$N(1) = \int_0^1 \frac{dN}{dt}(t)dt. \tag{4.1}$$

As an explicit assessment, in §4.8 we integrate an exponential spectrum. Nonetheless, the methods work equally well for integrating a table of numbers.

4.2 MODEL: QUADRATURE, SUMMING BOXES

The integration of a function may require some cleverness to do analytically, but it is relatively straightforward on a computer. The Riemann definition of an integral is the limit of the sum over boxes as the width h of the box

Table 4.1 Elementary weights for uniform-step integration rules

Name	Degree	Elementary Weights
Trapezoid	1	$(h/2, h/2)$
Simpson's	2	$(h/3, 4h/3, h/3)$
$\frac{3}{8}$	3	$(3h/8, 9h/8, 9h/8, 3h/8)$
Milne	4	$(14h/45, 64h/45, 24h/45, 64h/45, 14h/45)$

approaches zero:

$$\int_a^b f(x)dx = \lim_{h \to 0} \left[h \sum_{i=1}^{(b-a)/h} f(x_i) \right]. \tag{4.2}$$

For our problem, the function f is the spectrum as a function of time, $f(x) = dN(t)/dt$.

A traditional way to measure the area numerically is to take a piece of graph paper and count the number of boxes or *quadrilaterals* lying below the curve of the function. For this reason numerical integration is also called *numerical quadrature*, even when it gets beyond the box-counting stage.

The integral of a function $f(x)$ is approximated numerically by the equivalent of a sum over boxes:

$$\boxed{\int_a^b f(x)dx \approx \sum_{i=1}^{N} f(x_i)w_i.} \tag{4.3}$$

In this example, f is evaluated at N points in the interval $[a, b]$, and the function $f_i \equiv f(x_i)$ are summed with each term in the sum weighted by w_i. While, in general, the sum in (4.3) will give the exact integral only when $N \to \infty$, for certain classes of functions it may be exact for finite N.

The different integration algorithms amount to different ways of choosing the points and weights. Generally, the precision increases as N gets larger, with roundoff error eventually limiting the increase. Because the "best" approximation depends on the specific behavior of the function $f(x)$, there is no universally best approximation. In fact, some of the automated integration schemes found in subroutine libraries will switch from one method to another until they find one that works well.

In the simplest integration schemes, the integrand is approximated by a few terms in the Taylor series expansion of f, and the terms are integrated. Unless the integrand has some very unusual behaviors in each interval, successive terms should yield higher and higher precision. In these so-called Newton-Cotes methods, the total interval is divided into equal subintervals as

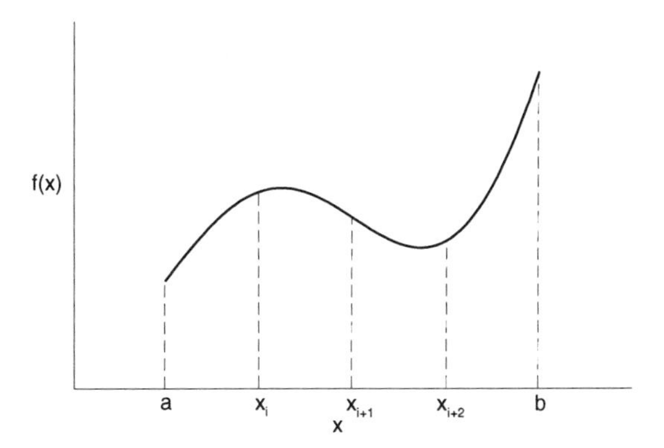

Fig. 4.1 The integral $\int_a^b f(x)dx$ is the area under the graph of $f(x)$ from a to b. Here we break up the area into four regions of equal widths.

shown in Fig. 4.1, with the integrand evaluated at equally spaced points x_i. These algorithms include the trapezoid rule (first-order) and the Simpson rule (second-order). The corresponding weights are given in Table 4.1. These rules are easy to apply and are the logical choice if the integrand is tabulated along evenly spaced points (or if it is tabulated along points that can be mapped onto equally spaced points).

More accurate integration rules are possible if the points are not constrained to be equally spaced. *Gaussian quadrature* methods have the ingenious ability to exactly[1] integrate the product of a function times a $(2N-1)$ degree polynomial, with only N function evaluations. This is no harder than evaluating the sum (4.3), and explains why so many computational scientists carry their Gaussian quadrature routines with them wherever they go.

In general, results with Gaussian quadrature will be superior to results with equally spaced points as long as there are no singularities in the integrand or its derivative. In the latter cases, the use of Simpson's rule may help avoid a catastrophe, but

> *you would be better off to remove singularities analytically from your integration before attempting any numerical quadrature.*

You may be able to do this by breaking the interval down into several subintervals, so the singularity is at an endpoint where a Gauss point never falls, or by a change of variable. For example

$$\int_{-1}^{1} |x| f(x)dx = \int_{-1}^{0} f(-x)dx + \int_{0}^{1} f(x)dx, \tag{4.4}$$

[1] "Exactly," except for roundoff error.

$$\int_0^1 x^{1/3}dx = \int_0^1 3y^3 dy, \qquad (y = x^{1/3}), \tag{4.5}$$

$$\int_0^1 \frac{f(x)dx}{\sqrt{1-x^2}} = 2\int_0^1 \frac{f(1-y^2)dy}{\sqrt{2-y^2}}, \qquad (y^2 = 1-x). \tag{4.6}$$

Likewise, if your integrand has a very slow variation in some region, you can speed up the integration by changing to a variable that compresses that region and places few points there. Conversely, if your integrand has a very rapid variation in some region, you may want to change to variables that expand that region to ensure that no oscillations are missed.

4.3 METHOD: TRAPEZOID RULE

The trapezoid and Simpson integration rules use values of $f(x)$ at evenly spaced values of x. They use N points x_i, $(i = 1, N)$, evenly spaced at a distance h apart throughout the integration region $[a, b]$ and *include the endpoints.* This means that there are $N-1$ intervals of length h:

$$h = \frac{b-a}{N-1}, \tag{4.7}$$

$$x_i = a + (i-1)h, \quad i = 1, N. \tag{4.8}$$

Notice that we start our counting at $i = 1$, and that Simpson's rule requires an *odd* number of points N.

In Fig. 4.2 we see that in the trapezoid rule we take the integration interval i and construct a trapezoid of width h in it. This approximates $f(x)$ by a straight line in that interval i, and uses the average height $(f_i + f_{i+1})/2$ as the value for f. The area of a single trapezoid is in this way

$$\int_{x_i}^{x_i+h} f(x)dx \simeq \frac{h(f_i + f_{i+1})}{2} = \tfrac{1}{2}hf_i + \tfrac{1}{2}hf_{i+1}. \tag{4.9}$$

In terms of our standard integration formula (4.3), the "rule" in (4.9) is for $N = 2$ points with weight $w_i \equiv \frac{1}{2}$.

In order to apply the trapezoid rule to the entire region $[a, b]$, we add the contributions from each subinterval:

$$\boxed{\int_a^b f(x)dx \approx \frac{h}{2}f_1 + hf_2 + hf_3 + \cdots hf_{N-1} + \frac{h}{2}f_N.} \tag{4.10}$$

You will notice that because each internal point gets counted twice, it has a weight of h, whereas the endpoints get counted just once and on that account have weights of only $h/2$. In terms of our standard integration rule (4.39), we

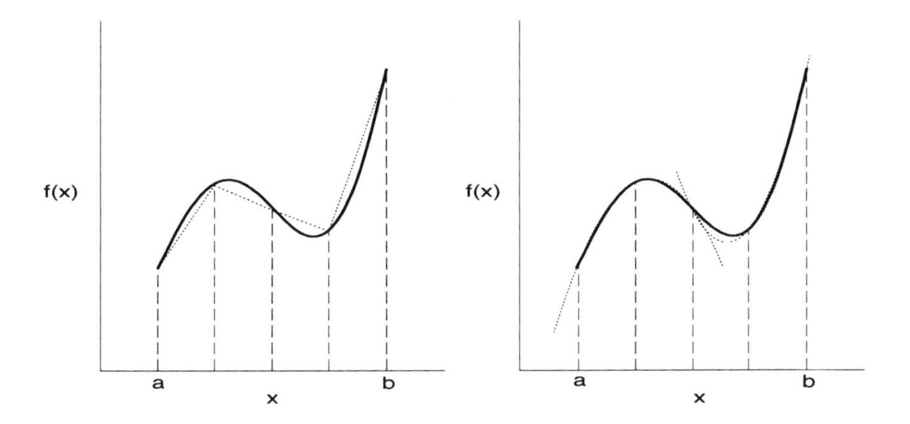

Fig. 4.2 (*Left*) straight-line sections used for the trapezoid rule; (*right*) the paraboli used in Simpson's rule.

have

$$w_i = \left\{ \frac{h}{2}, h, \ldots, h, \frac{h}{2} \right\}. \tag{4.11}$$

4.4 METHOD: SIMPSON'S RULE

In Simpson's rule we approximate the function $f(x)$ by a parabola

$$f(x) \approx \alpha x^2 + \beta x + \gamma, \tag{4.12}$$

for each interval, still keeping the intervals equally spaced. The area of each section is then the integral of this parabola

$$\int_{x_i}^{x_i+h} (\alpha x^2 + \beta x + \gamma) dx = \frac{\alpha x^3}{3} + \frac{\beta x^2}{2} + \gamma x \Big|_{x_i}^{x_i+h}. \tag{4.13}$$

This is equivalent to integrating the Taylor series up to the quadratic term. In order to relate the parameters α, β, and γ to the function, we consider an interval from -1 to $+1$, in which case

$$\int_{-1}^{1} (\alpha x^2 + \beta x + \gamma) dx = \frac{2\alpha}{3} + 2\gamma. \tag{4.14}$$

But we notice that

$$\begin{aligned} & f(-1) = \alpha - \beta + \gamma, \qquad f(0) = \gamma, \qquad f(1) = \alpha + \beta + \gamma, \\ \Rightarrow \quad & \alpha = \tfrac{f(1)+f(-1)}{2} - f(0), \quad \beta = \tfrac{f(1)+f(-1)}{2}, \qquad \gamma = f(0). \end{aligned} \tag{4.15}$$

In this way we can express the integral as the weighted sum over the values of the function at 3 points:

$$\int_{-1}^{1} (\alpha x^2 + \beta x + \gamma) dx = \frac{f(-1)}{3} + \frac{4f(0)}{3} + \frac{f(1)}{3}. \tag{4.16}$$

Because 3 values of the function are needed, we generalize this result to our problem by evaluating the integral over two adjacent intervals, in which case we evaluate the function at the two endpoints and the middle:

$$\int_{x_i-h}^{x_i+h} f(x)dx = \int_{x_i}^{x_i+h} f(x)dx + \int_{x_i-h}^{x_i} f(x)dx \simeq \frac{h}{3} f_{i-1} + \frac{4h}{3} f_i + \frac{h}{3} f_{i+1}. \tag{4.17}$$

Simpson's rule requires the elementary integration to be over *pairs* of intervals, and this requires that the total number of intervals be even or the number of points N be odd. In order to apply Simpson's rule to the entire interval, we add up the contributions from each pair of subintervals, counting all but the first and last endpoints twice:

$$\boxed{\int_a^b f(x)dx \approx \frac{h}{3} f_1 + \frac{4h}{3} f_2 + \frac{2h}{3} f_3 + \frac{4h}{3} f_4 + \cdots + \frac{4h}{3} f_{N-1} + \frac{h}{3} f_N.} \tag{4.18}$$

In terms of our standard integration rule (4.3), we have

$$w_i = \left\{ \frac{h}{3}, \frac{4h}{3}, \frac{2h}{3}, \frac{4h}{3}, \ldots \frac{4h}{3}, \frac{h}{3} \right\}. \tag{4.19}$$

The sum of these weights provides a useful check on your integration:

$$\sum_{i=1}^{N} w_i = (N-1)h. \tag{4.20}$$

Remember, N must be odd.

4.5 ASSESSMENT: INTEGRATION ERROR, ANALYTIC

In general, you want to choose an integration rule that gives an accurate answer using the least number of integration points. We obtain a feel for the absolute *approximation* or *algorithmic error* E and the relative error ϵ, by expanding $f(x)$ in a Taylor series around the midpoint of the integration interval. We then multiply that error by the number of intervals N to estimate the error for the entire region $[a, b]$. For the trapezoid and Simpson rules this yields

$$E_t = O\left(\frac{[b-a]^3}{N^2}\right) f^{(2)}, \tag{4.21}$$

$$E_s = O\left(\frac{[b-a]^5}{N^4}\right) f^{(4)}, \tag{4.22}$$

$$\epsilon_{t,s} = \frac{E_{t,s}}{f}. \tag{4.23}$$

We see that the third derivative term in Simpson's rule cancels (much like the central difference method in differentiation). The equations (4.21)–(4.22) are illuminating by showing how increasing the sophistication of an integration rule leads to an error that falls off with a higher inverse power of N, yet that is also proportional to higher derivatives of f. Consequently, for small intervals and $f(x)$ functions with well-behaved high derivatives, Simpson's rule should converge more rapidly than the trapezoid rule.

To be more specific, we assume that after N steps the *relative* roundoff error is random and of the form

$$\epsilon_{\text{ro}} \approx \sqrt{N}\epsilon_m, \tag{4.24}$$

where ϵ_m is the machine precision ($\sim 10^{-7}$ for single precision and $\sim 10^{-15}$ for double precision). We want to determine an N that minimizes the total error, that is, the sum of the approximation and roundoff error

$$\epsilon_{\text{tot}} = \epsilon_{\text{ro}} + \epsilon_{\text{approx}}. \tag{4.25}$$

This occurs, approximately, when the two relative errors are of equal magnitude, which we approximate even further by assuming that the two errors are equal

$$\epsilon_{\text{ro}} = \epsilon_{\text{approx}} = \frac{E_{\text{trap,simp}}}{f}. \tag{4.26}$$

To continue the search for optimum N for a general function f, we set the scale of function size by assuming

$$\frac{f^{(n)}}{f} \approx 1, \tag{4.27}$$

and the scale of length by assuming

$$b - a = 1 \quad \Rightarrow \quad h = \frac{1}{N}. \tag{4.28}$$

The estimate (4.26), when applied to the **trapezoid rule**, yields

$$\sqrt{N}\epsilon_m \approx \frac{f''(b-a)^3}{fN^2} = \frac{1}{N^2}, \tag{4.29}$$

$$\Rightarrow N \approx \frac{1}{(\epsilon_m)^{2/5}}. \tag{4.30}$$

Because the machine precision ϵ_m differs for single- and double-precision calculations, the optimum number of steps N for the **trapezoid rule** has the following two values:

$$N = \frac{1}{h} = \begin{cases} (1/10^{-7})^{2/5} = 631, & \text{for single precision,} \\ (1/10^{-15})^{2/5} = 10^6, & \text{for double precision.} \end{cases} \tag{4.31}$$

The corresponding errors are

$$\epsilon_{\text{ro}} \approx \sqrt{N}\epsilon_m = \begin{cases} 3 \times 10^{-6}, & \text{for single precision,} \\ 10^{-12}, & \text{for double precision.} \end{cases} \tag{4.32}$$

The estimate (4.26) when applied to **Simpson's rule** yields

$$\sqrt{N}\epsilon_m = \frac{f^{(4)}(b-a)^5}{fN^4} = \frac{1}{N^4}, \tag{4.33}$$

$$\Rightarrow N = \frac{1}{(\epsilon_m)^{2/9}}. \tag{4.34}$$

For single and double precision, this now yields

$$N = \frac{1}{h} = \begin{cases} (1/10^{-7})^{2/9} = 36, & \text{for single precision,} \\ (1/10^{-15})^{2/9} = 2154, & \text{for double precision.} \end{cases} \tag{4.35}$$

The corresponding errors are

$$\epsilon_{\text{ro}} \approx \sqrt{N}\epsilon_m = \begin{cases} 6 \times 10^{-7}, & \text{for single precision,} \\ 5 \times 10^{-14}, & \text{for double precision.} \end{cases} \tag{4.36}$$

These results are illuminating because they show that

1. Simpson's rule is an improvement over the trapezoid rule.

2. It is possible to obtain an error close to machine precision with Simpson's rule (and with other higher-order integration algorithms).

3. Obtaining the best numerical approximation to an integral is not obtained by letting $N \to \infty$, but with a relatively small $N \leq 1000$.

4.6 METHOD: GAUSSIAN QUADRATURE

It is often useful to rewrite the basic integration formula (4.3) such that we separate a weighting function $W(x)$ from the integrand:

$$\int_{\alpha}^{\beta} f(x)dx \equiv \int_{\alpha}^{\beta} W(x)F(x)dx \approx \sum_{i=1}^{N} w_i F(x_i). \tag{4.37}$$

In the Gaussian quadrature approach to integration, the N weights w_i are chosen to make the approximation error actually vanish if $f(x)$ was a $2N-1$ degree polynomial. To obtain this incredible optimization, the points x_i end up having a very specific distribution over $[a, b]$. (If f is only given by an equally spaced table, then a *Simpson* or *trapezoid rule* is simpler, although any integration algorithm can be used if the table is interpolated.[2])

In general, if $f(x)$ is smooth, or can be made smooth by factoring out some $W(x)$, Gaussian algorithms produce higher accuracy than lower-order ones, or conversely, the same accuracy with a fewer number of points. If the function being integrated is not smooth (for example, if it has some noise in it), then using a higher-order method such as Gaussian quadrature may well lead to lower accuracy. Sometimes the function may not be smooth because it has different behaviors in different regions. In these cases it makes sense to integrate each region separately with a low-order quadrature rule and then add the answers together. In fact, some of the "smart" integration subroutines will decide for themselves how many intervals to use and what rule to use in each interval.

All the rules indicated in Table 4.2 are Gaussian with the general form (4.37). We can see that in one case the weighting function is an exponential, in another a Gaussian, and in several an integrable singularity. In contrast to the equally-spaced rules, there is never an integration point at the extremes of the intervals, yet all of the points and weights change as the number of points N changes.

The values of α and β in (4.37) help define each integration scheme. If the integral you need to evaluate is over the range (a, b) and this differs from (α, β), you must *map* your (a, b) onto (α, β), possibly using one of the methods discussed below.

Although we will leave it to the references on numerical methods for the derivation of the Gauss points and weights, we note here that for ordinary Gaussian (Gauss–Legendre) integration, the points x_i turn out to be the N

[2]Interpolation schemes will be described soon.

Table 4.2 Types of Gaussian integration rules

Integral	Name
$\int_{-1}^{1} f(x)dx$	Gauss
$\int_{-1}^{1} \frac{F(x)}{\sqrt{1-x^2}}dx$	Gauss–Chebyshev (rational)
$\int_{-\infty}^{\infty} e^{-x^2} F(x)dx$	Gauss–Hermite
$\int_{0}^{\infty} e^{-x} F(x)dx$	Gauss–Laguerre
$\int_{0}^{\infty} \frac{e^{-x}}{\sqrt{x}} F(x)dx$	Associated Gauss–Laguerre

Table 4.3 Points and weights for 4-point Gaussian quadrature (points repeated for negative x)

$\pm x_i$	w_i
0.33998 10435 84856	0.65214 51548 62546
0.86113 63115 94053	0.34785 48451 37454

zeros of the Legendre polynomials, with the weights related to the derivatives:

$$P_N(x_i) = 0, \quad w_i = \frac{2}{(1 - x_i^2)[P_N'(x_i)]^2}. \tag{4.38}$$

Subroutines to generate these points and weights are standard in mathematical function libraries, are found in tables such as those in [A&S 64], or can be computed. The *gauss* subroutines we provide on the diskette and the Web also scales the points to a specified region. As a check that your points are correct, you may want to compare them to the four-point set we give in Table 4.3.

4.6.1 Scaling with Integration Rules

Our standard convention (4.3) for the general interval $[a, b]$ is

$$\int_a^b f(x)dx \approx \sum_{i=1}^{N} f(x_i)w_i. \tag{4.39}$$

But when the points and weights, $(y_i, w_i^{'})$, are for a fixed integration range (such as Gaussian), the programmer must *scale* the Gaussian interval to $[a, b]$. Here are some mappings we have found useful in our work (they are in the subroutine *gauss* given on the diskette and Web).

In the scalings below, $(y_i, w_i^{'})$ are the elementary Gaussian points and weights for the interval $[-1, 1]$, and we want to scale to x.

$[-1,1] \rightarrow [A,B]$ **uniformly,** $\frac{A+B}{2}$ **= midpoint:**

$$x_i = \frac{B+A}{2} + \frac{B-A}{2} y_i \tag{4.40}$$

$$\Rightarrow \int_A^B f(x)dx = \frac{B-A}{2} \int_{-1}^{1} f[x(y)]dy \tag{4.41}$$

$$\Rightarrow w_i = \frac{B-A}{2} w_i'. \tag{4.42}$$

$[0 \rightarrow \infty]$, A **= midpoint:**

$$x_i = A\frac{1+y_i}{1-y_i}, \tag{4.43}$$

$$w_i = \frac{2A}{(1-y_i)^2} w_i'. \tag{4.44}$$

$[-\infty \rightarrow \infty]$, **scale set by** A**:**

$$x_i = A\frac{y_i}{1-y_i^2}, \tag{4.45}$$

$$w_i = \frac{A(1+y_i^2)}{(1-y_i^2)^2} w_i'. \tag{4.46}$$

$[B \rightarrow \infty]$, $A+2B$ **= midpoint:**

$$x_i = \frac{A+2B+Ay_i}{1-y_i}, \tag{4.47}$$

$$w_i = \frac{2(B+A)}{(1-y_i)^2} w_i'. \tag{4.48}$$

$[0 \rightarrow B]$, $AB/(B+A)$ **= midpoint:**

$$x_i = \frac{AB(1+y_i)}{B+A-(B-A)y_i}, \tag{4.49}$$

$$w_i = \frac{2AB^2}{(B+A-(B-A)y_i)^2} w_i'. \tag{4.50}$$

You can see, that if your integration range extends out to infinity, there will be points at large but not infinite x. As you keep increasing the number of grid points N, the largest x_i moves farther and farther out.

4.7 IMPLEMENTATION: INTEGRATION, INTEG.F (.C)

Write a program to integrate an arbitrary function numerically using the trapezoid rule, the Simpson rule, and Gaussian quadrature. Use single precision in order to show more quickly the effects of error. (This may not be possible if your quadrature routines, such as the ones we supply on the diskette and the Web, are written in only double precision.)

For our **problem** we assume exponential decay so that there actually is an analytic answer:

$$\frac{dN(t)}{dt} = e^{-t} \tag{4.51}$$

$$\Rightarrow \quad N(1) = \int_0^1 e^{-t} dt = 1 - e^{-1}. \tag{4.52}$$

4.8 ASSESSMENT: EMPIRICAL ERROR ESTIMATE

Compare the relative error

$$\epsilon = \left| \frac{\text{numeric-exact}}{\text{exact}} \right|, \tag{4.53}$$

for the trapezoid rule, Simpson's rule, and Gaussian quadrature.

1. Make a table of the form

N	h	T-Rule	S-Rule	G-Quad	ϵ_T	ϵ_S	ϵ_G
10	0.1111	⋮	⋮	⋮	⋮		

   Try N values of 2, 10, 20, 40, 80, 160, ...

2. Make a plot like Fig. 4.3 of $\log_{10} \epsilon$ versus $\log_{10} h$. Note that the ordinate is effectively the number of decimal places of precision.

3. Use your plot to determine the approximate functional dependence of the error on the number of points N. (Notice that you may not be able to reach the roundoff error regime for the trapezoid rule because the approximation error is so large.)

4. (*Optional*) If possible, see how your answers change for double precision.

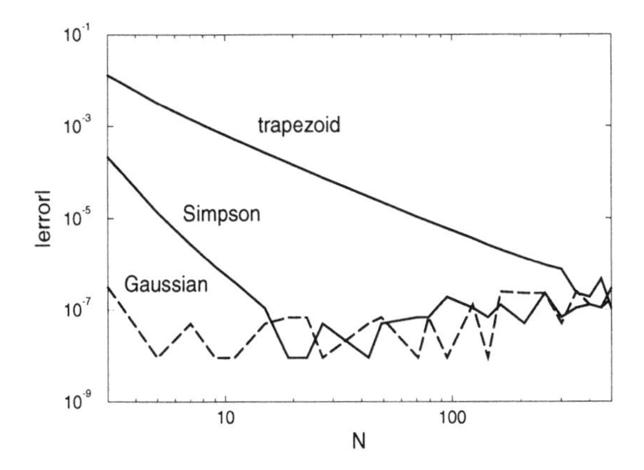

Fig. 4.3 Log–log plot of the error in integration of exponential decay using trapezoid rule, Simpson's rule, and Gaussian quadrature, versus number of integration points N. Approximately 7 decimal places of precision are attainable with single precision (shown here) and 15 places with double precision.

4.9 ASSESSMENT: EXPERIMENTATION

Try two integrals for which the answers are less obvious:

$$F_1 = \int_0^{2\pi} \sin(100x)dx, \quad F_2 = \int_0^{2\pi} \sin^x(100x)dx. \tag{4.54}$$

Explain why the computer may have trouble with these integrals.

4.10 METHOD: ROMBERG EXTRAPOLATION

As in the case of numerical differentiation, we can use the known functional dependence of the error on interval size h to reduce the error in integration. For simple rules like the trapezoid and Simpson's, we have the analytic estimates (4.26), while for others you may have to experiment to determine the h dependence. For example, if $A(h)$ and $A(h/2)$ are the values of the integral determined with the trapezoid rule using interval h and $h/2$, respectively, we know the integrals have expansions with a leading error term proportional to h^2:

$$A(h) \approx \int_a^b f(x)dx + \alpha h^2 + \beta h^4 + \ldots, \tag{4.55}$$

$$A(\tfrac{h}{2}) \approx \int_a^b f(x)dx + \frac{\alpha h^2}{4} + \frac{\beta h^4}{16} + \cdots. \tag{4.56}$$

Consequently, we make the h^2 term vanish by computing the combination

$$\tfrac{4}{3}A(\tfrac{h}{2}) - \tfrac{1}{3}A(h) \approx \int_a^b f(x)dx - \frac{\beta h^4}{4} + \cdots. \qquad (4.57)$$

Clearly this particular trick (Romberg's extrapolation) works only if the h^2 term dominates the error, and then only if the derivatives of the function are well behaved. An analogous extrapolation can also be made for other algorithms.

4.10.1 Other Closed Newton–Cotes Formulas

In Table 4.1 we gave the weights for several equal-interval rules. Whereas the Simpson rule used two intervals, the $\frac{3}{8}$ rule uses versus and the Milne[3] rule four. (These are single-interval rules and must be strung together to obtain a rule *extended* over the entire integration range. This means that the points that end one interval and begin the next get weighted twice.) You can easily determine the number of elementary intervals integrated over, and check whether you and we have written the weights right, by summing the weights for any rule. The sum is the integral of $f(x) = 1$ and must equal h times the number of intervals (which, in turn, equals $b - a$):

$$\sum_{i=1}^{N} w_i = h \times N_{\text{intervals}} = b - a. \qquad (4.58)$$

[3]There is, not coincidently, a Milne Computer Center at Oregon State University.

Part II

APPLICATIONS

5

Data Fitting

Data fitting is an art worthy of serious study. In this chapter we just scratch the surface. We examine how to interpolate within a table of numbers and how to do a least-squares fit for linear functions. If a least-squares fit is needed for nonlinear functions, then some of the search routines obtained from scientific subroutine libraries may be used. We describe how to do that in Chapter 15, *Matrix Computing and Subroutine Libraries* and on the Web. In some recent work [Penn 94], *simulated annealing* (which we describe in Chapter 22, *Thermodynamic Simulations: The Ising Model*) is used to assist the search in least-squares fitting.

5.1 PROBLEM: FITTING AN EXPERIMENTAL SPECTRUM

The cross section measured for the resonant scattering of a neutron from a nucleus is given in the Table 5.1. The first row is the energy; the second row, the experimental cross section; and the third row, the cross section as predicted by a friendly theoretical physicist (most are, you know). Your **problem** is to determine values for the cross section at values of energy between those measured by experiment. The theoretical prediction is for pedagogical purposes only.

Table 5.1 A scattering cross section as a function of energy.

$E(MeV)$	0	25	50	75	100	125	150	175	200
σ_{exp}(mb)	10.6	16.0	45.0	83.5	52.8	19.9	10.8	8.25	4.7
σ_{th}(mb)	9.34	17.9	41.5	85.5	51.5	21.5	10.8	6.29	4.09

5.2 THEORY: CURVE FITTING

You can view your **problem** in a number of ways. The most direct is to numerically interpolate the values of σ_{exp} given in Table 5.1. This is direct and easy, but it ignores the possibility of there being experimental noise in the data. It assumes, at least over some small range, that the data can be represented as a polynomial in the independent variable E (as is done in the algorithms used for differentiation and integration). This is more precisely "interpolation" and not data "fitting." We will discuss it first.

In a later section we discuss another way to view this problem, namely, to start with what one believes to be the "correct" theoretical description of the data, and then adjust whatever parameters may be present in that theory to obtain a *best fit* to the data.[1] This is a best fit in a statistical sense, but may not in fact go through all (or any) of the data points. For an easy, yet effective, introduction to statistical data analysis, we recommend [B&R 92].

As part of our **problem**, let's say that a friendly theoretical physicist told us that a resonant cross section as a function of energy should be described by the Breit–Wigner function

$$\sigma = \frac{\sigma_0}{(E - E_r)^2 + \gamma^2/4}, \tag{5.1}$$

where σ_0, E_r, and γ are constants to be determined by the fitting.

Interpolation and least-squares fitting are powerful tools that let you treat tables of numbers as if they were analytic functions, and sometimes let you deduce statistically meaningful constants or conclusions from measurements. In general, you can view data fitting as *global* or *local*. In global fits, a single function in x is used to represent the entire set of numbers in a table such as 5.1. While it may be spiritually satisfying to find a single function that passes through all the data points, that function may also oscillate rapidly between them in an unphysical or otherwise unreasonable manner (especially if the data points contain some sort of error or if the analytic function is not appropriate for the data). For this reason, global fits need to be looked at with a jaundiced eye.

[1] In Chapter 12 we discuss yet another way to fit data, namely, using Fourier series and Fourier transforms.

Table 5.2 A table of numbers fit with one polynomial using Lagrange interpolation

i	1	2	3	4
x_i	0	1	2	4
$f_i = f(x_i)$	-12	-12	-24	-60

5.3 METHOD: LAGRANGE INTERPOLATION

Consider Table 5.2 as the standard form for data we wish to fit. We interpolate among the entries in this table by assuming that in small regions, $f(x)$ can be approximated as a polynomial of low degree $(n-1)$ in x:

$$f_i(x) \simeq a_0 + a_1 x + a_2 x^2 + \cdots + a_{n-1} x^{n-1}, \quad (x \simeq x_i). \tag{5.2}$$

Because the fit is local, there is a different polynomial; that is, a different set of a_i values, for each region of the table. While the polynomial is of low degree, there are many polynomials that are needed to cover the entire table. If some care is taken, the set of polynomials so obtained behaves well enough to be used in further calculations without introducing much unwanted noise or discontinuities.

It might be tempting to use one higher-degree polynomial to fit all the data, but unless there is some good reason to believe that one polynomial may fit all of the data, this global fit might be a very bad representation of the data. Nevertheless, we ask you to do just that!

The classic of interpolation formulas was created by Lagrange. He figured out a closed-form one that directly fits the $(n-1)$-order polynomial (5.2) to n values of the function $f(x)$ evaluated at the points x_i. The formula is written as the sum of polynomials:

$$f(x) \simeq f_1\lambda_1(x) + f_2\lambda_2(x) + \cdots + f_n\lambda_n(x), \tag{5.3}$$

$$\lambda_i(x) = \prod_{j(\neq i)=1}^{n} \frac{x - x_j}{x_i - x_j} = \frac{x - x_1}{x_i - x_1}\frac{x - x_2}{x_i - x_2} \cdots \frac{x - x_n}{x_i - x_n}. \tag{5.4}$$

For 3 points, (5.3) provides a second-degree polynomial, while for eight points it gives a seventh-degree polynomial. Notice that Lagrange interpolation makes no restriction that the points in the table be evenly spaced. As a check, it is also worth noting that the sum of the Lagrange multipliers equals one:

$$\sum_{i=1}^{n} \lambda_i = 1. \tag{5.5}$$

Usually the Lagrange fit is made to only a small region of the table with a small value of n, even though the formula works perfectly well for fitting a high-degree polynomial to the entire table. The difference between the value of the polynomial evaluated at some x and that of the actual function is equal to the *remainder*

$$R_n \approx \frac{(x - x_1)(x - x_2) \cdots (x - x_n)}{n!} f^{(n)}(\zeta), \tag{5.6}$$

where ζ lies somewhere in the interval of interpolation, but is otherwise undetermined. This shows that if many high derivatives exist in $f(x)$, then $f(x)$ may not be approximated well as a polynomial. In particular, if $f(x)$ is a function or a table of numbers that has been determined experimentally, then it is likely to contain noise. In that case it might be very bad idea to fit a curve through all the data points.

5.3.1 Example

Consider again Table 5.2. We wish to use a four-point Lagrange interpolation to determine a third-order polynomial that reproduces each of the tabulated values:

$$\begin{aligned} f(x) &= \frac{(x-1)(x-2)(x-4)}{(0-1)(0-2)(0-4)}(-12) + \frac{x(x-2)(x-4)}{(1-0)(1-2)(1-4)}(-12) \\ &\quad + \frac{x(x-1)(x-4)}{(2-0)(2-1)(2-4)}(-24) + \frac{x(x-1)(x-2)}{(4-0)(4-1)(4-2)}(-60), \\ \Rightarrow f(x) &= x^3 - 9x^2 + 8x - 12. \end{aligned} \tag{5.7}$$

As a check we see that

$$f(4) = 4^3 - 94^2 + 32 - 12 = -60, \qquad f(0.5) = -10.125. \tag{5.8}$$

If the data contain little noise, the polynomial (5.7) can be used with some confidence within the range of data, but with great risk beyond the range of data.

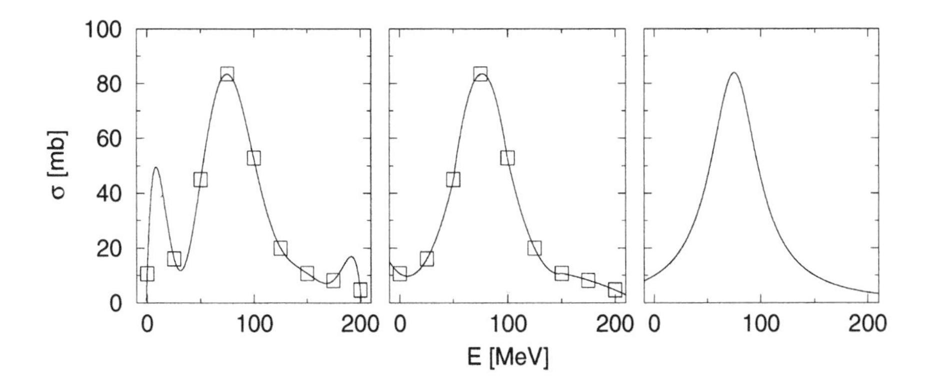

Fig. 5.1 From left to right: the fit to the cross section using an eighth-degree polynomial, the fit to the cross section using a series of third-degree polynomials, and the theoretical shape of the cross section.

5.4 IMPLEMENTATION: LAGRANGE INTERPOLATION, LAGRANGE.F (.C)

Write a subroutine to perform an n-point Lagrange interpolation using (5.3). Treat n as an arbitrary input parameter. (You can also do this exercise with a spline fit using subroutines from the libraries discussed in Chapter 15, *Matrix Computing and Subroutine Libraries*. We discuss the spline implementation in §5.8.)

5.5 ASSESSMENT: INTERPOLATING A RESONANT SPECTRUM

Consider the experimental neutron scattering data in Table 5.1. The expected theoretical functional form which describes these data is (5.1), and our empirical fits to these data are shown in Fig. 5.1.

1. Use the Lagrange interpolation formula to fit the entire experimental spectrum with one polynomial. (This means that you want to fit all nine data points with an eight degree polynomial.) Then use this fit to plot the cross section in steps of 5 MeV.

2. Use your graph to deduce the resonance energy E_r (your peak position) and γ (the full width at half maximum). Compare your results with those predicted by our theorist friend, $(E_r, \gamma) = (78, 55)$.

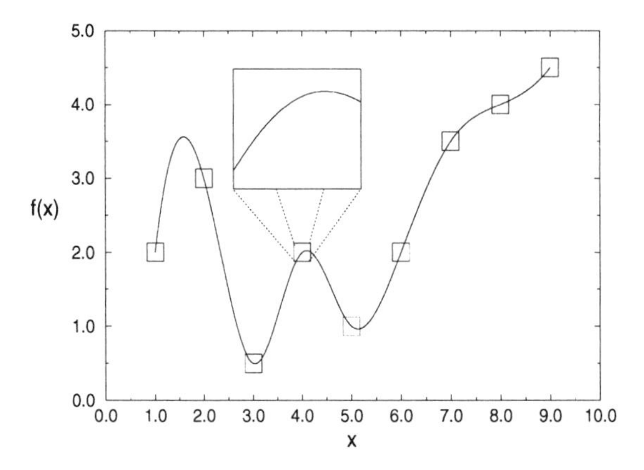

Fig. 5.2 Splines fit to a function in each interval. Notice in the insert how the splines are adjusted for a continuous derivative.

3. A more realistic use of Lagrange interpolation is for local interpolation with a small number of points, such as three. Interpolate the preceding cross-section data in 5 MeV steps using three-point Lagrange interpolation. (Note, the end intervals may be special cases.)

This example shows how easy it is to go wrong with a high-degree polynomial fit. Although the polynomial is guaranteed to actually pass through all the data points, the representation of the function away from these points can be quite unrealistic. Using a low-order interpolation formula, say, $n = 2$ or 3, in each interval usually eliminates the wild oscillations. If these local fits are then matched together, as we discuss in the next section, a rather continuous curve results. Nonetheless, you must recall that if the data contain errors, a curve that actually passes through them may lead you astray. We discuss this in §5.13.

5.6 ASSESSMENT: EXPLORATION

We deliberately have not discussed *extrapolation* of data because it can lead to serious *systematic* errors; the answer you get may well depend more on the function you assume than on the data you input. Add some adventure to your life and use the programs you have written to extrapolate. Compare your results to the theoretical Breit–Wigner shape (5.1).

5.7 METHOD: CUBIC SPLINES

If you tried to interpolate the resonant cross section with Lagrange interpolation, then you saw that fitting paraboli (three-point interpolation) within a table may avoid the erroneous and possibly catastrophic deviations of a high-order formula. (Two-point interpolation, which connects the points with straight lines, may not lead you far astray, but it is rarely pleasing to the eye or precise.) A sophisticated variation of $n = 4$ interpolation, known as *cubic splines*, often leads to surprising eye-pleasing fits. In this approach, illustrated in Fig. 5.2, cubic polynomials are fit to the function in each interval, with the additional constraint that the first and second derivatives of the polynomials must be continuous from one interval to the next. This continuity of slope and curvature makes the spline fit particularly eye-pleasing. It is analogous to what happens when you use the flexible drafting tool from which the method draws its name.

The series of cubic polynomials obtained by spline-fitting a table can be integrated and differentiated, and is guaranteed to have well-behaved derivatives. The existence of meaningful derivatives is an important consideration. For example, if the interpolated function is a potential, you can take the derivative to obtain the force.

The complexity of simultaneously matching polynomials and their derivatives over all the interpolation points leads to many simultaneous, linear equations to be solved. This makes splines unattractive for hand calculations, yet easy for computers. Splines have made recent gains in popularity and applicability in both calculations and graphics. For example, the smooth curves connecting points in most "draw" programs are usually splines.

The basic approximation of splines is the representation of the function $f(x)$ in the subinterval $[x_i, x_{i+1}]$ with a cubic polynomial:

$$f(x) \simeq f_i(x), \quad \text{for } x_i \le x \le x_{i+1}, \tag{5.9}$$

$$f_i(x) = f_i + f_i^{(1)}(x - x_i) + \tfrac{1}{2} f_i^{(2)}(x - x_i)^2 + \tfrac{1}{6} f_i^{(3)}(x - x_i)^3. \tag{5.10}$$

As written, the coefficients in the polynomial are manifestly related to the values of $f(x)$ and to its first, second, and third derivatives at x_i (derivatives beyond the third vanish in cubics). The computational chore is to determine these derivatives in terms of the N tabulated values f_i.

The matching of f_i from one interval to the next (at the *nodes*) provides the equations:

$$f_i(x_{i+1}) = f_{i+1}(x_{i+1}), \qquad i = 1, N-1. \tag{5.11}$$

Further equations are provided by requiring the first *and* second derivatives to be continuous at each subinterval's boundary:

$$f_{i-1}^{(1)}(x_i) = f_i^{(1)}(x_i), \qquad f_{i-1}^{(2)}(x_i) = f_i^{(2)}(x_i). \tag{5.12}$$

To provide the additional equations needed to determine all constants, the third derivatives at adjacent nodes are matched. Values for the third derivatives are found by approximating them in terms of the second derivatives:

$$f_i^{(3)} \simeq \frac{f_{i+1}^{(2)} - f_i^{(2)}}{x_{i+1} - x_i}. \tag{5.13}$$

[If you have read Chapter 8 on differentiation, then, yes, you know that a *central difference approximation* would be better for the forward difference; (5.13) keeps the equations simpler.]

5.7.1 Cubic Spline Boundary Conditions

It is straightforward but complicated to solve for all the parameters in (5.10). We leave that to other reference sources [Thom 92, Pres 94]. We can see, however, that matching at the boundaries of the intervals results in only $N-2$ linear equations for N unknowns. Further input is required. It usually is taken to be the boundary conditions at the endpoints $a = x_1$ and $b = x_N$, specifically, the second derivatives $f^{(2)}(a)$, and $f^{(2)}(b)$. There are several ways to determine these second derivatives:

natural spline: Set $f^{(2)}(a) = f^{(2)}(b) = 0$, that is, permit the function to have a slope at the endpoints but no curvature. This is "natural" because the derivative vanishes for the flexible spline drafting tool (its ends being free). The natural spline is often desirable because it produces a fit with the minimum integrated curvature, $\int dx f^{(2)}(x)^2$, and thus is likely to avoid erroneous oscillations. Nonetheless, the minimum curvature of the natural spline makes it a *stiff* fit, and if the actual function does not have vanishing curvature at the endpoints, the fit will be inaccurate near there.

input values for $f^{(1)}$ at boundaries: The computer uses $f^{(1)}(a)$ to approximate $f^{(2)}(a)$. If you do not know the first derivatives, you can calculate them from the table of f_i values with the techniques discussed in Chapter 8. For example, at the beginning of the table we can only form a forward difference, and this yields

$$f^{(1)}(x) \simeq \frac{f(x_2) - f(x_1)}{x_2 - x_1}. \tag{5.14}$$

input values for $f^{(2)}$ at boundaries: This is, of course, best because it is most realistic, but it requires the most input. If the values of $f^{(2)}$ are not known, the tabulated values of f can be used with the forward difference method:

$$f^{(2)}(x) \simeq \frac{[f(x_3) - f(x_2)]/[x_3 - x_2] - [f(x_2) - f(x_1)]/[x_2 - x_1]}{[x_3 - x_1]/2}. \tag{5.15}$$

5.7.2 Exploration: Cubic Spline Quadrature

A nice integration scheme is to use the spline fit to integrate f. If $f(x)$ is known only at its tabulated values, then this is about as good an integration scheme as possible; if you have the ability to actually calculate the function for arbitrary x, Gaussian quadrature may be preferable.

We know that the spline fit to f in each interval is the cubic (5.10)

$$f(x) \simeq f_i + f_i^{(1)}(x - x_i) + \tfrac{1}{2} f_i^{(2)}(x - x_i)^2 + \tfrac{1}{6} f_i^{(3)}(x - x_i)^3. \tag{5.16}$$

It is easy to integrate this to obtain the integral of f for this interval and then to sum over all intervals:

$$\int_{x_i}^{x_{i+1}} f(x)dx \simeq \left(f_i x + \tfrac{1}{2} f_i^{(1)} x_i^2 + \tfrac{1}{6} f_i^{(2)} x^3 + \tfrac{1}{24} f_i^{(3)} x^4 \right)\Big|_{x_i}^{x_{i+1}}, \tag{5.17}$$

$$\int_{x_j}^{x_k} f(x)dx = \sum_{i=j}^{k} \int_{x_i}^{x_{i+1}} f(x)dx. \tag{5.18}$$

Making the intervals smaller does not necessarily increase precision as subtractive cancellations may get large.

5.8 IMPLEMENTATION: PACKAGED SPLINE SUBPROGRAM, SPLINE.F

We recommend that you *not* write your own spline programs but instead get them from an advanced library such as IMSL or SLATEC. The procedures to do that can be found in Chapter 15, *Matrix Computing and Subroutine Libraries*.

5.9 ASSESSMENT: SPLINE FIT OF RESONANT CROSS SECTION

Carry out the assessment of §5.5 using cubic spline interpolation rather than Lagrange interpolation.

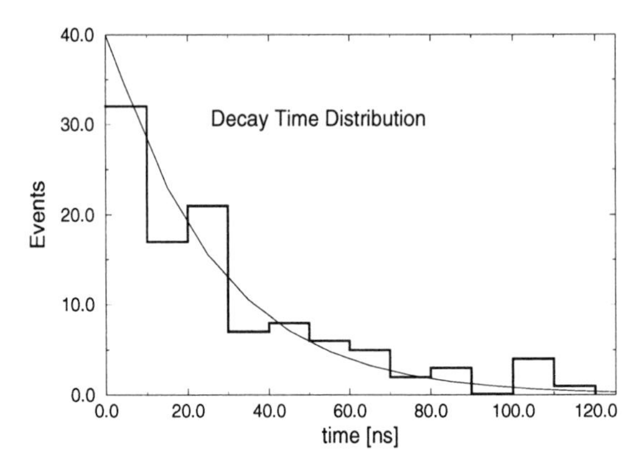

Fig. 5.3 A reproduction of the experimental measurement by [Stez 73] of the number of decays of a π meson as a function of time. Measurements are made during time intervals of 10-ns length.

5.10 PROBLEM: FITTING EXPONENTIAL DECAY

Fig. 5.3 presents experimental data of [Stez 73] on the number of decays ΔN of the π meson as a function of time. Notice that the time has been "binned" into $\Delta t = 10$ ns intervals and that the smooth curve gives the theoretical exponential decay law. The **problem** is to deduce the lifetime τ of the π meson from these data (the tabulated lifetime of the pion is 2.6×10^{-8} s).

5.11 MODEL: EXPONENTIAL DECAY

We start with a number N_0 of radioactive particles at time $t = 0$ that can decay to other particles.[2] If we wait a short time Δt, then a small number ΔN of the particles will decay spontaneously; that is, with no external influences. This decay is a stochastic process, which means that there is an element of chance involved and fluctuations are to be expected. The basic law of nature for spontaneous decay is that the number of decays ΔN in time interval Δt is proportional to the number of particles present at that time $N(t)$ and to the time interval

$$\Delta N(t) = -\frac{1}{\tau} N(t) \Delta t, \tag{5.19}$$

[2]Spontaneous decay is discussed further and simulated in §7.1.

where τ is the *lifetime* of the particle. This equation can be arranged into an equation for the average decay *rate*

$$\frac{\Delta N(t)}{\Delta t} = -\lambda N(t). \tag{5.20}$$

If the number of decays ΔN is very small compared to the number of particles N, and if we look at vanishingly small time intervals, then the difference equation (5.20) becomes the differential equation

$$\frac{dN(t)}{dt} \simeq -\lambda N(t). \tag{5.21}$$

This differential equation has an exponential solution for the number

$$N(t) = N_0 e^{-t/\tau}, \tag{5.22}$$

as well as an exponential solution for the decay rate

$$\frac{dN(t)}{dt} = -\frac{N_0}{\tau} e^{-t/\tau} = \frac{dN}{dt}(0) e^{-t/\tau}. \tag{5.23}$$

Equation (5.23) is the theoretical formula we wish to "fit" to the data in Fig. 5.3. The output of such a fit is a "best value" for the lifetime τ. Before we discuss how to carry out such a *least-squares fit*, we give some background information on probability theory. The reader familiar with probability theory may wish to skip ahead to §5.13.

5.12 THEORY: PROBABILITY THEORY

The field of statistics is an attempt to use mathematics to describe events, such as coin flips, in which there is an element of chance or randomness. A basic building block of statistics is the *binomial distribution* function

$$P_B(x) = \binom{N}{x} p^x (1-p)^{N-x} = \frac{N!}{(N-x)!x!} p^x (1-p)^{N-x}, \tag{5.24}$$

where $P_B(x)$ is the probability that the independent event (heads) will occur x times in the N trials. Here p is the probability of an individual event occurring; for example, the probability of "heads" in any one toss is $p = \frac{1}{2}$. The variable N is the number of *trials* or experiments in which that event can occur; for example, the number of times we flip the coin. For coin flipping, the probability of success p and the probability of failure $(1-p)$ are both $\frac{1}{2}$, but in the general case p can be any number between 0 and 1.

For calculational convenience, the factorials in the binomial distribution (5.24) are usually eliminated by considering the limit in which the number of

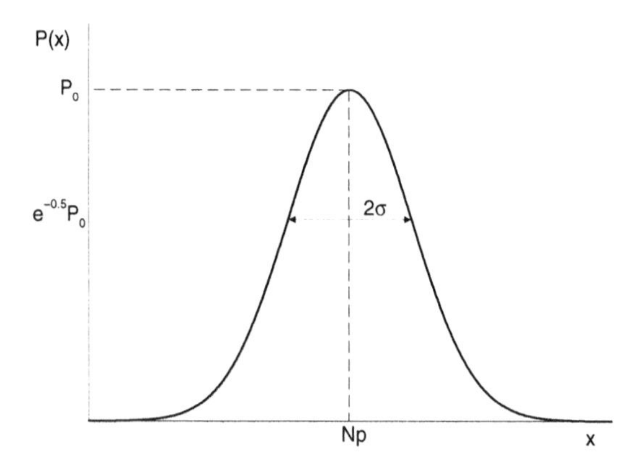

Fig. 5.4 A Gaussian distribution of m successes in N tries, each with probability p.

trials $N \to \infty$. In **Gaussian** or **normal** statistics, the probability p of an individual event (heads) remains finite as $N \to \infty$:

$$P_G(x) = \lim_{N\to\infty, p\neq 0} P_B(x) = \frac{1}{\sqrt{2\pi}\sigma} \exp\left[-\frac{(x-\mu)^2}{2\sigma^2}\right]. \tag{5.25}$$

This produces the function shown in Fig. 5.4, where $\mu \equiv \overline{x}$ is the mean and σ is the variance. These constants are related to the others by

$$\mu = Np\,, \quad \sigma = \sqrt{Np(1-p)}. \tag{5.26}$$

The Gaussian distribution is generally a very good approximation to the binomial distribution even for N as small as 10. To repeat, it describes an experiment in which N measurements of the variable x are made. The average of these measurements is μ and the "error" or uncertainty in μ is σ. As an example, in N =1000 coin flips, the probability of a head is $p = \frac{1}{2}$ and the average number of heads μ should be $Np = N/2 = 500$.

As shown in Fig. 5.4, the Gaussian distribution has a width

$$\sigma = \sqrt{Np(1-p)} \propto \sqrt{N}, \tag{5.27}$$

so that the distribution actually gets wider and wider as more measurements are made. Yet the *relative width,* whose inverse gives us an indication of the probability of obtaining the average μ, decreases with N:

$$\frac{\text{width}}{N} \propto \frac{\sqrt{N}}{N} = \frac{1}{\sqrt{N}} \to 0, \quad (N \to \infty). \tag{5.28}$$

Another limit of the binomial distribution is the **Poisson** distribution. In

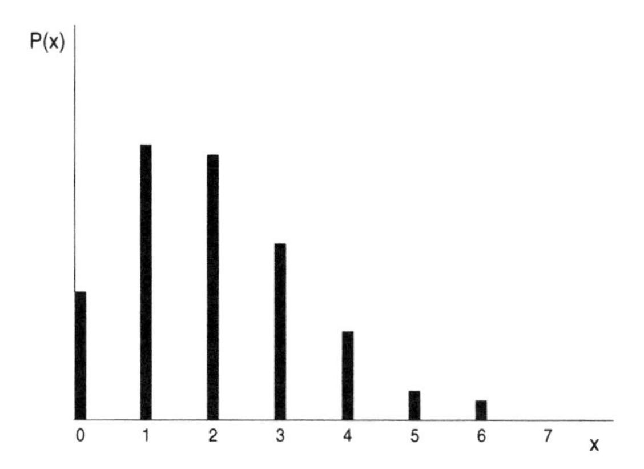

Fig. 5.5 A Poisson distribution for m successes with total success $a = 2$.

the Poisson distribution, the number of trials $N \to \infty$, yet the probability of an individual success $p \to 0$ in such a way that the product Np remains finite:

$$P_P(x) = \lim_{N\to\infty, p\to 0} P_B(x) = \frac{\mu^x e^{-\mu}}{x!}. \tag{5.29}$$

A **Poisson** distribution describes radioactive decay experiments or telephone interchanges where there may be a very large number of trials (each microsecond when the counter is on), but a low probability of an event (a decay or phone call) occurring in this one microsecond. As we see in Fig. 5.5, the Poisson distribution is quite asymmetric for small μ, and in this way quite different from a Gaussian distribution. For $\mu \gg 1$, the Poisson distribution approaches a Gaussian distribution.

5.13 METHOD: LEAST-SQUARES FITTING

Books have been written and careers have been spent discussing what is meant by a good fit to experimental data. We cannot do justice to the subject here and refer the reader to several other sources [B&R 92, Pres 94, M&W 65, Thom 92]. However, we will emphasize two points:

- If the data being fit contain errors, then the "best" fit in a statistical sense will not necessarily pass through the data points.
- Only for the simplest case of a linear, least-squares fit, can we write down a closed-form solution to evaluate and obtain the fit. More realistic problems are usually solved by *trial-and-error* search procedures using

library subroutines such as SLATEC.

Imagine that you have measured N_D values of the independent variable y as a function of the dependent variable x:

$$(x_i, y_i \pm \sigma_i), \quad i = 1, N_D, \tag{5.30}$$

where $\pm\sigma_i$ is the uncertainty in the ith value of y. (For simplicity we assume that all the errors σ_i occur in the dependent variable, although this is hardly ever true [Thom 92]). For our problem, y is the number of decays as a function of time.

Our goal is to determine how well a mathematical function $y = f(x)$ (also-called *theory*) can describe these data. Alternatively, if the theory contains some parameters or constants, our goal can be viewed as determining best values for these parameters. We assume that the model function $f(x)$ contains, in addition to the functional dependence on x, an additional dependence upon M_P parameters $\{a_1, a_2, \ldots, a_{M_P}\}$. For the exponential decay function (5.23), the parameter is the lifetime τ. We indicate this as

$$f(x) = f(x; \{a_1, a_2, \ldots, a_{M_P}\}) = f(x; \{a_m\}). \tag{5.31}$$

Notice that the parameters $\{a_m\}$ are not variables, in the sense of numbers read from a meter, but rather are parts of the theoretical model such as the size of a box, the mass of a particle, or the depth of a potential well.

We take the chi-squared (χ^2) measure as a gauge of how well a theoretical function f reproduces data

$$\boxed{\chi^2 \stackrel{\text{def}}{=} \sum_{i=1}^{N_D} \left(\frac{y_i - f(x_i; \{a_m\})}{\sigma_i} \right)^2,} \tag{5.32}$$

where the sum is over the N_D experimental points $(x_i, y_i \pm \sigma_i)$. The definition (5.32) is such that smaller values of χ^2 are better fits, with $\chi^2 = 0$ occurring if the theoretical curve went through the center of every data point. Notice also that the $1/\sigma_i^2$ weighting means that measurements with larger errors[3] contribute less to χ^2.

Least-squares fitting refers to adjusting the theory until a minimum in χ^2 is found; that is, finding a curve that produces the least value for the summed squares of the deviations of the data from the function $f(x)$. In general, this is the best fit possible or the best way to determine the parameters in a theory.

[3] If you are not given the errors, you can guess them on the basis of the apparent deviation of the data from a smooth curve, or you can weigh all points equally by setting $\sigma_i \equiv 1$ and continue with the fitting.

The M_P parameters $\{a_m, m = 1, M_P\}$ that make χ^2 an extremum are found by solving the M_P equations:

$$\frac{\partial \chi^2}{\partial a_m} = 0, \quad (m = 1, M_P), \tag{5.33}$$

$$\Rightarrow \quad \sum_{i=1}^{N_D} \frac{y_i - f(x_i)}{\sigma_i^2} \frac{\partial f(x; \{a_m\})}{\partial a_m} = 0, \quad (m = 1, M_P). \tag{5.34}$$

More usually, the function $f(x; \{a_m\})$ has a sufficiently complicated dependence on the a_m values for (5.34) to produce M_P simultaneous, nonlinear equations in the a_m values. In these cases, solutions are found by a trial-and-error search through the M_P-dimensional parameter space. To be safe, when such a search is completed you need to check that the minimum χ^2 you found is *global* and not *local.* One way to do that is to repeat the search for a whole grid of starting values, and if different minima are found, to pick the one with the lowest χ^2.

5.14 THEORY: GOODNESS OF FIT

When the deviations from theory are due to random errors and when these errors are described by a Gaussian distribution, there are some useful rules of thumb to remember [B&R 92]. You know that your fit is good if the value of χ^2 calculated via the definition (5.32) is approximately equal to the number of degrees of freedom

$$\chi^2 \approx N_D - M_P, \tag{5.35}$$

where N_D is the number of data points and M_P the number of parameters in the theoretical function. If your χ^2 is much less than (5.35), it doesn't mean that you have a "great" theory or a really precise measurement; instead, you probably have too many parameters or have assigned errors (σ_i values) that are too large. In fact, too small a χ^2 may indicate that you are fitting the random scatter in the data rather than missing $\sim \frac{1}{3}$ of the error bars (as expected for Gaussian statistics). If your χ^2 is significantly greater than (5.35), the theory may not be good, you may have significantly underestimated your errors, or you may have errors which are not random.

If you think you obtained a good fit to the data, but cannot determine the χ^2 because you did not have values for the experimental errors (maybe you assumed $\sigma_i \equiv 1$), you can get an approximate σ^2 for use in calculating χ^2. First you fit the data using an arbitrary value for σ_i. Then you calculate the *variance* of your data,

$$\sigma_{\text{exp}}^2 \stackrel{\text{def}}{=} \frac{1}{N_D} \sum_{i=1}^{N_D} [y_i - f(x_i)]^2 . \tag{5.36}$$

Finally, you use σ_{exp} as an approximation to σ_i and apply (5.32) to obtain a meaningful χ^2.

5.15 IMPLEMENTATION: LEAST-SQUARES FITS, FIT.F (.C)

The M_P simultaneous equations (5.34) simplify considerably if the functions $f(x;\{a_m\})$ depend *linearly* on the a values. This happens, for example, when the theory function $f(x)$ is linear:

$$f(x;\{a_1,a_2\}) = a_1 + a_2 x. \tag{5.37}$$

In this case (also known as *linear regression*) there are $M_P = 2$ parameters, the slope a_2 and the y intercept a_1. Notice that while there are only two parameters to determine, there still may be an arbitrary number N_D of data points to fit. Remember, a unique solution is not possible unless the number of data points is equal to or greater than the number of parameters.

For the linear case, the χ^2 minimization equations (5.34) become two in number, and determine the parameters in terms of all the data points [Pres 94]:

$$a_1 = \frac{S_{xx}S_y - S_x S_{xy}}{\Delta}, \quad a_2 = \frac{SS_{xy} - S_x S_y}{\Delta}, \tag{5.38}$$

$$S = \sum_{i=1}^{N_D} \frac{1}{\sigma_i^2}, \quad S_x = \sum_{i=1}^{N_D} \frac{x_i}{\sigma_i^2}, \tag{5.39}$$

$$S_y = \sum_{i=1}^{N_D} \frac{y_i}{\sigma_i^2}, \quad S_{xx} = \sum_{i=1}^{N_D} \frac{x_i^2}{\sigma_i^2}, \tag{5.40}$$

$$S_{xy} = \sum_{i=1}^{N_D} \frac{x_i y_i}{\sigma_i^2}, \quad \Delta = SS_{xx} - S_x^2. \tag{5.41}$$

If you know the errors σ_i in your experimental measurements of the y_i, or have determined an approximate σ from the sample variance from your fitted function, the theory then gives you an expression for the *variance* or uncertainty in the deduced parameters:

$$\sigma_{a_1}^2 = \frac{S_{xx}}{\Delta}, \quad \sigma_{a_2}^2 = \frac{S}{\Delta}. \tag{5.42}$$

This is a measure of the uncertainties in the values of the fitted parameters arising from the uncertainties σ_i in the measured y_i values.

A measure of the dependence of the parameters on each other is given by the *correlation coefficient*:

$$\rho(a_1, a_2) = \frac{\mathrm{cov}(a_1, a_2)}{\sigma_{a_1}\sigma_{a_2}}, \tag{5.43}$$

$$\mathrm{cov}(a_1, a_2) = \frac{-S_x}{\Delta}. \tag{5.44}$$

Here $\mathrm{cov}(a_1, a_2)$ is the *covariance* of a_1 and a_2 and vanishes if a_1 and a_2 are independent. The correlation coefficient $\rho(a_1, a_2)$ lies in the range $-1 \le \rho \le 1$. Positive ρ indicates that the errors in a_1 and a_2 are likely to have the same sign; negative ρ indicates opposite signs.

The preceding analytic solutions for the parameters are of the form found in statistics books, but are not optimal for numerical calculations because subtractive cancellation can make the answers unstable. As discussed in Chapter 3, *Errors and Uncertainties in Computations*, a rearrangement of the equations can decrease this type of error. For example, [Thom 92] gives improved expressions that measure the data relative to their averages:

$$a_1 = y_{\mathrm{av}} - a_2 x_{\mathrm{av}}, \quad a_2 = \frac{S_{xy}}{S_{xx}}, \tag{5.45}$$

$$S_{xy} = \sum_{i=1}^{N_d} \frac{(x_i - x_{\mathrm{av}})(y_i - y_{\mathrm{av}})}{\sigma_i^2}, \quad S_{xx} = \sum_{i=1}^{N_d} \frac{(x_i - x_{\mathrm{av}})^2}{\sigma_i^2}, \tag{5.46}$$

$$x_{\mathrm{av}} = \frac{1}{N}\sum_{i=1}^{N_d} x_i, \quad y_{\mathrm{av}} = \frac{1}{N}\sum_{i=1}^{N_d} y_i. \tag{5.47}$$

5.16 ASSESSMENT: FITTING EXPONENTIAL DECAY

Fit the exponential decay law (5.23) to the data in Fig. 5.3. This means finding a value of τ that provides a best fit to the data, and then judging how good the fit is.

1. Construct a table $(dN/dt_i,\, t_i)$, for $i = 1, N_D$ from Fig. 5.3. Notice that because time was measured in bins, t_i should correspond to the middle of a bin.

2. Add an estimate of the error σ_i to obtain a table of the form $(dN/dt_i \pm \sigma_i,\, t_i)$. You can estimate the errors by eye, say, by estimating how much the histogram values appear to fluctuate about a smooth curve, or you can take $\sigma_i \simeq \sqrt{\mathrm{Events}}$. (This last approximation is reasonable for large numbers, which this is not.)

3. In the limit of very large numbers, we would expect that a plot of $\ln dN/dt$ versus t is a straight line:

$$\ln \frac{dN(t)}{dt} \simeq \ln \frac{dN_0}{dt} - \frac{1}{\tau}t. \tag{5.48}$$

(While we do not have truly large numbers here, this result should be good on average.) This means that if we treat $\ln \frac{dN(t)}{dt}$ as the dependent variable and time t as the independent variable, we can use our linear fit results. Plot $\ln dN/dt$ versus t.

4. Make a least-squares fit of a straight line to your data and use it to determine the lifetime τ of the π meson. Compare your deduction to the tabulated lifetime of 2.6×10^{-8} s and comment on the difference.

5. Plot your best fit on the same graph as the data and comment on the agreement.

6. Use the formulas from statistics to deduce the goodness of fit of your straight line and the approximate error in your deduced lifetime. Do these agree with what your "eye" tells you?

5.17 ASSESSMENT: FITTING HEAT FLOW

Here is a table that gives the temperature T along a metal rod whose ends are kept at fixed constant temperatures. The temperature is a function of the distance x along the rod.

Position x_i (cm)	Temperature T_i (°C)
1.0	14.6
2.0	18.5
3.0	36.6
4.0	30.8
5.0	59.2
6.0	60.1
7.0	62.2
8.0	79.4
9.0	99.9

1. Plot up the data to verify the appropriateness of a linear relation

$$T = a + bx. \tag{5.49}$$

2. Because you are not given the errors for each measurement, assume that the least-significant figure has been rounded off and so $\sigma \simeq 0.05$. Use that to compute a least-square, straight-line fit to these data.

3. Plot your best $a + bx$ on the curve with the data.

4. After fitting the data, compute the variance and compare it to the deviation of your fit from the data. Verify that about one-third of the points miss the σ error band (that's what is expected for a normal distribution of errors).

5. Use your computed variance to determine the χ^2 of the fit. Comment on the value obtained.

6. Determine the variances σ_a and σ_b and check to see if they make sense as the errors in a and b.

7. What correlation is expected between a and b?

5.18 IMPLEMENTATION: LINEAR QUADRATIC FITS

As indicated earlier, as long as the function being fit depends *linearly* on the parameters a_i, the condition of minimum χ^2 leads to a set of simultaneous linear equations. These can be solved directly on the computer. For example, suppose we want to fit the experimental measurements $(x_i, y_i, i = 1, N_D)$ to the quadratic polynomial

$$y(x) = b_0 + b_1 x + b_2 x^2. \tag{5.50}$$

The χ^2 will be a minimum with respect to variation of these parameters (in other words, there will be maximum likelihood that these are the correct parameters describing the measurements), when we satisfy the three simultaneous linear equations:

$$\begin{aligned} N_D b_0 + S_x b_1 + S_{xx} b_2 &= S_y, & (5.51)\\ S_x b_0 + S_{xx} b_1 + S_{xxx} b_2 &= S_{xy}, & (5.52)\\ S_{xx} b_0 + S_{xxx} b_1 + S_{xxxx} b_2 &= S_{xxy}. & (5.53) \end{aligned}$$

Here the definitions of the S's are simple extensions of those used in (5.38). These equations can be written in matrix form:

$$\begin{bmatrix} N_D & S_x & S_{xx} \\ S_x & S_{xx} & S_{xxx} \\ S_{xx} & S_{xxx} & S_{xxxx} \end{bmatrix} \begin{bmatrix} b_0 \\ b_1 \\ b_2 \end{bmatrix} = \begin{bmatrix} S_y \\ S_{xy} \\ S_{xxy} \end{bmatrix}. \tag{5.54}$$

The solution follows after finding the inverse of the S matrix:

$$[S]\mathbf{b} = \mathbf{s} \tag{5.55}$$
$$\mathbf{b} = [S]^{-1}\mathbf{s}. \tag{5.56}$$

The inversion can be accomplished with the techniques discussed in Chapter 15, *Matrix Computing and Subroutine Libraries.*

5.19 ASSESSMENT: QUADRATIC FIT

Fit a quadratic to the following data sets [given as $(x_1, y_1), (x_2, y_2), \ldots$]. In each case determine the solution to these equations, the number of degrees of freedom in the problem, *and* the value of χ^2.

1. $(0, 1)$
2. $(0, 1), (1, 3)$
3. $(0, 1), (1, 3), (2, 7)$
4. $(0, 1), (1, 3), (2, 7), (3, 15)$

5.20 METHOD: NONLINEAR LEAST-SQUARES FITTING

An example of a subroutine for conducting a nonlinear search is *snls1* from SLATEC.

5.21 ASSESSMENT: NONLINEAR FITTING

Return to Table 5.1, which gives the scattering cross section versus energy. Determine what values for the parameters E_r, σ_0, and γ in the Breit–Wigner formula (5.1) provide a best fit to the data in the table (that is, minimize χ^2).

6

Deterministic Randomness

6.1 PROBLEM: DETERMINISTIC RANDOMNESS

Some people are attracted to computing by its deterministic nature; it's nice to have something in life where nothing is left to chance. Barring random machine errors or undefined variables, you should get the same output every time you feed your program the same input. Nevertheless, many computer cycles are used for *Monte Carlo* calculations that at their very core strive to be random. These are calculations in which random numbers generated by the computer are used to *simulate* naturally random processes, such as thermal motion or radioactive decay, or to solve equations on the average. Indeed, much of the recognition of computational physics as a specialty has come about from the ability of computers to solve previously intractable thermodynamic and quantum mechanics problems using Monte Carlo techniques.

The **problem** in this chapter is explore how computers can generate random numbers and how well they can do it. To check whether it really works, you *simulate* some simple physical processes and evaluate some multidimensional integrals. Other applications, such as radioactive decay, magnetism, and lattice quantum mechanics, are considered in latter chapters.

6.2 THEORY: RANDOM SEQUENCES

We define a sequence of numbers $r_1, r_2, \ldots$ as *random* if there are no correlations among the numbers in the sequence. Yet randomness does not necessarily mean all numbers in the sequence are equally likely to occur. If all numbers in a sequence are equally likely to occur, then the sequence is *uniform*. For example, 1, 2, 3, 4, ... is uniform but not random, while 3, 1, 4, 2, 3, 1, 3, 2, 4, ... may be random but does not appear uniform. In addition, it is possible to have a sequence of numbers that, in some sense, are random but have very short-range correlations, for example; $r_1(1-r_1)r_2(1-r_2)r_3(1-r_3)\cdots$.

Mathematically, the likelihood of a random number occurring is described by a distribution function $P(r)$. This means the probability of finding r_i in the interval $[r, r+dr]$ is $P(x)dx$. The standard random-number generator on computers generates uniform distributions ($P = 1$) between 0 and 1. In other words, the standard random-number generator outputs numbers in this interval, each with an equal probability yet each independent of the previous number. As we shall see, numbers can also be generated nonuniformly and still be random.

By their very construction we know computers are deterministic and so they cannot truly create a random sequence. Although it may be a bit of work, if we know r_m and its preceding elements, it is always possible to figure out r_{m+1}. For this reason, computers generate *"pseudo"-random numbers*. By the very nature of their creation, computed random numbers must contain correlations and in this way are not truly random. (Yet with our incurable laziness we won't bother saying "pseudo" all the time.) While the more sophisticated generators do a better job at hiding the correlations, experience shows that if you look hard enough, or use these numbers enough you will notice correlations. A primitive alternative to generating random numbers is to read in a table of true random numbers; that is, numbers determined by naturally random processes such as radioactive decay. While not an attractive way to spend one's time, it may provide a valuable comparison.

6.3 METHOD: PSEUDO-RANDOM-NUMBER GENERATORS

The *linear congruent* or *power residue* method is the most common way of generating a random sequence of numbers, $\{r_1, r_2, \ldots, r_k\}$ over the interval $[0, M-1]$. You multiply the previous random number r_{i-1} by the constant a, add on another constant c, take the *modulus* by M, and then keep just the fractional part (remainder)[1] as the next random number r_i:

$$r_i \stackrel{\text{def}}{=} (ar_{i-1} + c)\text{mod}M \tag{6.1}$$

$$= \text{remainder}\left(\frac{ar_{i-1} + c}{M}\right). \tag{6.2}$$

[1] You may obtain the same result for the modulus operation by subtracting M until any further subtractions would leave a negative number, what remains is the *remainder*.

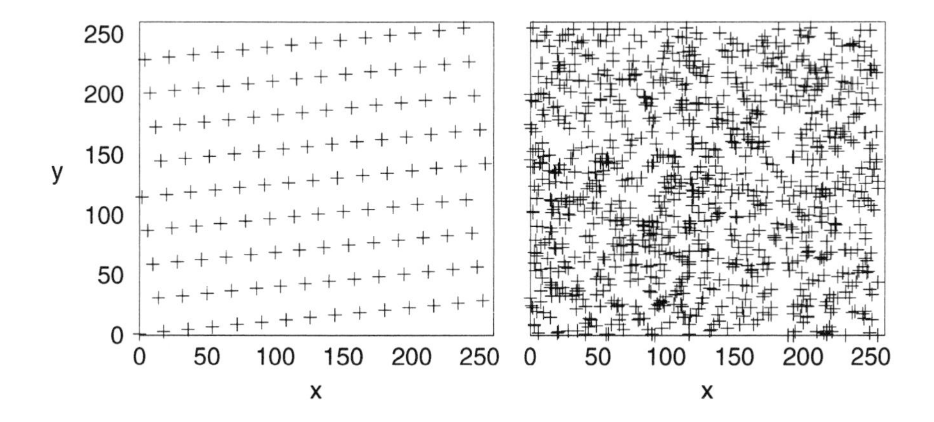

Fig. 6.1 (*Left*) A plot of successive random numbers $(x, y) = (r_i, r_{i+1})$, generated with a deliberately "bad" generator; (*right*) a plot with the library routine *drand48*.

The value for r_1 (the *seed*) is frequently supplied by the user, and *mod* is a built-in function on your computer for *remaindering* (it may be called *amod* or *dmod*). This is essentially a bit-shift operation that ends up with the least-significant part of the input number, and thus counts on the randomness of roundoff errors to produce a random sequence.

As an example, if $c = 1, a = 4, M = 9$, and you supply $r_1 = 3$, then you obtain the sequence

$$r_1 = 3, \tag{6.3}$$
$$r_2 = (4 \times 3 + 1)\text{mod}9 = 13\text{mod}9 = \text{rem}\tfrac{13}{9} = 4, \tag{6.4}$$
$$r_3 = (4 \times 4 + 1)\text{mod}9 = 17\text{mod}9 = \text{rem}\tfrac{17}{9} = 8, \tag{6.5}$$
$$r_4 = (4 \times 8 + 1)\text{mod}9 = 33\text{mod}9 = \text{rem}\tfrac{33}{9} = 6, \tag{6.6}$$
$$r_{5-10} = 7, 2, 0, 1, 5, 3. \tag{6.7}$$

We get a sequence of length $M = 9$, after which the entire sequence repeats. If we want the numbers in the range $[0, 1]$, we would divide these r values by $M = 9$. (It would still be a sequence of length 9, but no longer of integers.)

As a general operating procedure

> *Before using a random number generator in your programs, you may check its range and that it is producing numbers that "look" random.*

Although this is not a strict mathematical test, your visual cortex is quite refined at recognizing patterns, and, in any case, it's easy to perform. For example, in Fig. 6.1 we show results from "good" and "bad" generators; it's really quite easy to tell them apart.

The rule (6.1) produces integers in the range $[0, M-1]$, but not necessarily every integer. When a particular integer comes up a second time, the whole cycle repeats. In order to obtain a longer sequence, a and M should be large numbers, but not so large that ar_{i-1} overflows. On a scientific computer using 48-bit integer arithmetic, the built-in random number generator may use M values as large as $2^{48} \simeq 3\times 10^{14}$. A 32-bit machine may use $M = 2^{31} \simeq 2\times 10^9$. If your program uses approximately this many random numbers, you may need to reseed the sequence during intermediate steps to avoid repetitions.

Your computer probably has random-number generators that are better than the one you will compute with the power residue method. You may check this out in the manual or help pages (try the *man* command in Unix) and then test the generated sequence. These routines may have names like *rand*, *rn*, *random*, *srand*, *erand*, *drand*, or *drand48*.

We recommend *drand48* as a random-number generator. It generates random numbers in the range [0, 1] with good spectral properties by using 48-bit integer arithmetic with the parameters

$$M = 2^{48}, \quad c = B \text{ (base16)} = 13 \text{ (base8)}, \tag{6.8}$$

$$a = 5DEECE66D \text{ (base16)} = 273673163155 \text{ (base8)}. \tag{6.9}$$

To use it properly, you need to call the subroutine *srand48* to plant your seed.[2]

The definition (6.1) will generate r_i values in the range $[0, M]$ or $[0, 1]$ if you divide by M. If random numbers in the range $[A, B]$ are needed, you need to only **scale**; for example

$$x_i = A + (B - A)r_i, \quad 0 \leq r_i \leq 1, \tag{6.10}$$

$$\text{produces} \quad A \leq x_i \leq B. \tag{6.11}$$

6.4 ASSESSMENT: RANDOM SEQUENCES

For scientific work we recommend using an industrial-strength random-number generator. To see why, here we assess how *bad* a careless application of the power residue method can be.

1. Write a simple program to generate random numbers using the linear congruent method (6.1).

2. For pedagogical purposes, try the unwise choice: $(a, c, M, r_1) = (57, 1, 256, 10)$. Determine the *period*, that is, how many numbers are generated before the sequence repeats.

[2]Unless you know how to do 48-bit arithmetic and how to input numbers in different bases, we do not recommend that you try these numbers yourself. For pedagogical purposes, large numbers like $M = 112233$ and $a = 9999$ work well.

3. Take the pedagogical sequence of random numbers and look for correlations by observing clustering on a plot of successive pairs $(x_i, y_i) = (r_{2i-1}, r_{2i})$, $i = 1, 2, \ldots$ (Do *not* connect the points with lines.) As in Fig. 6.1, you may "see" correlations, which means that you should not use this sequence for serious work.

4. Test the built-in random number generator on your computer for correlations by plotting the same pairs as above. (This should be good for serious work.)

5. Test the linear congruent method again with reasonable constants like those in (6.8)–(6.9). Compare the scatterplot you obtain with that of the built-in random-number generator. (This, too, should be good for serious work.)

6.5 IMPLEMENTATION: SIMPLE AND NOT [RANDOM.F (.C); CALL.F (.C)]

6.6 ASSESSMENT: RANDOMNESS AND UNIFORMITY

Because the computer's random numbers are generated according to a definite rule, the numbers in the sequence must be correlated to each other. This can affect a simulation that assumes random events. Therefore, it is wise for you to test a random-number generator before you stake your scientific reputation on results obtained with it. In fact, some tests are simple enough that you may make it a habit to run them simultaneously with your simulation.

In the examples to follow, we test for either randomness or uniformity.

1. One simple test of uniformity evaluates the kth moment of the random-number distribution:

$$\langle x^k \rangle = \frac{1}{N} \sum_{i=1}^{N} x_i^k. \tag{6.12}$$

If the random numbers are distributed with a *uniform* probability distribution $P(x)$, then (6.12) is approximately the moment of $P(x)$:

$$\frac{1}{N} \sum_{i=1}^{N} x_i^k \simeq \int_0^1 dx x^k P(x) + O(1/\sqrt{N}) \tag{6.13}$$

$$\simeq \frac{1}{k+1}. \tag{6.14}$$

If (6.14) holds for your generator, then you know that the distribution is uniform. If the deviation from (6.14) varies as $1/\sqrt{N}$, then you *also* know that the distribution is random.

2. Another simple test determines the near-neighbor correlation in your random sequence by taking sums of products for small k:

$$C(k) = \frac{1}{N}\sum_{i=1}^{N} x_i x_{i+k}, \qquad (k = 1, 2, \ldots). \tag{6.15}$$

If your random numbers x_i and x_{i+k} are distributed with the joint probability distribution $P(x_i, x_{i+k})$ and are independent and uniform, then (6.15) is approximately the integral:

$$\frac{1}{N}\sum_{i=1}^{N} x_i x_{i+k} \simeq \int_0^1 dx \int_0^1 dy xy P(x, y) \tag{6.16}$$

$$= \tfrac{1}{4}. \tag{6.17}$$

If (6.17) holds for your random numbers, then you know that they are not correlated. If the deviation from (6.17) varies as $1/\sqrt{N}$, then you *also* know that the distribution is random.

3. An effective test for randomness is performed visually by making a scatterplot of $(x_i = r_{2i}, y_i = r_{2i+1})$ for many i values. If your points have noticeable regularity, the sequence is not random. If the points are random, they should uniformly fill a square with no discernible pattern (a cloud). Such a test is shown in Fig. 6.1.

4. Another test is to run your calculation or simulation with the sequence r_1, r_2, $r_3, \ldots$, and then again with the sequence $(1 - r_1)$, $(1 - r_2)$, $(1 - r_3), \ldots$. Because both sequences should be random, if your results differ beyond statistics, then your sequence is probably not random.

5. Yet another test is to run your simulation with a sequence of true random numbers from a table and compare it to the results with the pseudo-random-number generator. In order to be practical, you may need to reduce the number of trials being made.

6.7 ASSESSMENT: TESTS FOR RANDOMNESS AND UNIFORMITY

1. Test your random-number generator with (6.14) for k =1, 3, 7, and N =100, 10,000, 100,000. In each case print out

$$\sqrt{N}\left|\frac{1}{N}\sum_{i=1}^{N} x_i^k - \frac{1}{k+1}\right| \tag{6.18}$$

to check that it is of order 1.

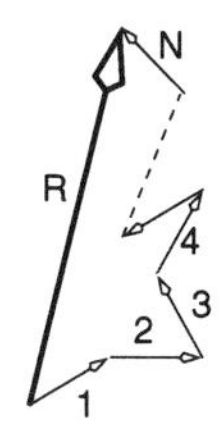

Fig. 6.2 Some of the N steps in a random walk that end up a distance R from the origin.

2. Test the mildly correlated series $r_1(1-r_1)r_2(1-r_2)r_3(1-r_3)\cdots$ with (6.17) for N =100, 10,000, 100,000. Again print out the deviation from the expected result and divide the deviation by $1/\sqrt{N}$ to check that it is of order 1.

6.8 PROBLEM: A RANDOM WALK

There are many physical processes, such as Brownian motion and electron transport through metals, in which a particle appears to move randomly. For example, consider a perfume atom released in the middle of a classroom. It collides randomly with other atoms in the air and eventually reaches the instructor's nose. The **problem** is to determine how many collisions, on average, the atom must make to travel a radial distance of R.

6.9 MODEL: RANDOM WALK SIMULATION

In a random-walk simulation, such as that in Fig. 6.2, an artificial *walker* takes many steps, usually with the *direction* of each step *independent* from the direction of the previous one. This is illustrated in Fig. 6.2. For our model, we start at the origin and take N steps in the $x-y$ plane of *lengths* (not coordinates)

$$(\Delta x_1, \Delta y_1), (\Delta x_2, \Delta y_2), (\Delta x_3, \Delta y_3), \ldots, (\Delta x_n, \Delta y_N). \tag{6.19}$$

The radial distance R from the starting point traveled after N steps is

$$\begin{aligned} R^2 &= (\Delta x_1 + \Delta x_2 + \cdots + \Delta x_N)^2 + (\Delta y_1 + \Delta y_2 + \cdots + \boldsymbol{\Delta} y_N)^2 \\ &= \Delta x_1^2 + \Delta x_2^2 + \cdots + \Delta x_N^2 + 2\Delta x_1 \Delta x_2 + 2\Delta x_1 \Delta x_3 + 2\Delta x_2 \Delta x_1 + \cdots \\ &\quad +(x \rightarrow y). \end{aligned} \tag{6.20}$$

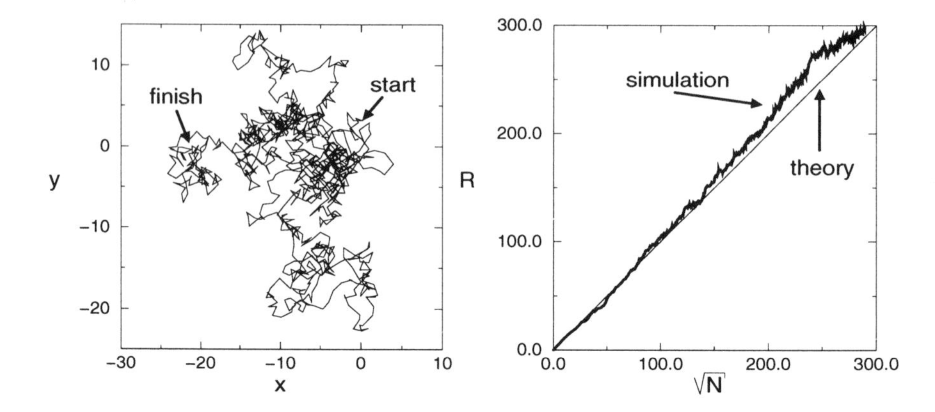

Fig. 6.3 (*Left*) A computer simulation of a random walk; (*right*) the distance covered in a simulated random walk of N steps compared to the theoretical prediction (6.22).

Equation (6.20) is valid for any walk. If it is random, the particle is equally likely to in any direction in each step. *On average*, for a large number of random steps, all the cross terms in (6.20) will vanish and we will be left with

$$\begin{aligned} R^2 &\simeq \Delta x_1^2 + \Delta x_2^2 + \cdots + \Delta x_N^2 + \Delta y_1^2 + \Delta y_2^2 + \cdots + \Delta y_N^2 \\ &= N\langle r^2 \rangle, \qquad (6.21) \\ \Rightarrow \quad R &\simeq \sqrt{N} r_{\rm rms}. \qquad (6.22) \end{aligned}$$

Here $r_{\rm rms}$ is the average (*root-mean-squared*) step size. Notice that that the same result obtains for a three-dimensional walk. According to (6.22), even though the total distance walked after N steps is $Nr_{\rm rms}$, on average, the radial distance from the starting point is only $\sqrt{N} r_{\rm rms}$.

6.10 METHOD: NUMERICAL RANDOM WALK

When using your computer to simulate a random walk, you should only expect to obtain (6.22) as the average displacement after many trials, not necessarily as the answer for each trial. Furthermore, you may get different answers depending on how you take your random steps. Below we give several ways to generate 2-D random walks. The results of our simulation are shown in Fig. 6.3.

We obtained the best results with the second method.

1. Choose a random angle θ in the range $[0, 2\pi]$. Set $\Delta x = \cos\theta$ and $\Delta y = \sin\theta$. (While this seems very reasonable, we have found some problems

with it which may be due to the trigonometric functions distorting the uniformity of the random distribution.)

2. Choose a random Δx in the range $[-\sqrt{2}, \sqrt{2}]$ and a separate random Δy in the range $[-\sqrt{2}, \sqrt{2}]$. In this way positive and negative steps in each direction are equally likely.

3. Choose random values for Δx in the range $[-1, 1]$ and $\Delta y = \pm\sqrt{1 - \Delta x^2}$ (choose the sign randomly, too).

4. Choose the directions (N, E, S, W) randomly as the step directions (no trigonometric functions are then needed). Notice that choosing one of four directions is equivalent to choosing a random *integer* in [1,4].

5. Choose the directions (N, NE, E, SE, S, SW, W, NW) randomly as the step directions (no trigonometric functions are then needed). This is equivalent to choosing a random *integer* in [1,8].

6.11 IMPLEMENTATION: RANDOM WALK, WALK.F (.C)

6.12 ASSESSMENT: DIFFERENT RANDOM WALKERS

Start at the origin and use your computer to take a two-dimensional random walk with unit steps.

1. If you have your walker taking N steps in a single trial, then conduct a total of $K \simeq \sqrt{N}$ trials each with N steps and each with a different seed.

2. Take the average of the results from your K trials as your output:

$$\langle R(N) \rangle = \frac{1}{K} \sum_{k=1}^{K} R^{(k)}(N) \approx \sqrt{N}. \tag{6.23}$$

3. After each step, calculate and output R, the distance from the origin.

4. Plot R as a function of $\sqrt{N}$ for one trial as well as the average after K trials. Notice how many steps must be made before Gaussian statistics start to be valid; that is, before $R \approx \sqrt{N}$.

5. Check to see if you obtained the expected statistical "error band" of $\pm\sqrt{R} \simeq (N)^{1/4}$ for large values of N.

6. Make up a scatterplot showing the quadrant in which the walker ends for each trial. If your walk is random, there should be no preferred direction.

7

Monte Carlo Applications

7.1 PROBLEM: RADIOACTIVE DECAY

We have already encountered spontaneous radioactive decay in Chapter 5, *Data Fitting*, where we fit an exponential function to a decay spectrum. Your **problem** now is to simulate how a small number of radioactive particles decay. In particular, you are to determine when radioactive decay looks exponential and when it looks *stochastic* (that is, determined by chance). As seen in Fig. 7.1, because the exponential decay law is only a large-number approximation to the natural process, our simulation should be closer to nature than the exponential decay law. In fact, if you go to the Web and "listen" to the output of the codes developed here, what you hear sounds very much like a Geiger counter: a convincing demonstration of how realistic the simulations are.

7.2 THEORY: SPONTANEOUS DECAY

Spontaneous decay is a natural process in which a particle, with no external stimulation, and at one instant in time, decays into other particles. Because the exact moment when any one particle decays is random, it does not matter how long the particle has been around or what is happening to the other particles. In other words, the probability $\mathcal{P}$ of any one particle decaying per unit time is a constant, and when that particle decays, it is gone forever. Of course, as the number of particles decreases with time, so will the number of decays. Nonetheless, the probability of any one particle decaying in some

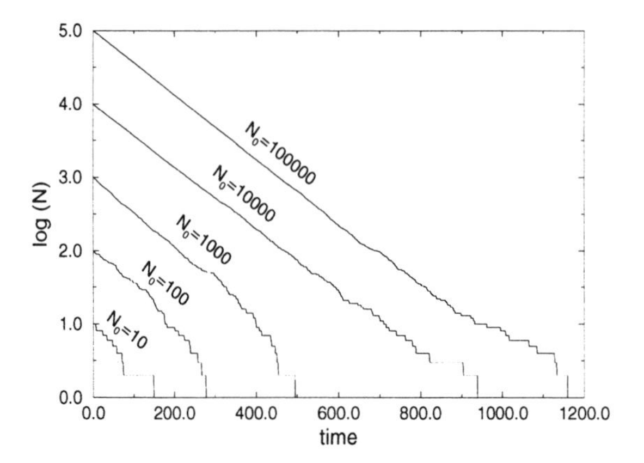

Fig. 7.1 Semilog plots of the results from several decay simulations. Notice how the decay appears exponential (like a straight line) when the number of nuclei is large, but stochastic for $\log N \leq 2.0$.

time interval is always the same constant as long as the particle still exists.

7.3 MODEL: DISCRETE DECAY

Imagine having $N(t)$ radioactive particles at time t. Let ΔN be the number of particles that decay in some small time interval Δt. We convert the statement "the probability $\mathcal{P}$ of any one particle decaying per unit time is a constant" into an equation by noting that the decay probability per particle, $\Delta N/N$, is proportional to the length of the time interval over which we observe the particle

$$\mathcal{P} = \frac{\Delta N(t)}{N(t)} = -\lambda \Delta t, \tag{7.1}$$

where λ is a constant. If we divide both sides of (7.1) by Δt, we obtain the constant decay rate per particle:

$$\frac{\Delta N(t)}{N(t)\Delta t} = -\lambda. \tag{7.2}$$

Sure enough, (7.2) says that the probability of any one particle decaying per unit time is a constant.

Equation (7.1) is a *finite-difference equation* in which $\Delta N(t)$ and Δt are experimental observables. Although it cannot be integrated the way a differential equation can, it can be solved numerically or algebraically. Because the decay process is random, we cannot predict an exact value for $\Delta N(t)$. Instead, we may think of $\Delta N(t)$ as the average number of decays when observations are made of many identical systems of N radioactive particles.

We convert (7.2) into a finite-difference equation for the decay rate by multiplying both sides by $N(t)$:

$$\frac{\Delta N(t)}{\Delta t} = -\lambda N(t). \tag{7.3}$$

The absolute decay rate $\Delta N(t)/\Delta t$ is called the *activity*, and because it is proportional to the number of particles present (which decreases in time), it, too, decreases in time.

7.4 MODEL: CONTINUOUS DECAY

When the number of particles $N \to \infty$ and the observation time interval approaches zero, an approximate form of the radioactive decay law (7.3) results:

$$\frac{\Delta N(t)}{\Delta t} \longrightarrow \frac{dN(t)}{dt} = -\lambda N(t). \tag{7.4}$$

This can be integrated to obtain the exponential decay law for the number and for the activity:

$$N(t) = N(0)e^{-\lambda t} = N(0)e^{-t/\tau}, \tag{7.5}$$

$$\frac{dN}{dt}(t) = -\lambda N(0)e^{-\lambda t} = \frac{dN}{dt}(0)e^{-\lambda t}. \tag{7.6}$$

In this limit we get exponential decay. We identify the decay rate λ with the inverse of the lifetime:

$$\lambda = \frac{1}{\tau}. \tag{7.7}$$

We see from its derivation that exponential decay is a good description of nature only *on the average*, and only for a large number of particles. The basic law of nature (7.2) is always valid, but, as we will see in the simulation, (7.6) becomes less and less accurate as the number of particles gets smaller and smaller.

7.5 METHOD: DECAY SIMULATION

The simulation program for radioactive decay is surprisingly simple, but not without its subtleties. To help you understand and write it, we give some pseudocode for it:

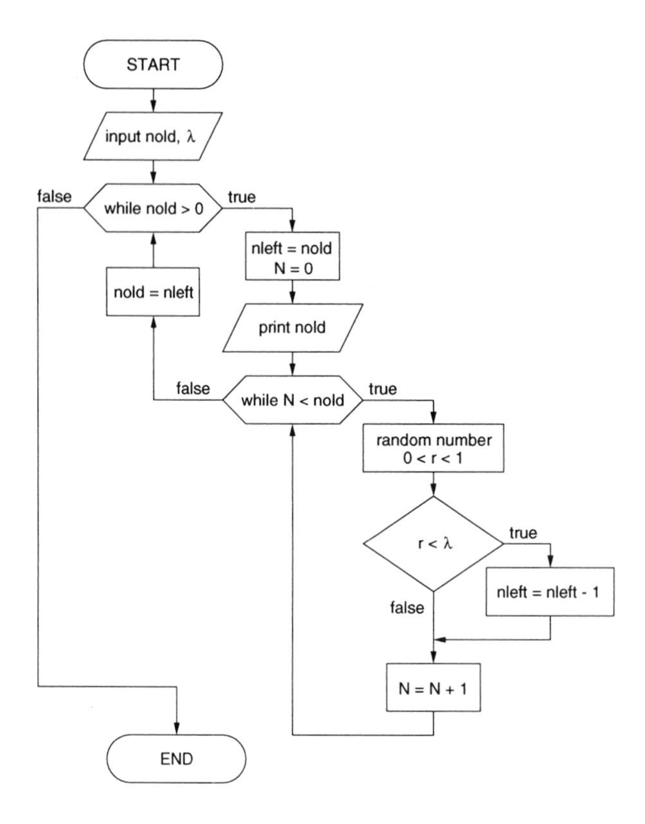

Fig. 7.2 A flowchart describing the program used in the simulation of radioactive decay.

```
Input N(0), λ
N = Nleft = N(0)                                              Initialize
Do till Nleft = 0                                  Do until all nuclei gone
  Do for N                      Go through all nuclei for one time period
    0 ≤ r ≤ 1                                       Generate random number
    if r ≤ λ,  Nleft = Nleft -1                      THIS IS THE ALGORITHM
  Endo
  print N, λ*(N-Nleft)                  Print number left and decay rate
  N = Nleft
Endo
```

As an alternative mode of understanding, in Fig.7.2 we give a flowchart of the decay simulation program. Many scientists find that crude flowcharts and pseudocodes are helpful in writing programs (yet, detailed ones may not work as well, in addition to being very time consuming).

The decay constant λ sets the time scale for the simulation, that is, the scale for Δt. This means that the decay rate is λ times the number of decays in each cycle. A good value to start with is $\lambda \simeq 0.3$.

7.6 IMPLEMENTATION: RADIOACTIVE DECAY, DECAY.F (.C)

Write a program to simulate radioactive decay. You should obtain results like those in Fig. 7.1.

7.7 ASSESSMENT: DECAY VISUALIZATION

1. Try several values of N increasing from a small one like 10, which should show large statistical fluctuations, to a large one like 10^5, which should exhibit exponential decay for a long time.
2. Plot the logarithm of the number left $\ln N$ and the logarithm of the decay rate $\ln(\lambda \Delta N)$ versus time (λ^{-1} times the generation number).
3. Check that for large $N(0)$ you obtain exponential decay, and that for small $N(0)$ you get a stochastic process.
4. Verify that the slopes of your curves (which should be proportional to λ) are *independent* of the value used for $N(0)$.
5. Verify that, within statistical errors, $\ln N$ and $\ln(\lambda \Delta N)$ have the same time dependence.
6. Verify that the time scale is set by the value of the decay rate λ.

7.8 PROBLEM: MEASURING BY STONE THROWING

Imagine yourself as a farmer walking to your furthermost field to add chemicals to a pond having an algae explosion. You get there, only to read the instructions and discover that you need to know the area of the pond to get the correct concentration. Your **problem** is to measure the area of this irregularly shaped pond with just the materials at hand [G&T 96].

7.9 THEORY: INTEGRATION BY REJECTION

It is hard to believe that Monte Carlo techniques could be used to evaluate integrals. After all, we do not want to gamble on their values! While it is true that other methods are preferable for single and double integrals, when the number of integrations required gets large, Monte Carlo techniques are best!

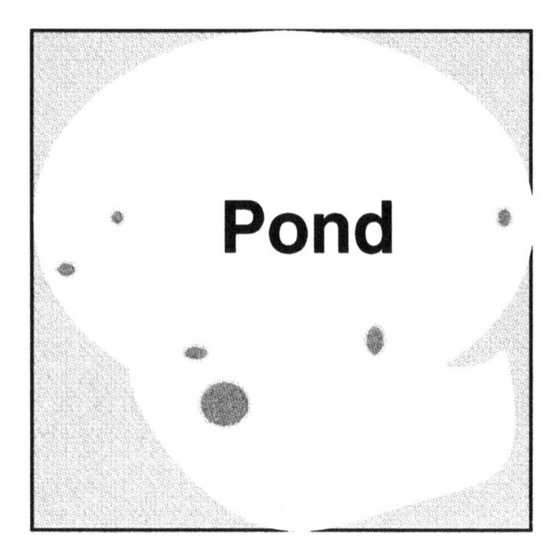

Fig. 7.3 Throwing stones in a pond as a technique for measuring its area. There is a Web tutorial of this where you can see the actual "splashes" (the dark spots) used in an integration.

For our pond problem, we will use the *sampling* technique illustrated in Fig. 7.3:

1. Walk off a box that completely encloses the pond and remove any pebbles within the box.

2. Measure the lengths of the sides in natural units like *feet*. This tells you the area of the enclosing box A_{box}.

3. Grab a bunch of pebbles and throw them up in the air in random directions.

4. Count the number of splashes in the pond N_{pond} and the number of pebbles lying on the ground within your box N_{box}.

5. Assuming that you threw the pebbles uniformly and randomly, the number of pebbles falling into the pond should be proportional to the area of the pond A_{pond}. You determine that area from the simple ratio

$$\frac{N_{\text{pond}}}{N_{\text{pond}} + N_{\text{box}}} = \frac{A_{\text{pond}}}{A_{\text{box}}} \tag{7.8}$$

$$\Rightarrow \quad A_{\text{pond}} = \frac{N_{\text{pond}}}{N_{\text{pond}} + N_{\text{box}}} A_{\text{box}}. \tag{7.9}$$

7.10 IMPLEMENTATION: STONE THROWING, POND.F (.C)

Use sampling, as illustrated in Fig. 7.3, to perform a 2-D integration and thereby determine π:

1. Imagine a circular pond centered at the origin and enclosed in a square of side 2.

2. We know the analytic result

$$\oint dA = \pi. \tag{7.10}$$

3. Generate a sequence of random numbers $\{r_i\}$ in $[-1, 1]$.

4. For $i = 1$ to N, pick $(x_i, y_i) = (r_{2i-1}, r_{2i})$.

5. If $x_i^2 + y_i^2 < 1$, let $N_{\text{pond}} = N_{\text{pond}} + 1$, else let $N_{\text{box}} = N_{\text{box}} + 1$.

6. Use (7.9) to calculate the area and in this way π.

Try increasing N until you get π to three significant figures (we don't ask much; that's only slide-rule accuracy).

7.11 PROBLEM: HIGH-DIMENSIONAL INTEGRATION

Let's say that we want to calculate some properties of a small atom such as magnesium with 12 electrons. To do that, we need to integrate some function over the three coordinates of each electron. This amounts to a $3 \times 12 = 36$-dimensional integral. If we use 64 points for each integration, this requires some $64^{36} \simeq 10^{65}$ evaluations of the integrand. If the computer were fast and could evaluate the integrand a million times per second, this would take some 10^{59} seconds, which is significantly longer than the age of the universe ($\sim 10^{17}$ seconds).

Your **problem** is to find a way to perform multidimensional integrations so that you are still alive to savor the answers. Specifically, evaluate the 10-D integral

$$I = \int_0^1 dx_1 \int_0^1 dx_2 \cdots \int_0^1 dx_{10} \, (x_1 + x_2 + \cdots + x_{10})^2. \tag{7.11}$$

Check your numerical answer against the analytic one, $\frac{155}{6}$.

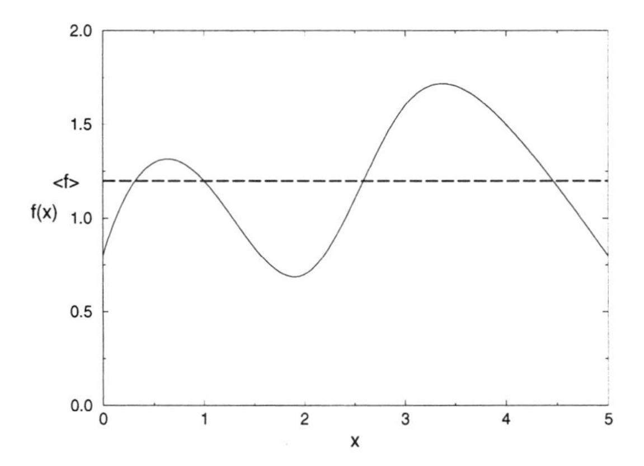

Fig. 7.4 The area under the curve $f(x)$ is the same as that under the line $\langle f \rangle$.

7.12 METHOD: INTEGRATION BY MEAN VALUE

The standard Monte Carlo technique for integration is based on the *mean-value theorem* (presumably familiar from elementary calculus):

$$I = \int_a^b dx f(x) = (b-a)\langle f \rangle. \tag{7.12}$$

As shown in Fig. 7.4, the theorem states the obvious if you think of integrals as areas: the value of the integral of some function $f(x)$ between a and b equals the length of the interval $(b-a)$ times the average value of the function over that interval $\langle f \rangle$. The integration algorithm uses Monte Carlo techniques to evaluate the mean in (7.12). With a sequence x_i of N uniform random numbers in $[a, b]$, we want to determine the *sample mean* by *sampling* the function $f(x)$ at these points:

$$\langle f \rangle \simeq \frac{1}{N} \sum_{i=1}^{N} f(x_i). \tag{7.13}$$

This gives us the very simple integration rule:

$$\boxed{\int_a^b dx f(x) \simeq (b-a) \frac{1}{N} \sum_{i=1}^{N} f(x_i) = (b-a)\langle f \rangle.} \tag{7.14}$$

Equation (7.14) looks much like our standard algorithm for integration (4.3), with the "points" x_i chosen randomly and with constant weights $w_i = (b-a)/N$. Because no attempt has been made to get the best answer for a given value of N, this is by no means an optimized way to evaluate integrals; but you will admit it is simple. If we let the number of samples of $f(x)$ approach infinity $N \to \infty$, or we keep the number of samples finite and take the average of infinitely many runs, the laws of statistics assure us that, barring roundoff errors, (7.14) approaches the correct answer.

For those readers who are somewhat familiar with statistics, we remind you that the uncertainty in the value obtained for the integral I after N samples of $f(x)$ is measured by the standard deviation σ_I. The standard deviation of the integrand f in the sampling is an intrinsic property of the function $f(x)$; that is, something we do not change by taking more samples. For normal distributions, the two are related by

$$\sigma_I \approx \frac{1}{\sqrt{N}} \sigma_f. \tag{7.15}$$

This means that for large N, the error in the value of the integral always decreases as $1/\sqrt{N}$.

7.12.1 Multidimensional Monte Carlo

It is easy to generalize mean-value integration to many dimensions by picking random points in a multidimensional space. For example

$$\int_a^b dx \int_c^d dy f(x,y) \simeq (b-a)(c-d)\frac{1}{N}\sum_i^N f(\mathbf{x}_i) = (b-a)(c-d)\langle f \rangle. \tag{7.16}$$

7.13 ASSESSMENT: ERROR IN MULTIDIMENSIONAL INTEGRATION

When we perform a multidimensional integration, the error in the Monte Carlo technique, being statistical, decreases as $1/\sqrt{N}$. This is valid even if the N points are distributed over D dimensions. In contrast, when we use these same N points to perform a D-dimensional integration as D one-dimensional integrals, we use N/D points for each integration. For fixed N, this means that the number of points used for each integration decreases as the number of dimensions D increases, and so the error in each integration increases with D. Furthermore, the total error will be approximately N times the error in each integral. If we put these trends together and look at a particular integration rule, we would find that at a value of $D \simeq 3-4$ the error in Monte Carlo integration is similar to that of conventional schemes. For larger values of D, the Monte Carlo method is more accurate!

7.14 IMPLEMENTATION: MONTE CARLO 10-D INTEGRATION, INT_10D.F (.C)

Use a built-in random-number generator to perform the 10-dimensional Monte Carlo integration in (7.11).

1. Conduct 16 trials and take the average as your answer.
2. Try sample sizes of $N = 2, 4, 8, \ldots, 8192$.
3. Plot the absolute value of the error versus $1/\sqrt{N}$ and try to identify linear behavior.

7.15 PROBLEM: MC INTEGRATION OF RAPIDLY VARYING FUNCTIONS$\odot$

It is common in many physical applications to integrate a function with an approximately Gaussian dependence on x. The rapid falloff of the integrand means that our Monte Carlo integration technique would require an incredibly large number of integrations points to obtain even modest accuracy. Your **problem** is to make Monte Carlo integration more efficient for rapidly varying integrands.

7.16 METHOD: VARIANCE REDUCTION$\odot$

If the function being integrated never differs much from its average value, then the standard Monte Carlo mean-value method (7.14) should work well with a not too ridiculously large number of points. Yet for a function with a large *variance* (i.e., one that is not "flat"), many of the random evaluations of the function may occur where the function makes a slight contribution to the integral; this is, basically, a waste of time. The method can be improved by mapping the function f into a function g that has a smaller variance over the interval. We indicate two methods here and refer you to the References (at the end of this book) for more details.

The first method is a *variance reduction* or *subtraction technique* in which we devise a flatter function on which to apply the Monte Carlo technique. Suppose we construct a function $g(x)$ with the following properties on $[a, b]$:

$$|f(x) - g(x)| \leq \epsilon, \tag{7.17}$$

$$\int_a^b dx\; g(x) = J. \tag{7.18}$$

We now evaluate the integral of $f(x) - g(x)$ and add the result to J to obtain the required integral

$$\int_a^b dx\ f(x) = \int_a^b dx\,\{f(x) - g(x)\} + J. \tag{7.19}$$

If we are clever enough to find a simple $g(x)$ that makes the variance of $f(x) - g(x)$ less than that of $f(x)$, and that we can integrate analytically, we obtain more accurate answers in less time.

7.17 METHOD: IMPORTANCE SAMPLING⊙

A second method to improve Monte Carlo integration is called *importance sampling* because it lets us sample the integrand in the most important regions. It derives from expressing the integral in the form

$$I = \int_a^b dx\ f(x) = \int_a^b dx\ w(x)\frac{f(x)}{w(x)}. \tag{7.20}$$

If we now use $w(x)$ as the *weighting function* or *probability distribution* for our random numbers, the integral can be approximated as

$$I = \left\langle \frac{f}{w} \right\rangle \simeq \frac{1}{N}\sum_{i=1}^{N} \frac{f(x_i)}{w(x_i)}. \tag{7.21}$$

The improvement with (7.21) is that a judicious choice of the weighting function $w(x)$ makes $f(x)/w(x)$ a rather constant function and consequently easier to integrate.

7.18 IMPLEMENTATION: NONUNIFORM RANDOMNESS⊙

In order for $w(x)$ to be the weighting function for random numbers over $[a, b]$, we want it with the properties

$$\int_a^b dx\ w(x) = 1, \quad [w(x) > 0], \tag{7.22}$$

$$d\mathcal{P}(x \to x + dx) = w(x)dx, \tag{7.23}$$

where $d\mathcal{P}$ is the probability of obtaining an x in the range $x \to x + dx$. For the uniform distribution over $[a, b]$, $w(x) = 1/(b - a)$.

7.18.1 Inverse Transform/Change of Variable Method

Let us consider a change of variables that take our original integral I (7.20) to the form

$$I = \int_a^b dx\, f(x) = \int_0^1 dW\, \frac{f[x(W)]}{w[x(W)]}. \tag{7.24}$$

Our aim is to make this transformation such that there are equal contributions from all parts of the range in W; that is, we want to use a uniform sequence of random numbers for W. To determine the new variable, we start with $u(r)$, the uniform distribution over $[0, 1]$,

$$u(r) = \begin{cases} 1, & \text{for } 0 \le r \le 1, \\ 0, & \text{otherwise.} \end{cases} \tag{7.25}$$

We want to find a mapping $r \leftrightarrow x$ or probability function $w(x)$ for which probability is conserved:

$$w(x)dx = u(r)dr, \quad \Rightarrow w(x) = \left|\frac{dr}{dx}\right| u(r). \tag{7.26}$$

This means that even though x and r are related by some (possibly) complicated mapping, x is also random with the probability of x lying in $x \to x + dx$ equal to that of r lying in $r \to r + dr$.

To find the mapping between x and r (the tricky part), we change variables to $W(x)$ defined by the integral

$$W(x) = \int_{-\infty}^{x} dx'\; w(x'). \tag{7.27}$$

We recognize $W(x)$ as the (incomplete) integral of the probability density $u(r)$ up to some point x. It is in this way another type of distribution function, the integrated probability of finding a random number less than the value x. The function $W(x)$ is on that account called a *cumulative distribution function* and can also be thought of as the area to the left of $r = x$ on the plot of $u(r)$ versus r. It follows immediately from the definition (7.27) that $W(x)$ has the properties

$$W(-\infty) = 0; \qquad W(\infty) = 1. \tag{7.28}$$

$$\frac{dW(x)}{dx} = w(x); \qquad dW(x) = w(x)dx = u(r)dr. \tag{7.29}$$

Consequently, $W_i = \{r_i\}$ is a uniform sequence of random numbers and we invert (7.27) to obtain x values distributed with probability $w(x)$.

The crux of this technique is being able to invert (7.27) to obtain x:

$$x = W^{-1}(r). \tag{7.30}$$

Let us look at some analytic examples to get a feel for these steps (numerical inversion is possible and frequent in realistic cases).

7.18.2 Uniform Weight Function w

We start off with our old friend, the uniform distribution:

$$w(x) = \begin{cases} \frac{1}{b-a}, & \text{if } a \le x \le b, \\ 0, & \text{otherwise.} \end{cases} \tag{7.31}$$

By following the rules this then leads to

$$W(x) = \int_a^x dx' \, \frac{1}{b-a} = \frac{x-a}{b-a} \tag{7.32}$$

$$\Rightarrow \quad x = a + (b-a)W \tag{7.33}$$

$$\Rightarrow \quad W^{-1}(r) = a + (b-a)r, \tag{7.34}$$

where $W(x)$ is always taken as uniform. In this way we generate uniform random $r : [0,1]$ and uniform random $x = a + (b-a)r : [a,b]$.

7.18.3 Exponential Weight

We want to generate points with the exponential distribution:

$$w(x) = \begin{cases} \frac{1}{\lambda} e^{-x/\lambda}, & \text{for } x > 0, \\ 0, & \text{for } x < 0, \end{cases} \tag{7.35}$$

$$W(x) = \int_0^x dx' \, \frac{1}{\lambda} e^{-x'/\lambda} = 1 - e^{-x/\lambda}, \tag{7.36}$$

$$\Rightarrow \quad x = -\lambda \ln(1-W) \equiv -\lambda \ln(1-r). \tag{7.37}$$

In this way we generate uniform random $r : [0,1]$ and obtain $x = -\lambda \ln(1-r)$ distributed with an exponential probability distribution for $x > 0$.

Note that according to our prescription (7.20)–(7.21), we could use $w(x) = e^{-x/\lambda}/\lambda$ to remove the exponential-like behavior out of an integrand and place it into the weights. Because the resulting integrand will vary less, the error will decrease. So, for example

$$\int_0^\infty dx \, e^{-x/\lambda} f(x) \simeq \frac{1}{N} \sum_{i=1}^N f(x_i), \tag{7.38}$$

$$x_i = -\lambda \ln(1-r_i) r_N, \quad (0 \le x_i \le \infty), \tag{7.39}$$

where $f(x)$ can have any functional behavior, but the error will be decreased if it is exponential-like.

7.18.4 Gaussian (Normal) Distribution

We want to generate points with a normal distribution:

$$w(x') = \frac{1}{\sqrt{2\pi}\sigma} e^{-(x'-\overline{x})^2/2\sigma^2}. \tag{7.40}$$

This by itself is rather hard, but it is made easier by generating uniform distributions in angles and then using trigonometric relations to convert these to a Gaussian distribution. But before doing that, we keep things simple by considering the distribution

$$w(x) = \frac{1}{\sqrt{2\pi}} e^{-x^2/2} dx. \tag{7.41}$$

To obtain (7.40) with a mean $\overline{x}$ and standard deviation σ, we need to only scale

$$x' = \sigma x + \overline{x}. \tag{7.42}$$

We start by generalizing the statement of probability conservation for two different distributions (7.26) to two dimensions [Pres 94]:

$$p(x,y)dxdy = u(r_1,r_2)dr_1 dr_2 \quad \Rightarrow p(x,y) = u(r_1,r_2)\left|\frac{\partial(r_1,r_2)}{\partial(x,y)}\right|. \tag{7.43}$$

We recognize the term in vertical bars as the Jacobian determinant:

$$J = \left|\frac{\partial(r_1,r_2)}{\partial(x,y)}\right| \stackrel{\text{def}}{=} \frac{\partial r_1}{\partial x}\frac{\partial r_2}{\partial y} - \frac{\partial r_2}{\partial x}\frac{\partial r_1}{\partial y}. \tag{7.44}$$

To specialize to a Gaussian distribution, we consider $2\pi r$ as angles obtained from a uniform random distribution r, and x and y as Cartesian coordinates that will have a Gaussian distribution. The two are related by

$$x = \sqrt{-2\ln r_1}\cos 2\pi r_2, \quad y = \sqrt{-2\ln r_1}\sin 2\pi r_2. \tag{7.45}$$

The inversion of this mapping produces the Gaussian distribution

$$r_1 = e^{-(x^2+y^2)/2}, \tag{7.46}$$

$$r_2 = \frac{1}{2\pi}\tan^{-1}\frac{y}{x}, \tag{7.47}$$

$$J = -\frac{e^{-(x^2+y^2)/2}}{2\pi}. \tag{7.48}$$

The solution to our problem is at hand. We use (7.45) with r_1 and r_2 uniform random distributions, and x and y will then be Gaussian random distributions centered around $x = 0$.

7.18.5 Alternate Gaussian Distribution

The central limit theorem can be used to deduce a Gaussian distribution via a simple summation. The theorem states, under rather general conditions, that if $\{r_i\}$ is a sequence of mutually independent random numbers, then the sum

$$x_N = \sum_{i=1}^{N} r_i \tag{7.49}$$

is distributed normally. This means that the generated x values have the distribution

$$P_N(x) = \frac{1}{\sqrt{2\pi\sigma^2}} \exp\left[-\frac{(x-\mu)^2}{2\sigma^2}\right], \tag{7.50}$$

$$\text{where} \quad \mu = N\langle r\rangle, \quad \sigma^2 = N\left(\langle r^2\rangle - \langle r\rangle^2\right). \tag{7.51}$$

7.19 METHOD: VON NEUMANN REJECTION⊙

A simple and ingenious method for generating random points with a probability distribution $w(x)$ was deduced by von Neumann. This method is essentially the same as the rejection or sampling method used to guess the area of the pond, only now the pond is replaced by the weighting function $w(x)$ and the arbitrary box around the lake by the arbitrary constant W_0. Imagine a graph of $w(x)$ versus x as visualized in Fig. 7.5. Walk off your box by placing the line $W = W_0$ on the graph, with the only condition being $W_0 \geq w(x)$. We next "throw stones" at this graph and count only those that fall into the $w(x)$ pond. That is, we generate uniform distributions in x and $y \equiv W$ with the maximum y value equal to the width of the box W_0:

$$(x_i, W_i) = (r_{2i-1}, W_0 r_{2i}). \tag{7.52}$$

We then reject all x_i that do not fall into the pond:

$$\text{if } W_i < w(x_i), \qquad \text{accept}, \tag{7.53}$$

$$\text{if } W_i > w(x_i), \qquad \text{reject}. \tag{7.54}$$

The x_i values so accepted will have the weighting $w(x)$. As is clear from Fig. 7.5, the largest acceptance occurs where $w(x)$ is large, in this case for midrange x.

In Chapter 22, *Thermodynamic Simulations: The Ising Model,* we describe

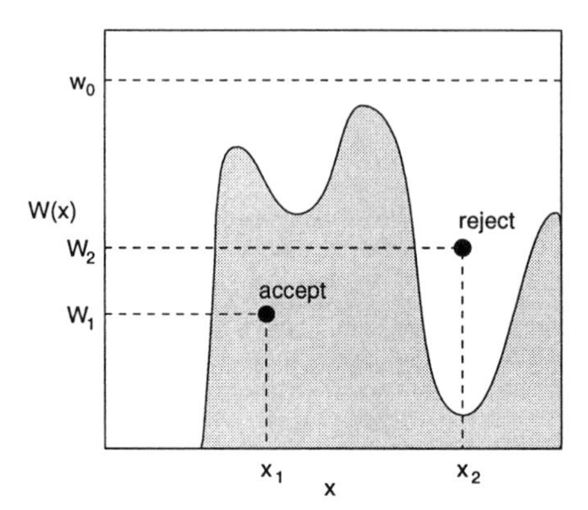

Fig. 7.5 The von Neumann rejection technique for generating random points with weight $w(x)$.

and apply a variation of the rejection technique known as the *Metropolis algorithm.* That algorithm has now become the cornerstone of computation thermodynamics.

7.20 ASSESSMENT⊙

Use the von Neumann rejection technique to generate a normal distribution of standard deviation 1, and compare to the preceding *Box–Muller* method.

8 Differentiation

8.1 PROBLEM 1: NUMERICAL LIMITS

A particle is moving through space, and you record its position as a function of time $x(t)$ in a table. Your **problem** is to determine its velocity $v(t) = dx/dt$ when all you have is this table of x versus t.

8.2 METHOD: NUMERIC

You probably did rather well in your first calculus course and feel competent at taking derivatives. However, you probably did not take derivatives of a table of numbers using the elementary definition:

$$\frac{df(x)}{dx} \stackrel{\text{def}}{=} \lim_{h \to 0} \frac{f(x+h) - f(x)}{h}. \tag{8.1}$$

In fact, even a computer runs into errors with this kind of limit because it is wrought with subtractive cancellation; the computer's finite word length causes the numerator to fluctuate between 0 and the machine precision ϵ_m as the denominator approaches zero.

8.2.1 Method: Forward Difference

The most direct method for numerical differentiation of a function starts by expanding it in a Taylor series. This series advances the function one small

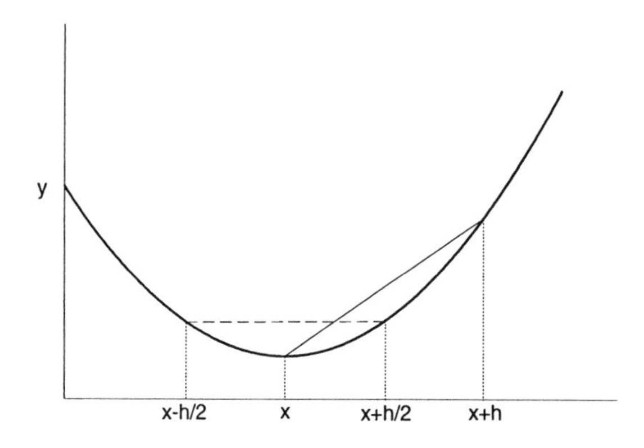

Fig. 8.1 Forward difference (solid line) and central-difference (dashed line) methods for numerical first derivative.

step forward:

$$f(x+h) = f(x) + hf'(x) + \tfrac{h^2}{2}f''(x) + \tfrac{h^3}{6}f^{(3)}(x) + \ldots, \tag{8.2}$$

where h is the *step size.* This is illustrated in Fig. 8.1. We obtain the *forward-difference* derivative algorithm by solving (8.2) for $f'(x)$:

$$f_c'(x) \simeq \frac{f(x+h) - f(x)}{h}, \tag{8.3}$$

$$\simeq f'(x) + \frac{h}{2}f''(x) + \cdots, \tag{8.4}$$

where the subscript c denotes a computed expression. You can think of this approximation as using two points to represent the function by a straight line in the interval from x to $x + h$.

The approximation (8.3) has an error proportional to h (unless the "ultra-user" looks down kindly upon you and makes f'' vanish). We can make the approximation error smaller and smaller by making h smaller and smaller. For too small an h, however, all precision will be lost through the subtractive cancellation on the LHS of (8.4), and the decreased approximation error becomes irrelevant.

For example, consider a case where $f(x)$ is a simple, analytic polynomial:

$$f(x) = a + bx^2. \tag{8.5}$$

The exact derivative is

$$f' = 2bx. \tag{8.6}$$

The computed derivative is

$$f'_c(x) \approx \frac{f(x+h) - f(x)}{h} = 2bx + bh. \tag{8.7}$$

This clearly becomes a good approximation only for small h ($h \ll 2x$).

8.2.2 Method: Central Difference

An improved approximation to the derivative starts with the basic definition (8.1). Rather than making a single step of h forward, we form a *central difference* by stepping forward by $h/2$ and backward by $h/2$:

$$\boxed{f'_c(x) \approx \frac{f(x+h/2) - f(x-h/2)}{h} \stackrel{\text{def}}{=} D_c f(x,h).} \tag{8.8}$$

Here D_c is the symbol for the central difference. The central-difference approximation is illustrated in Fig. 8.1.

When the Taylor series for $f(x \pm h/2)$ are substituted into (8.8), we obtain

$$f'_c(x) \simeq f'(x) + \tfrac{1}{24} h^2 f^{(3)}(x) + \cdots. \tag{8.9}$$

The important difference from (8.3) is that when $f(x - h/2)$ is subtracted from $f(x + h/2)$, all terms containing an odd power of h in the Taylor series cancel. Therefore, the central-difference algorithm becomes accurate to one order higher in h; that is, h^2. If the function is well behaved; that is, if $(f^{(3)}h^2)/24 \ll (f^{(2)}h)/2$, then you can expect the error with the central-difference method to be smaller than with the forward difference (8.4).

If we now return to our polynomial example (8.5), we find that for a parabola, the central difference gives the exact answer regardless of the size of h:

$$f'_c(x) \approx \frac{f(x+h/2) - f(x-h/2)}{h} = 2bx. \tag{8.10}$$

8.2.3 Method: Extrapolated Difference

Because the Taylor series provides an analytic expression for the error, we can be even more clever. While the central difference (8.8) makes the error term proportional to h vanish, we can also make the term proportional to h^2 vanish by algebraically *extrapolating* from relatively large h (and, because of this, small roundoff error) to $h \to 0$:

$$f'_c(x) \simeq \lim_{h \to 0} D_c f(x,h). \tag{8.11}$$

We introduce the additional information by forming the central difference with step size $h/2$:

$$D_c f(x, h/2) \stackrel{\text{def}}{=} \frac{f(x + h/4) - f(x - h/4)}{h/2} \tag{8.12}$$

$$\approx f'(x) + \frac{h^2 f^{(3)}(x)}{96} + \cdots. \tag{8.13}$$

We now eliminate the quadratic error term as well as the linear error term in (8.9) with the combination

$$f'_c(x) \approx \frac{4D_c f(x, h/2) - D_c f(x, h)}{3} \tag{8.14}$$

$$\approx f'(x) - \frac{h^4 f^{(3)}(x)}{1920 \times 4 \times 16} + \cdots. \tag{8.15}$$

If $h = 0.4$, there is only one place of roundoff error, and if $f^{(3)} = O(1)$, the truncation error in (8.15) is now the same size as machine precision ϵ_m, which is the best you can hope for.

A good way of computing (8.14) is to group the terms as

$$f'_c(x) = \tfrac{1}{3h} \left\{8\left[f(x + \tfrac{h}{4}) - f(x - \tfrac{h}{4})\right] - \left[f(x + \tfrac{h}{2}) - f(x - \tfrac{h}{2})\right]\right\}. \tag{8.16}$$

The advantage to (8.16) is that it reduces the loss of precision that occurs when large and small numbers are added together, only to be subtracted from other large numbers; it is better to first subtract the large numbers from each other and then add the difference to the small numbers.

When working with these and similar higher-order methods, it is important to remember that while they may work very well for well-behaved functions, they may fail badly for computed or measured functions containing noise. In these difficult cases it may be better to first fit the data with some analytic function using the techniques of Chapter 5, *Data Fitting*, and then differentiate the fit.

But regardless of the algorithm, you may remember that evaluating the derivative of $f(x)$ at x requires you to know the values of f surrounding x. We shall use this same idea in Chapter 9, *Differential Equations and Oscillations*.

8.3 ASSESSMENT: ERROR ANALYSIS

The approximation errors in numerical differentiation decrease with decreasing step size h while roundoff errors increase with a smaller step size (you have to take more steps and do more calculations). We know from our discussion in Chapter 3, *Errors and Uncertainties in Computations*, that the best approximation occurs for an h that makes the total error $\epsilon_{\text{approx}} + \epsilon_{\text{ro}}$ a minimum.

As a rough guide, this occurs when $\epsilon_{\rm ro} \approx \epsilon_{\rm approx}$.

Because differentiation subtracts two numbers very close in value, the limit of roundoff error is essentially machine precision:

$$f' \approx \frac{f(x+h)-f(x)}{h} \approx \frac{\epsilon_m}{h}, \tag{8.17}$$

$$\Rightarrow \quad \epsilon_{\rm ro} \approx \frac{\epsilon_m}{h}. \tag{8.18}$$

The approximation error with the forward-difference algorithm (8.4) is an $\mathcal{O}(h)$ term, while that with the central-difference algorithm (8.9) is an $\mathcal{O}(h^2)$ term;

$$\epsilon^{\rm fd}_{\rm approx} \approx \frac{f^{(2)}h}{2}, \tag{8.19}$$

$$\epsilon^{\rm cd}_{\rm approx} \approx \frac{f^{(3)}h^2}{24}. \tag{8.20}$$

The h value for which roundoff and approximation errors are equal is therefore

$$\frac{\epsilon_m}{h} \approx \epsilon^{\rm fd}_{\rm approx} = \frac{f^{(2)}h}{2}, \tag{8.21}$$

$$\frac{\epsilon_m}{h} \approx \epsilon^{\rm cd}_{\rm approx} = \frac{f^{(3)}h^2}{24}, \tag{8.22}$$

$$\Rightarrow \quad h^2_{\rm fd} = \frac{2\epsilon_m}{f^{(2)}}, \qquad h^3_{\rm cd} = \frac{24\epsilon_m}{f^{(3)}}. \tag{8.23}$$

We take $f' \approx f^{(2)} \approx f^{(3)}$ (which is crude, although not really so crude for e^x or $\cos x$), and assume a single precision calculation $\epsilon_m \approx 10^{-7}$. In this case we obtain

$$h_{\rm fd} \approx 0.0005, \tag{8.24}$$

$$h_{\rm cd} \approx 0.01. \tag{8.25}$$

This may seem backward because the better algorithm leads to a larger h value. It is not. The ability to use a large h means that the error in the central-difference method is some 20 times smaller than the error in the forward-difference method. For the forward-difference algorithm, approximately half of the machine precision is lost.

8.4 IMPLEMENTATION: DIFFERENTIATION, DIFF.F (.C)

8.5 ASSESSMENT: ERROR ANALYSIS, NUMERICAL

1. Differentiate the functions $\cos x$ and e^x at $x = 0.1$, 1., and 100 using single-precision forward-, central-, and extrapolated-difference algorithms.

   (a) Print out the derivative and its relative error $\mathcal{E}$ as a function of h. Reduce the step size h until it equals machine precision $h \approx \epsilon_m$.

   (b) Plot $\log_{10} |\mathcal{E}|$ versus $\log_{10} h$ and check whether the number of decimal places obtained agrees with the estimates in the text.

   (c) See if you can identify truncation error at large h and roundoff error at small h in your plot. Do the slopes agree with our predictions?

8.6 PROBLEM 2: SECOND DERIVATIVES

Let's say that you have measured the position versus time $x(t)$ for a particle. Your **problem** is to determine the force on the particle.

8.7 THEORY: NEWTON II

Newton's second law tells us that the force and acceleration are linearly related:

$$F = ma, \tag{8.26}$$

where F is the force, m is the particle's mass, and a is the acceleration. So if we can determine the acceleration $a(t) = d^2x/dt^2$ from the $x(t)$ table, we can determine the force.

8.8 METHOD: NUMERICAL SECOND DERIVATIVES

The concern about errors we expressed for first derivatives is even more of a concern for second derivatives because the additional subtractions lead to more cancellations. Let's go back to the central-difference method:

$$f'(x) \simeq \frac{f(x+h/2) - f(x-h/2)}{h}. \tag{8.27}$$

This algorithm gives the derivative at x by moving forward and backward from x by $h/2$. The second derivative $f^{(2)}(x)$ is taken to be the central difference of the first derivative:

$$f^{(2)}(x) \simeq \frac{f'(x+h/2) - f'(x-h/2)}{h}, \tag{8.28}$$

$$f^{(2)}(x) \simeq \frac{[f(x+h) - f(x)] - [f(x) - f(x-h)]}{h^2}. \tag{8.29}$$

As was true for first derivatives, by evaluating a function in the region surrounding x, it is possible to determine the second derivative at x. But some care is required to preserve the grouping in (8.29) and not convert it to the neater form:

$$f^{(2)}(x) \simeq \frac{f(x+h) + f(x-h) - 2f(x)}{h^2}. \tag{8.30}$$

This latter form increases subtractive cancellation because the computer first stores the "large" number $f(x+h) + f(x-h)$ [well, if f does not change sign, then the sum is larger than either $f(x+h)$ or $f(x-h)$] only to subtract another large number $2f(x)$ from it.

8.9 ASSESSMENT: NUMERICAL SECOND DERIVATIVES

Write a program to calculate the second derivative of $\cos x$ using the central-difference algorithm (8.29). Test it over four cycles. Start with $h \approx \pi/10$ and keep reducing h until you reach machine precision.

9

Differential Equations and Oscillations

9.1 PROBLEM: A FORCED NONLINEAR OSCILLATOR

In Fig. 9.1 there is a mass m attached to a spring and pushed by an external force. The properties of the spring are completely general; it need not be elastic and it need not be linear (harmonic). Your **problem** is to solve for the motion of the mass as a function of time. You may assume the motion is constrained to one dimension with the spring keeping the mass near the origin.

Part of the fascination of computational physics is that we can solve arbitrary problems of this sort rather easily. Furthermore, while most traditional treatments are restricted to small oscillations and *linear*, or nearly linear, oscillations, we will break those bonds and go exploring.[1]

9.2 THEORY, PHYSICS: NEWTON'S LAWS

This is a problem in classical mechanics. Newton's second law of motion provides us with the equation of motion

$$F_k(x) + F_{\text{ext}}(x,t) = m\frac{d^2x}{dt^2}, \tag{9.1}$$

[1]Some special properties of nonlinear equations are discussed in Chapters 13 and 14 and Part V, *Nonlinear PDEs*.

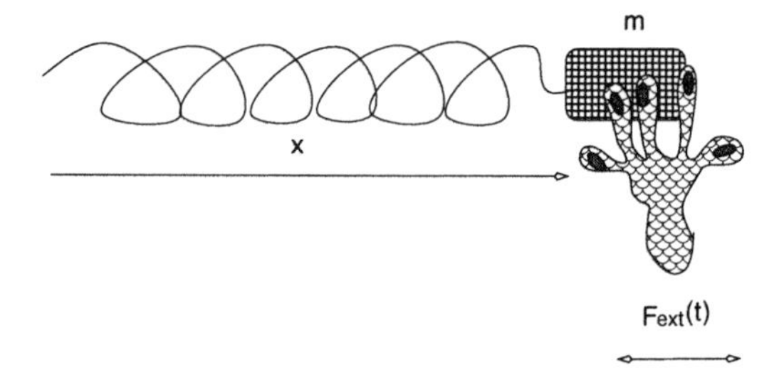

Fig. 9.1 A mass m attached to a spring with arbitrary properties. A hand subjects the mass to an external, time-dependent driving force $F_{\text{ext}}(t)$.

where $F_k(x)$ is the force exerted by the spring and $F_{ext}(x,t)$ is the external force. Equation (9.1) is the differential equation we must solve for arbitrary forces.[2]

9.3 MODEL: NONLINEAR OSCILLATOR

We model the general spring by assuming that the potential energy stored in it is proportional to some arbitrary power p of the displacement x from equilibrium:

$$V(x) = \frac{1}{p} k|x|^p. \tag{9.2}$$

Here the absolute value signs guarantees that the resulting force is toward $x = 0$.[3] The corresponding force felt by the particle due to the spring is the negative derivative of the potential:

$$F_k(x) = -\frac{dV(x)}{dx} = -k|x|^{p-1}\frac{x}{|x|} = \begin{cases} -kx^{p-1}, & \text{for } x > 0, \\ +kx^{p-1}, & \text{for } x < 0. \end{cases} \tag{9.3}$$

We display the characteristics of this potential in Fig. 9.2. For $p = 1$, the mass effectively slides up and down the inclined planes to the right and left of the origin with a discontinuous "bump" at the bottom (a bouncing ball);

[2]The differential nature of many physical laws, such as Newton's, may be a reflection of our use of continuous variables such as position and probability to describe nature. The use of differential equations may be a reflection of the traditional training in science. Modern advances in computational and theoretical physics may be changing all that and making integral equations as friendly as differential ones (see Chapter 17, *Quantum Scattering via Integral Equations*, and Chapter 23, *Functional Integration on Quantum Paths*).
[3]Another useful potential is the harmonic oscillator with an anharmonic correction, $V(x) = \frac{1}{2}kx^2(1 - \frac{2}{3}\alpha x)$. As α is made larger and larger, this potential becomes less and less like a harmonic oscillator. This is discussed in Chapter 11, *Anharmonic Oscillations*.

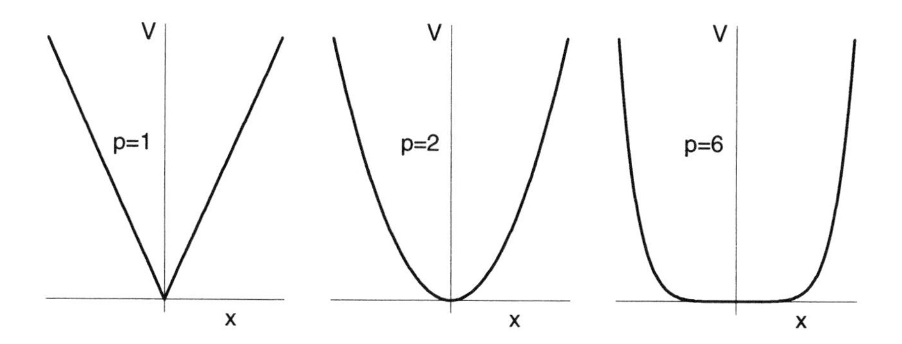

Fig. 9.2 The shape of the potential energy function $V(x) \propto |x|^p$ for different p values.

for $p = 2$, we have the familiar harmonic oscillator; for $p = 3$, we get an anharmonic oscillator:

$$V(x) = \begin{cases} kx, \\ \frac{1}{2}kx^2, \\ \frac{1}{3}kx^3, \end{cases} \quad F(x) = \begin{cases} -k\frac{x}{|x|}, & \text{for } p = 1, \\ -kx, & \text{for } p = 2, \\ -kx^2, & \text{for } p = 3. \end{cases} \tag{9.4}$$

We also notice in Fig. 9.2 that for $p \geq 6$, the mass is effectively in a box and moves almost freely until it hits the wall at $x \approx \pm 1$.

Application of Newton's law (9.1) to the combination of this potential with the external driving force produces the second-order differential equation we need to solve:

$$F_{\text{ext}}(x,t) - k|x|^{p-1}\frac{x}{|x|} = m\frac{d^2x}{dt^2}. \tag{9.5}$$

9.4 THEORY, MATHEMATICS: TYPES OF DIFFERENTIAL EQUATIONS

(The background material in this section is presented to avoid confusion over semantics. The well-versed student may want to skim or skip it.)

9.4.1 Order

A general form for a *first order* differential equation is

$$\frac{dy}{dt} = f(t,y), \tag{9.6}$$

where the "order" refers to the degree of the derivative on the LHS. The derivative or force function $f(t,y)$ on the RHS, is arbitrary. For example,

$f(t, y)$ may be a nasty function of y and t:

$$\frac{dy}{dt} = -3t^2 y + t^9 + y^7, \tag{9.7}$$

yet this is still first order in the derivative.

A general form for a *second order* differential equation is

$$\frac{d^2y}{dt^2} = f\left(t, \frac{dy}{dt}, y\right). \tag{9.8}$$

The derivative function f on the RHS is arbitrary and may involve any power of the first derivative as well. For example

$$\frac{d^2y}{dt^2} = -3t^2 \left(\frac{dy}{dt}\right)^4 + t^9 y(t), \tag{9.9}$$

is a second-order differential equation. Newton's law (9.1) is also a second-order differential equation.

In the differential equations (9.6) and (9.8), the time t is the *independent* variable and the position y is the *dependent* variable. This means that we are free to vary the time at which we want a solution, but not the value of y at that time. (In other applications, the position may be the independent variable.)

9.4.2 Ordinary and Partial

Differential equations such as (9.1) and (9.6) are *ordinary* differential equations because they contain only *one* independent variable, in this case x. In contrast, equations such as wave equations contain several independent variables, and this makes them *partial differential equations.* An example is the Schrödinger equation

$$i\frac{\partial\psi(\boldsymbol{x},t)}{\partial t} = -\frac{1}{2m}\left[\frac{\partial^2\psi(\boldsymbol{x},t)}{\partial x^2} + \frac{\partial^2\psi(\boldsymbol{x},t)}{\partial y^2} + \frac{\partial^2\psi(\boldsymbol{x},t)}{\partial z^2}\right] + V(\boldsymbol{x})\psi(\boldsymbol{x},t), \tag{9.10}$$

where the partial derivative symbol is used to indicate that the dependent variable ψ depends simultaneously on several independent variables.

In the early parts of this book we limit ourselves to ordinary differential equations (ODEs). In Part IV we will examine a variety of partial differential equations (PDEs).

9.4.3 Linear and Nonlinear

Part of the liberation of computational science is that we are no longer limited to studying *linear equations.* A *linear* equation is one in which only the first power of y or dy/dt appears; a *nonlinear* equation may contain higher powers. For example

$$\frac{dy}{dt} = g^3(t)y(t), \quad \text{is linear,} \tag{9.11}$$

$$\frac{dy}{dt} = \lambda y(t) - \lambda^2 y^2(t), \quad \text{is nonlinear.} \tag{9.12}$$

An important property of linear equations is the *law of linear superposition,* which lets us add solutions together to form new ones. Explicitly, if $A(t)$ and $B(t)$ are solutions of the linear equation (9.11), then

$$y(t) = \alpha A(t) + \beta B(t) \tag{9.13}$$

is also a solution for arbitrary values of the constants α and β. In contrast, let us assume that we were clever enough to solve the nonlinear equation (9.12) for the solution

$$y(t) = \frac{a}{1 + be^{-\lambda t}}, \tag{9.14}$$

(which you can verify by substitution). If we now add together two such solutions

$$y_1(t) = \frac{a}{1 + be^{-\lambda t}} + \frac{a'}{1 + b'e^{-\lambda t}}, \tag{9.15}$$

we do not obtain a $y_1(t)$, which is a solution of (9.12) (which you can verify by substitution).

9.4.4 Initial and Boundary Conditions

It is mathematical fact that a solution of a first-order differential equation always contains one arbitrary constant, that a solution of a second-order differential equation contains two constants, and so forth. For any specific problem, these constants are fixed by the *initial conditions.* For a first-order equation, the initial condition is usually the position $y(t)$ at some time, while for a second-order equation, the conditions are usually position and velocity. Regardless of how powerful a computer you use, the mathematical fact still remains and you must know the initial conditions in order to solve the problem.

In addition to initial conditions, it is possible to further restrict solutions of differential equations. One such way is by *boundary values* that constrain the solution to have fixed value(s) in a region of space. Problems of this sort are more demanding, may require a *search* to find solutions (the *eigenvalue problem*), and may not always have solutions. In Chapter 10, *Quantum Eigenvalues; Zero-Finding and Matching*, we discuss how to extend the techniques of the present chapter to boundary-value problems.

9.5 THEORY, MATH, AND PHYSICS: THE DYNAMICAL FORM FOR ODES

It is useful in both classical dynamics and numerical methods to write differential equations in the standard form [Schk 94, Tab 89, Pres 94]:

$$\frac{d\mathbf{y}}{dt}(t) = \mathbf{f}(t, \mathbf{y}). \tag{9.16}$$

Here $\mathbf{y}$ and $\mathbf{f}$ are N-dimensional vectors:

$$\mathbf{y} = \begin{pmatrix} y^{(1)}(t) \\ y^{(2)}(t) \\ \vdots \\ y^{(N)}(t) \end{pmatrix}, \quad \mathbf{f} = \begin{pmatrix} f^{(1)}[t, \mathbf{y}] \\ f^{(2)}[t, \mathbf{y}] \\ \vdots \\ f^{(N)}[t, \mathbf{y}] \end{pmatrix}, \tag{9.17}$$

where each $f^{(i)}$ may depend on all the $y^{(i)}$ values and time, but not on the derivatives $dy^{(i)}/dt$. That being so, when we solve (9.16), we are actually solving the N simultaneous first-order ODEs:

$$\begin{aligned} \frac{dy^{(1)}}{dt}(t) &= f^{(1)}[t, \mathbf{y}], \\ \frac{dy^{(2)}}{dt}(t) &= f^{(2)}[t, \mathbf{y}], \\ &\vdots \\ \frac{dy^{(N)}}{dt}(t) &= f^{(N)}[t, \mathbf{y}]. \end{aligned} \tag{9.18}$$

9.5.1 Dynamical Form for Second-Order Equation

We wish to take a second-order differential equation, such as Newton's law

$$\frac{d^2x}{dt^2} = F\left(t, \frac{dx}{dt}, x\right), \tag{9.19}$$

and write it in the standard dynamical form (9.16) with no derivatives on the RHS and only first derivatives on the LHS. We do that by first defining the position x as the dependent variable:

$$y^{(1)}(t) \stackrel{\text{def}}{=} x(t). \tag{9.20}$$

The trick is to now define the velocity dx/dt as a new, dependent variable:

$$y^{(2)}(t) \stackrel{\text{def}}{=} \frac{dx}{dt} \equiv \frac{dy^{(1)}}{dt}. \tag{9.21}$$

The second-order ODE (9.19) can now be written as two simultaneous first-order ODEs,

$$\frac{dy^{(1)}}{dt}(t) = y^{(2)}(t), \tag{9.22}$$

$$\frac{dy^{(2)}}{dt}(t) = F(t, y^{(1)}, y^{(2)}). \tag{9.23}$$

Here we have expressed the acceleration [the second derivative in (9.16)] as the first derivative of the velocity [the $y^{(2)}$ variable]. These equations are clearly in the vector form (9.18) with the derivative or force function having the two components

$$f^{(1)} = y^{(2)}(t), \tag{9.24}$$

$$f^{(2)} = F(t, y^{(1)}, y^{(2)}). \tag{9.25}$$

Breaking a second-order differential equation into two first-order ones is not just an arcane mathematical maneuver. In classical dynamics it occurs when transforming the single Newtonian equation of motion involving position and acceleration, (9.1), into two *Hamiltonian* equations involving position and momentum:

$$\frac{dp_i}{dt} = F_i, \tag{9.26}$$

$$m\frac{dy_i}{dt} = p_i. \tag{9.27}$$

9.6 IMPLEMENTATION: DYNAMICAL FORM FOR OSCILLATOR

By applying these definitions to our spring problem (9.5), we obtain the coupled first-order equations:

$$\frac{dy^{(1)}}{dt}(t) = y^{(2)}(t), \tag{9.28}$$

$$\frac{dy^{(2)}}{dt}(t) = \frac{1}{m}\left[F_{\text{ext}}(x,t) - k|y^{(1)}(t)|^{p-1}\frac{y^{(1)}(t)}{|y^{(1)}(t)|}\right], \tag{9.29}$$

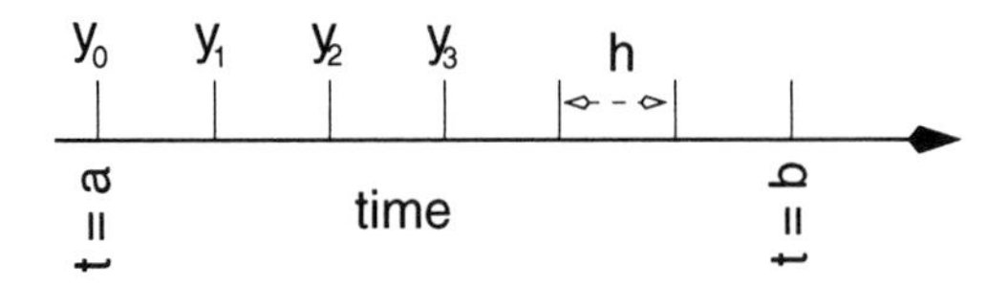

Fig. 9.3 The steps of length h taken in solving a differential equation. The solution starts at time $t = a$ and is integrated to $t = b$.

where $y^{(1)}(t)$ is the position of the mass at time t, and $y^{(2)}(t)$ is its velocity. In the standard form, the force or derivative function is

$$\boxed{\begin{aligned} f^{(1)}(t, \mathbf{y}) &= y^{(2)}(t), \\ f^{(2)}(t, \mathbf{y}) &= \frac{1}{m}\left[F_{\text{ext}}(x, t) - k|y^{(1)}|^{p-1}\frac{y^{(1)}}{|y^{(1)}|}\right]. \end{aligned}} \tag{9.30}$$

The *initial conditions* for the spring problem are its initial position and velocity:

$$y^{(1)}(0) = x_0, \quad y^{(2)}(0) = v_0. \tag{9.31}$$

9.7 NUMERICAL METHOD: DIFFERENTIAL EQUATION ALGORITHMS

The classic way to solve a differential equation is to start with the known initial value, $y_0 \equiv y(t = 0)$. Then you use the derivative function $f(t, y)$ to advance y_0 a small step $\Delta t = h$ forward in time to $y(t = h) \equiv y_1$. Then you take y_1 and advance it to y_2. This is continued until the solution at some large time $y(t = Nh) = y_N$ is obtained.[4] The process is illustrated in Fig. 9.3.

It is simplest if the time steps used throughout the integration remain constant in size, and that is what we shall do. Industrial-strength algorithms adapt the step size by making h larger in those regions where y varies slowly (this speeds up the integration and cuts down on roundoff error), and make h smaller in those regions where y varies rapidly (this provides better precision).

Error is always a concern when integrating differential equations. For one thing, we know that derivatives are prone to subtractive cancellations. More to the point, our stepping procedure for solving the differential equation is a continuous extrapolation of the initial conditions, each step building on a previous extrapolation. This is somewhat like a castle built on sand; in

[4]To avoid confusion, notice that $y^{(n)}$ is the nth component of the y vector, while y_n is the value of y after n time steps. (Yes, there is a price to pay for elegance in notation.)

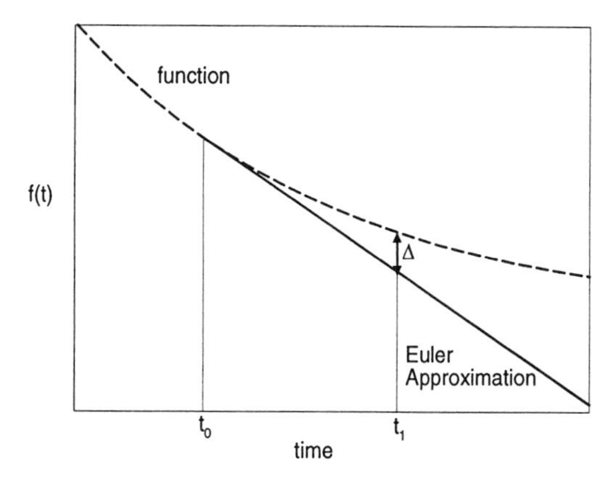

Fig. 9.4 Euler's algorithm for the forward integration of a differential equation by one time step. The linear extrapolation is seen to cause the error Δ.

contrast to interpolation, there are no tabulated values on which to anchor your solution.

9.8 METHOD (NUMERICAL): EULER'S ALGORITHM

A simple algorithm to integrate the differential equation (9.6) by one step is illustrated in Fig. 9.4. One simply substitutes the simplest algorithm for the derivative:

$$\frac{d\mathbf{y}}{dt}(t) = \mathbf{f}(t, \mathbf{y}), \tag{9.32}$$

$$\frac{\mathbf{y}(t_{n+1}) - \mathbf{y}(t_n)}{h} \simeq \mathbf{f}(t_n, \mathbf{y}), \tag{9.33}$$

$$\Rightarrow \quad \mathbf{y}_{n+1} \simeq \mathbf{y}_n + h\mathbf{f}(t_n, \mathbf{y}_n). \tag{9.34}$$

We know from our discussion of differentiation that the error in (9.34) is $\mathcal{O}(h^2)$; that is, the size of the next term in the Taylor series.

To indicate the simplicity of this algorithm, we apply it to our oscillator problem for one time step:

$$y_1^{(1)} = x_0 + v_0 h, \tag{9.35}$$

$$y_1^{(2)} = v_0 + h\frac{1}{m}\left[F_{\text{ext}}(t=0) + F_k(y=0)\right]. \tag{9.36}$$

Compare this to the projectile equations familiar from the first-year course in physics:

$$x = x_0 + v_0 h + \tfrac{1}{2} a h^2, \quad v = v_0 + ah. \tag{9.37}$$

We see that in (9.35) the acceleration does not contribute to the distance covered (no h^2 term), yet it does contribute to the velocity in (9.36).

The small values needed for h in order to make Euler's algorithm precise, increase roundoff error and may lead to instabilities.[5] Although we do not recommend Euler's algorithm for general use, it is commonly used to start some of the more sophisticated algorithms.

9.9 METHOD (NUMERICAL): SECOND-ORDER RUNGE–KUTTA

Although no one algorithm will work for all possible cases, the fourth-order Runge–Kutta method with adaptive step size has proved to be robust and capable of industrial-strength work. It is our recommended method. To understand this important tool we derive the simpler second-order method. The fourth-order Runge–Kutta is just more work and we will present it without derivation.

A Runge–Kutta algorithm for integrating a differential equation is based upon the formal integral of the differential equation:

$$\frac{dy}{dt}(t) = f(t, y) \quad \Rightarrow \quad y(t) = \int f(t, y) dt \tag{9.38}$$

$$\Rightarrow \quad y_{n+1} \quad = \quad y_n + \int_{t_n}^{t_{n+1}} f(t, y) dt. \tag{9.39}$$

The approximation enters by expanding $f(t \simeq t_n, y \simeq y_n)$ in a Taylor series about the *midpoint* of the integration interval:

$$f(t, y) \simeq f(t_{n+1/2}, y_{n+1/2}) + (t - t_{n+1/2}) \frac{df}{dt}(t_{n+1/2}) + \mathcal{O}(h^2). \tag{9.40}$$

When this is substituted into (9.39), the integral of $(t - t_{n+1/2})$ vanishes and we get a higher-order algorithm than Euler's, even though we use the same number of terms:

$$\int f(t, y) dt \quad \simeq \quad f(t_{n+1/2},\, y_{n+1/2}) h, \tag{9.41}$$

$$\Rightarrow \quad y_{n+1} \quad \simeq \quad y_n + h f(t_{n+1/2},\, y_{n+1/2}). \tag{9.42}$$

[5] Instability is often a problem when you integrate a $y(t)$ which decreases as the integration proceeds, analogous to upward recursion of spherical Bessel functions. In that case, and if you have a linear problem, you are best off integrating *inward* from large times to small times and then scaling the answer to agree with the initial conditions.

We see that the price for improved precision is having to evaluate the derivative at more than just the ends of the interval.[6]

The algorithm (9.40) cannot be applied immediately because it requires knowledge of $y_{n+1/2}$, which is not given by the initial conditions. Nonetheless, we can use Euler's algorithm to express $y_{n+1/2}$ in terms of the initial conditions:

$$y_{n+1/2} \simeq y_n + \frac{dy}{dt}\frac{h}{2} = y_n + \tfrac{1}{2}hf(t_n,\, y_n). \tag{9.43}$$

All together, and for simultaneous equations (the **bold** symbols), the second-order Runge–Kutta algorithm is

$$\mathbf{y}_{n+1} \simeq \mathbf{y}_n + \mathbf{k}_2, \tag{9.44}$$

$$\mathbf{k}_2 = h\mathbf{f}(t_n + \frac{h}{2},\, \mathbf{y}_n + \frac{\mathbf{k}_1}{2}), \quad \mathbf{k}_1 = h\mathbf{f}(t_n,\, \mathbf{y}_n). \tag{9.45}$$

We see that the known derivative function $\mathbf{f}$ is evaluated at the ends and midpoint of the interval, but that only the initial value of the unknown $\mathbf{y}$ is required. This means that the algorithm is self-starting.

As an example, we apply this algorithm to our spring problem:

$$\begin{aligned}
y_1^{(1)} &= y_0^{(1)} + hf^{(1)}(\frac{h}{2},\, y_0^{(1)} + k_1) \\
&\simeq x_0 + h[v_0 + \frac{h}{2}F_k(0)], \qquad \text{(Euler's rule for midpoint)} \\
y_1^{(2)} &= y_0^{(2)} + hf^{(2)}\left[\frac{h}{2},\, y_0 + \frac{h}{2}f(0,\, y_0)\right], \\
&\simeq v_0 + h\frac{1}{m}\left[F_{\text{ext}}(t = \frac{h}{2}) + F_k\left(y^{(1)}(0) + \frac{k_1}{2}\right)\right].
\end{aligned}$$

We see that the position now has an h^2 time dependence, which, at last, brings us up to the equation for free fall studied in first-year physics.

9.10 METHOD (NUMERICAL): FOURTH-ORDER RUNGE–KUTTA

The fourth-order Runge–Kutta method provides an excellent balance of power, precision, and programming simplicity. There are now four gradient (k) terms to provide a better approximation to $f(t, y)$ near the midpoint, and they can

[6]One can be even more clever and use the values of the derivatives in this one interval to estimate the function in the next interval. However, that would make integration of successive intervals dependent and would make adaptive step sizing difficult.

be determined with just four subroutine calls:

$$
\begin{aligned}
\mathbf{y}_{n+1} &= \mathbf{y}_n + \tfrac{1}{6}(\mathbf{k}_1 + 2\mathbf{k}_2 + 2\mathbf{k}_3 + \mathbf{k}_4), \\
\mathbf{k}_1 &= h\mathbf{f}(t_n,\ \mathbf{y}_n), \qquad \mathbf{k}_2 = h\mathbf{f}(t_n + \frac{h}{2},\ \mathbf{y}_n + \frac{\mathbf{k}_1}{2}), \\
\mathbf{k}_3 &= h\mathbf{f}(t_n + \frac{h}{2},\ \mathbf{y}_n + \frac{\mathbf{k}_2}{2}), \qquad \mathbf{k}_4 = h\mathbf{f}(t_n + h,\ \mathbf{y}_n + \mathbf{k}_3).
\end{aligned} \tag{9.46}
$$

9.11 IMPLEMENTATION: ODE SOLVER, RK4.F (.C)

1. Write a program to solve the equation of motion (9.5) or (9.29) using the fourth-order Runge–Kutta subroutine supplied on the diskette. [It is not hard to write your own based on (9.46), but the results will not be accurate unless you calculate every term perfectly.] Be sure to use double precision to help control subtractive cancellation.

2. Design your program to be general enough to solve other differential equations as well. In particular, make the derivative function $f(t, x)$ a subroutine.

9.12 ASSESSMENT: RK4 AND LINEAR OSCILLATIONS

In general, you may do a number of things to check that your differential equation solver is working well and that you have picked a reasonable value for the step size h. Start with a harmonic oscillator ($p = 2$) because in this case you know the analytic result:

$$x(t) = A\sin(\omega_0 t + \phi), \tag{9.47}$$

$$v(t) = \omega_0 A\cos(\omega_0 t + \phi), \tag{9.48}$$

$$\omega_0 = \frac{2\pi}{T} = \sqrt{\frac{k}{m}}. \tag{9.49}$$

1. Pick a value of k and m such that the period T in (9.49) is a nice number to work with (in other words, T =1).

2. Try out step sizes starting at $h \simeq T/5$ and make them smaller and smaller as needed. Your solution should look smooth and have a period that *never* changes even after many oscillations.

3. Plot your computed solution along with the analytic one. You may not be able to tell them apart.

4. Try different amplitudes A (that is, different initial conditions) and verify that a *harmonic* oscillator is *isochronous*; that is, its period does *not* change as the amplitude varies.

9.13 ASSESSMENT: RK4 AND NONLINEAR OSCILLATIONS

Test the Runge–Kutta method for anharmonic oscillations. Try powers in the range $p = 1$–11.

1. Observe your solutions and see if you need to decrease the step size h from the value used for the harmonic oscillator. As p gets larger, the potential changes more rapidly and so the forces and accelerations get larger. Consequently, a smaller step size may be needed to "keep up" with the changes. Likewise, the abrupt change in the potential for $p = 1$ may cause problems.

2. Check that the solution is periodic with a constant amplitude and period for a given initial condition and value of p. In particular, check that the maximum speed occurs at $x = 0$ and that the minimum speed occurs at maximum x.

3. Try different amplitudes A (that is, different initial conditions) and verify that for an anharmonic (*nonisochronous*) oscillator the period *does* indeed change as the amplitude varies.

4. Explain the difference in shape of the solution for different values of p.

9.14 EXPLORATION: ENERGY CONSERVATION

We have not explicitly built energy conservation into the solution of the differential equation. Nonetheless, unless you have included a frictional force, energy must be a constant (integral) of the motion for any value of p (it follows from Newton's second law). That being so, the constancy of energy is a demanding test of the accuracy of your solution.

1. Plot the potential energy, the kinetic energy, and the total energy for hundreds of periods. These are defined as

$$\mathrm{PE}(t) = V[x(t)], \quad \mathrm{KE}(t) = \tfrac{1}{2}mv^2(t), \quad E = \mathrm{KE}(t) + \mathrm{PE}(t). \tag{9.50}$$

   Notice the correlation between $\mathrm{PE}(t)$ and $\mathrm{KE}(t)$ arising from the constancy of E.

2. Check the long-term *stability* of your solution by plotting $\log[|E(t) - E(t=0)|/E(t=0)]$ for a large number of cycles. You may get 11 or more places of precision. If you do not, then you may need to decrease the value of h or look for bugs in your program.

3. Because a particle bound by a highly anharmonic oscillator is essentially "free" most of the time, its average kinetic energy (KE) should exceed its average potential energy (PE). This is actually a physical explanation of the Virial theorem:

$$\langle \mathrm{KE} \rangle = \frac{p}{2} \langle \mathrm{PE} \rangle. \tag{9.51}$$

   Verify that your solution satisfies the virial theorem.

10

Quantum Eigenvalues; Zero-Finding and Matching

10.1 PROBLEM: BINDING A QUANTUM PARTICLE

Your **problem** in this chapter is to determine whether the rules of quantum mechanics are applicable inside of a nucleus. More specifically, you are told that nuclei contain neutrons and protons ("nucleons") with mass

$$mc^2 \simeq 940\text{MeV}, \tag{10.1}$$

and that a nucleus has a size of about 2 fm.[1] Your explicit **problem** is to see if these experimental facts are compatible, first, with quantum mechanics and, second, with the observation that there is a typical spacing of several million electron volts (MeV) between the ground and excited states in nuclei.

We will solve this problem in coordinate space. We start with a semi-analytic treatment and then search with an ODE solver. In §16.3, we discuss how to solve the equivalent momentum-space eigenvalue problem as a matrix problem. In Chapter 30, *Confined Electronic Wave Packets*, we study the related (but more complicated) problem of the motion of a quantum wave packet confined to a potential well. Further discussion of the numerical bound-state problem is found in [Schd 87] and [Koon 86]. To broaden your programming experiences, you may want to try out the sample programs on the diskette and the Web.

[1] A fm, or fermi, equals 10^{-13} cm, and $\hbar c \simeq 197.32$ MeV fm.

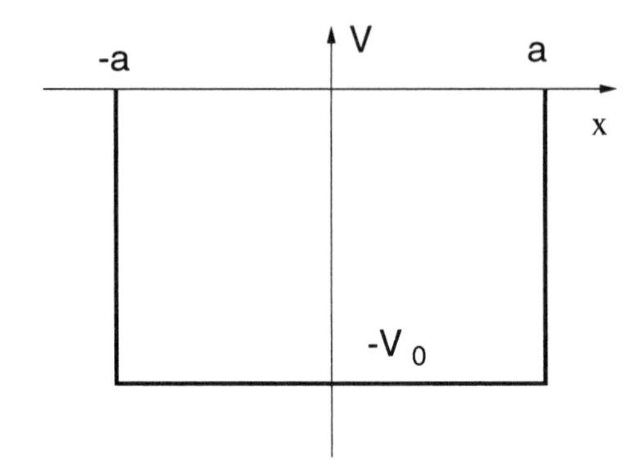

Fig. 10.1 A finite square well potential in one dimension. The depth is V_0 and the radius is a.

10.2 THEORY: QUANTUM WAVES

Quantum mechanics is the theory that describes phenomena at an atomic or subatomic scale of size (a nucleus is subatomic). It is a statistical theory in which the probability that a particle is located in the region of space from x to $x + dx$ is

$$\mathcal{P} = |\psi(x)|^2 dx, \tag{10.2}$$

where $\psi(x)$ is the *wave function.* When a particle is bound, it is in a *stationary state* with an $\exp(-iEt/\hbar)$ time dependence. In that case the wave function is determined by the time-independent form of the Schrödinger equation:[2]

$$\frac{-\hbar^2}{2m}\frac{d^2\psi(x)}{dx^2} + V(x)\psi(x) = E\psi(x), \tag{10.3}$$

where $V(x)$ is the potential experienced by the particle and E is its energy. This is the equation we must solve.

When a particle is bound, it is confined to some finite region of space. In quantum mechanics this means that the probability of finding the particle over all of space must be one,

$$\int_0^\infty dx|\psi(x)|^2 = 1. \tag{10.4}$$

So, in addition to solving (10.3), we require that $\psi(x)$ also satisfies (10.4). This extra condition turns the quantum bound-state problem into an *eigenvalue problem* in which solutions (*eigenvectors*) exist only for a few values of E (*eigenvalues*).

[2]The time-dependent equation requires the solution of a partial differential equation, as discussed in Chapter 30, *Confined Electronic Wave Packets.*

10.3 MODEL: PARTICLE IN A BOX

The numerical methods we will be developing are able to handle the most realistic situations. Yet to make a connection with the standard textbook case, and to permit some analytic checking, we will use a simple model in which the potential $V(x)$ in (10.3) is the finite square well pictured in Fig. 10.1:

$$V(x) = \begin{cases} -V_0 = -83\text{MeV}, & \text{for } |x| \leq a = 2 \text{ fm}, \\ 0, & \text{for } |x| > a = 2 \text{ fm}. \end{cases} \tag{10.5}$$

A depth of 83 MeV and radius of 2 fm are typical for nuclei, and therefore this problem will be solved with energies in millions of electron volts and lengths in fermis. The Schrödinger equation (10.3) is now

$$\frac{d^2\psi(x)}{dx^2} + \frac{2m}{\hbar^2}(E + V_0)\psi(x) = 0, \quad \text{for } |x| \leq a, \tag{10.6}$$

$$\frac{d^2\psi(x)}{dx^2} + \frac{2m}{\hbar^2}E\psi(x) = 0, \quad \text{for } |x| > a. \tag{10.7}$$

We write the constants in (10.6) with c^2 inserted in both the numerator and the denominator [L 96, Appendix A.1]:

$$\frac{2m}{\hbar^2} = \frac{2mc^2}{(\hbar c)^2} = \frac{2 \times 940\text{MeV}}{(197.32\text{MeV fm})^2} = 0.4829\text{MeV}^{-1}\text{fm}^{-2}. \tag{10.8}$$

10.4 SOLUTION: SEMIANALYTIC

For a general case we would solve (10.6) as an ordinary differential equation (ODE) eigenvalue problem. Before we do that in §10.6, we make connections with the standard textbook treatment of bound states in a square well, and introduce the *bisection algorithm*, which will be useful later.

The general form of the solution in the $|x| > a$ region (the "outer" wave function) depends on the sign of the energy E. If $E > 0$, the outer solution is a trigonometric function. Because a trigonometric function is not normalizable, it cannot meet the confinement requirements we have set on a bound-state wave function, and so there cannot be any $E > 0$ bound states.

If $E < 0$, the possible outer solutions are proportional to $\exp(\pm\beta x)$. Because a decaying exponential can be normalized, we use it to form[3] an accept-

[3]The form of the wave function given in (10.9) corresponds to choosing a positive-parity

able bound-state solution:

$$\psi(x) = \begin{cases} Ce^{\beta x}, & \text{for } -\infty < x < -a, \\ B\cos \alpha x, & \text{for } -a < x < a, \\ Ce^{-\beta x}, & \text{for } a < x < +\infty, \end{cases} \tag{10.9}$$

where

$$\beta = \sqrt{\frac{-2mE}{\hbar^2}} = \sqrt{-0.4829E}, \tag{10.10}$$

$$\alpha = \sqrt{\frac{2m(E+V_0)}{\hbar^2}} = \sqrt{0.4829(E+83)}, \tag{10.11}$$

and where energies are in MeV and are negative for bound states.

One way to solve our problem is to forget about a possible negative x solution and construct ψ to be symmetric about $x = 0$. We do not impose that symmetry externally, but rather solve for the wave function for both positive and negative x values, and then use the symmetry as a check.

In order for probability and current to be continuous at the edge of the well $x = \pm a$, the wave function and its first derivative must be continuous there. This demands that

$$B\cos \alpha a = Ce^{-\beta a}, \tag{10.12}$$

$$-\alpha B\sin \alpha a = -\beta Ce^{-\beta a}. \tag{10.13}$$

We remove the dependence on the normalization constants B and C by dividing these equations by each other:

$$\alpha a \tan \alpha a - \beta a = 0. \tag{10.14}$$

If we substitute for α and β, it becomes clear that (10.14) is a transcendental equation in the variable E:

$$\sqrt{2m(E+V_0)}\tan\sqrt{\frac{2m(E+V_0)}{\hbar^2}}a - \sqrt{-2mE} = 0. \tag{10.15}$$

The transcendental equation (10.15) has solutions only for certain values of E, the eigenvalues. In general, it can be solved only by numerical techniques. A classic way of finding solutions of (10.15) starts by converting it to an

wave function; that is, one that is symmetric about the origin. We make that choice because we are interested in the lowest energy states and they are usually the ones with the fewest wiggles.

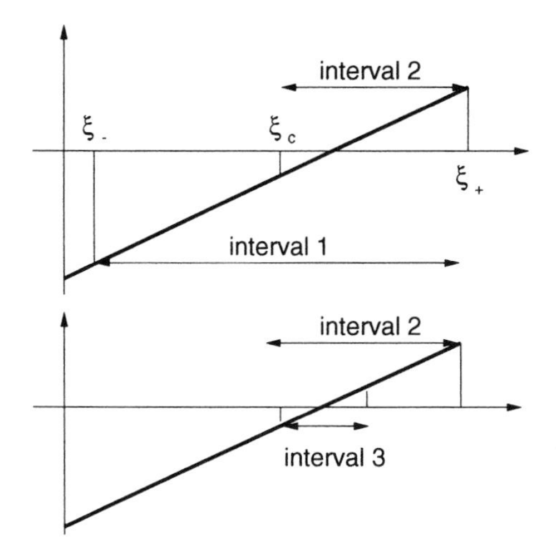

Fig. 10.2 The approximate behavior of the function $f(\xi)$ near its zero. The center point ξ_c is the approximate value of the root. The figure assumes that $f(\xi_+) > 0$ and that $f(\xi_-) < 0$, but only the relative sign matters. Observe how interval 3, resulting from the second iteration, is smaller than interval 2, resulting from the first iteration.

equation in dimensionless variables:

$$\xi \tan \xi - \eta = 0, \tag{10.16}$$

where

$$\xi = \alpha a, \qquad \eta = \beta a, \tag{10.17}$$

$$\xi^2 + \eta^2 = \frac{2mV_0 a^2}{\hbar^2} = 16.083. \tag{10.18}$$

We then look for a zero or root of the function $f(\xi)$:

$$f(\xi) = \xi \tan \xi - \eta = 0. \tag{10.19}$$

If $\xi \tan \xi$ and η are plotted (as in Fig. 10.2) as functions of ξ, the solutions of (10.19) are the intersections of the two curves. Alternatively, one can sit down with a calculator, evaluate $f(\xi = 0)$, and increase ξ in small steps until $f(\xi)$ is found to change signs. The locations of the sign changes are the zeros. In this way we found the approximate eigenvalues $\xi \simeq 1.25$, and $\xi \simeq 3.6$. You may check these with your calculator. To determine the eigenvalues more precisely, we use the bisection algorithm described next.

10.5 METHOD: FINDING ZERO VIA THE BISECTION ALGORITHM

The bisection algorithm, illustrated in Fig. 10.2, is a simple and reliable way to find a root of the equation $f(\xi) = 0$; it may also be the slowest. We assume that we have an interval, $\xi_- < \xi < \xi_+$, in which $f(\xi)$ changes sign; that is, for which

$$f(\xi_-)f(\xi_+) < 0. \tag{10.20}$$

The bisection algorithm keeps checking that the condition (10.20) is satisfied for ever closer values of ξ_- and ξ_+, and quits when the difference in ξ_- and ξ_+ is smaller than some tolerance ϵ. At that point we say that we have a solution. (The intermediate-value theorem of calculus guarantees that a solution must exist in the interval.) Explicitly, we define the center of the interval

$$\xi_c = \tfrac{1}{2}(\xi_- + \xi_+), \tag{10.21}$$

and use ξ_c as a trial root:

$$\text{if} \qquad f(\xi_c)f(\xi_+) < 0, \qquad \begin{cases} \text{then} & \xi_- = \xi_c, \\ \text{else} & \xi+ = \xi_c. \end{cases} \tag{10.22}$$

The process is continued, dividing the interval in half and analyzing which side the root is on until $|\xi_- - \xi_+| < \epsilon$.

10.6 METHOD: EIGENVALUES FROM AN ODE SOLVER

For more realistic potentials we would not be able to analytically solve the Schrödinger equation, but instead we would have to use an ODE solver to obtain a numerical wave function. In that case we would start with a guess for the energy. Then, as shown in Fig. 10.3, see how well the inner and outer wave functions match, and then guess a new energy. After two guesses for the energy we would use the bisection algorithm to predict an energy at which the wave functions match. The process continues until the difference in energy falls within some tolerance.

Generally, as discussed in Chapter 9, *Differential Equations and Oscillations*, we recommend the fourth-order Runge–Kutta method for solving ODEs. For the present eigenvalue problem we will use the *Numerov method* that is specialized for those ODEs not containing any first derivatives (the Schrödinger equation is such an equation). This algorithm provides one order more of accuracy than fourth-order Runge–Kutta, and does so with fewer computations (no partial steps). Nonetheless, it is not as general.

We start by rewriting the Schrödinger equation (10.6) as

$$\frac{d^2\psi}{dx^2} + k^2(x)\psi \quad = \quad 0, \tag{10.23}$$

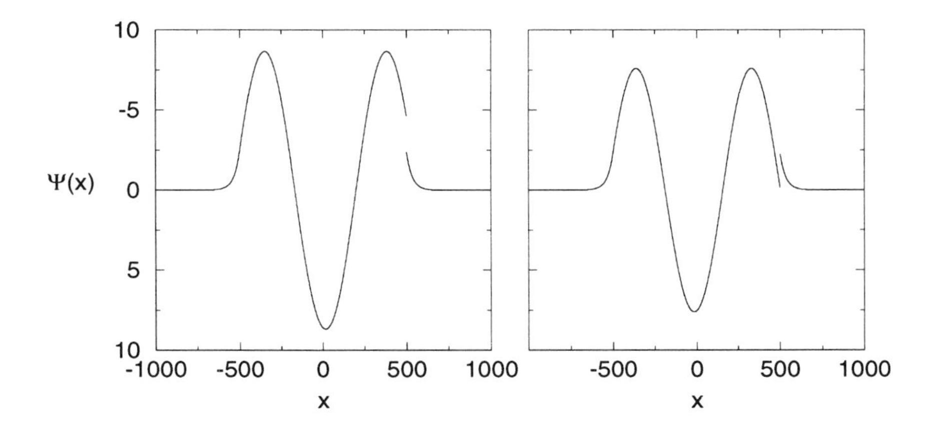

Fig. 10.3 (*Left*) A first guess at a wave function with an energy E that is 0.5% too low. We see that the left wave function does not vary rapidly enough to match the right one at $x = 500$. (*Right*) A second guess at a wave function with an energy E that is 0.5% too high. We see that, now, the left wave function varies too rapidly.

$$k^2(x) \stackrel{\text{def}}{=} \frac{2m}{\hbar^2} \begin{cases} E + V_0, & \text{for } |x| < a, \\ E, & \text{for } |x| > a. \end{cases} \tag{10.24}$$

We remind the student that $E < 0$ for bound states, and that in our units, $2m/\hbar^2$ =0.4829. Observe that the present functional form for $k^2(x)$ (10.24) is for a square well, but it can be generalized for a realistic potential. As usual, the algorithm integrates (10.23) for a single interval $x \to x + h$, in which the potential is assumed constant. The process is repeated until a wave function throughout all of space is found.

The trick in the Numerov method is to get an extra order of precision in the second derivative. We start with the Taylor expansion of the wave function,

$$\psi(x+h) = \psi(x)+h\psi^{(1)}(x)+\frac{h^2}{2}\psi^{(2)}(x)+\frac{h^3}{3!}\psi^{(3)}(x)+\frac{h^4}{4!}\psi^{(4)}(x)+\cdots, \tag{10.25}$$

where $\psi^{(n)}$ is used as shorthand for the nth derivative $d^n\psi/dx^n$. Because the corresponding Taylor expansion of $\psi(x - h)$ has odd powers of h appearing with negative signs, all odd powers cancel when we add $\psi(x+h)$ and $\psi(x-h)$:

$$\psi(h + h) + \psi(x - h) \simeq 2\psi(x) + h^2\psi^{(2)}(x) + \frac{h^4}{12}\psi^{(4)}(x) + \mathcal{O}(h^6). \tag{10.26}$$

We bring the $\psi^{(2)}$ term to the LHS to obtain a three-point algorithm for the second derivative:

$$\psi^{(2)}(x) \simeq \frac{\psi(x + h) + \psi(x - h) - 2\psi(x)}{h^2} - \frac{h^2}{12}\psi^{(4)}(x) + \mathcal{O}(h^6). \tag{10.27}$$

To eliminate fourth-derivative errors in the algorithm, we now apply the operator $1+\frac{h^2}{12}\frac{d^2}{dx^2}$ to the Schrödinger equation (10.23) to obtain a modified Schrödinger equation:

$$\psi^{(2)}(x)+\frac{h^2}{12}\psi^{(4)}(x)+k^2(x)\psi+\frac{h^2}{12}\frac{d^2}{dx^2}[k^2(x)\psi^{(4)}(x)]=0, \tag{10.28}$$

where the x dependence in $k^2(x)$ permits general potentials. Next we substitute the approximate expression for the second derivative (10.27) to obtain

$$\begin{aligned}&\frac{\psi(x+h)+\psi(x-h)-2\psi(x)}{h^2}\\&\quad-\frac{h^2}{12}\psi^{(4)}(x)+\frac{h^2}{12}\psi^{(4)}(x)+k^2(x)\psi(x)+\frac{h^2}{12}\frac{d^2}{dx^2}[k^2(x)\psi(x)]\simeq 0.\end{aligned} \tag{10.29}$$

We see that the $\psi^{(4)}$ terms now cancel each other, which means that the error is $\mathcal{O}(h^6)$.

To handle the general x dependence of k^2, we approximate[4] the second derivative of $k^2(x)\psi(x)$ as

$$\begin{aligned}&\frac{d^2[k^2(x)\psi(x)]}{dx^2}\simeq\\&\quad\frac{[k^2(x+h)\psi(x+h)-k^2(x)\psi(x)]+[k^2(x-h)\psi(x-h)-k^2(x)\psi(x)]}{h^2}.\end{aligned} \tag{10.30}$$

After making a substitution, we obtain our algorithm:

$$\psi(x+h)\simeq\frac{2\left[(1-\frac{5}{12}h^2k^2(x)\right]\psi(x)-\left[(1+\frac{h^2}{12}k^2(x-h)\right]\psi(x-h)}{1+\frac{h^2}{12}k^2(x+h)}. \tag{10.31}$$

This uses the values of ψ at two previous steps to move ψ forward one step (to step backward in x, we only need to reverse the sign of h). In terms of discrete indices, $x=ih$, the algorithm is simply

$$\boxed{\psi_{i+1}\simeq\frac{2\left(1-\frac{5}{12}h^2k_i^2\right)\psi_i-\left(1+\frac{h^2}{12}k_{i-1}^2\right)\psi_{i-1}}{1+\frac{h^2}{12}k_{i+1}^2}.} \tag{10.32}$$

[4]See equation (8.29) in Chapter 8, *Differentiation*.

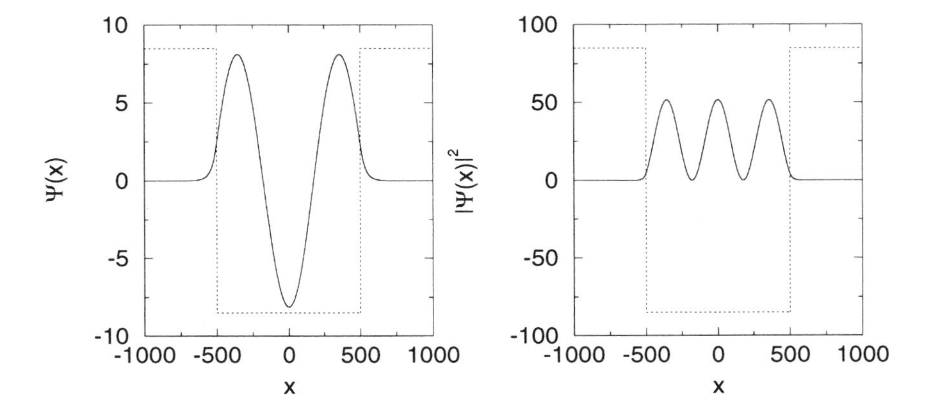

Fig. 10.4 (*Left*) The eigenfunction after the eigenenergy was found. The dashed curve shows the potential well. Observe the perfect symmetry. (*Right*) The probability density for this eigenfunction is clearly confined to the well.

10.6.1 Matching

The bound-state problem adds the extra requirement that the solution at large $|x|$ be the decaying exponential $\exp(-\beta|x|)$. We numerically impose that by integrating the ODE, starting at large positive x with $\psi = \exp(-\beta x)$, and then stepping inward (to the left) one h at a time. Next, we integrate the ODE, starting at large negative x with $\psi = \exp(\beta x)$, and step inward (to the right) one h at a time. At some arbitrary value of x, the *matching radius*, we test to see if the left and right values for ψ and $d\psi/dx$ agree and apply the bisection algorithm. When they agree, we have a solution.

Because the overall normalization of the wave function is arbitrary, there is really only one condition we must meet at the matching radius. Most usually, the normalization constraint is removed by requiring continuity of the logarithmic derivative ψ'/ψ. But in our case, we start with perfectly symmetric starting wave functions of the same magnitude on the right and left, and so we need to match only their magnitudes. If they do match, so will their derivatives.

The first two steps in the matching is shown in Fig. 10.3, and the final wave function is shown in Fig. 10.4. In this case, the wave function has a node, which means that it is an excited state rather than the ground state. To find the ground state, we lowered the energy at which we started the search.

10.7 IMPLEMENTATION: ODE EIGENVALUES, NUMEROV.C

1. Start with a step size $h = 0.04$. This means that there are 200 steps in the interval $x_{\min} = -4 \leq x \leq x_{\max} = 4$. Your ψ will be then an array of dimension 200, with the $i = 0, 1, \ldots$ elements representing the wave function to the left ψ_l, and the $i = 200, 199, \ldots$ elements representing ψ_r. (You will need to store separately the left and right values of the wave function at the matching radius.)

2. Write a subroutine which calculates the matching function

$$f(E, x) = \frac{\psi_l(x - h) - \psi_r(x - h)}{\psi_l(x)} \tag{10.33}$$

   as a function of energy and matching radius. This subroutine will be called by the bisection algorithm program to search for the energy at which $f(E, x = 2)$ vanishes.

3. As a first guess, take

$$E \simeq \xi^2 \frac{\hbar^2 c^2}{2mc^2 a^2} \simeq (3.6)^2 \frac{(197.32\text{MeV fm})^2}{2 \times 940\text{MeV}2(\text{fm})^2} \simeq 65\text{MeV}. \tag{10.34}$$

4. Determine the left and right wave functions by integrating the Schrödinger equation and using the Numerov algorithm (10.32). Start with

$$\psi_l \equiv \psi_{i=0} = \psi(x_{\min}) = e^{\beta(-4)}, \qquad \psi_1 = e^{\beta(-4+h)}, \tag{10.35}$$
$$\psi_r \equiv \psi_{200} = \psi(x_{\max}) = e^{\beta(-4)}, \qquad \psi_{199} = e^{\beta(-4+h)}. \tag{10.36}$$

5. Continue the step-by-step integration to $x_m = -2$ where you will match wave functions.

6. Renormalize the wave function so the left and right values match at x_m.

7. Write a main program that calls your subroutine for different energies (say, within ± 15% of your energy guess). Then use the bisection algorithm to zero in on the eigenenergy.

8. Print out the value of the energy for each iteration. This will give you a feel as to how well the procedure converges, as well as a measure of the precision obtained. Try different values for the tolerance until you are confident that you are obtaining three good decimal places in the energy.

9. Build in a limit to the number of energy iterations you permit, and print out (emphatically) when the iteration scheme fails.

10. Check that your computation is a solution of the transcendental equation (10.15).

11. Have you solved the **problem**? Is the spacing between levels on the order of MeV for a nucleon bound in a several-fm well?

10.8 ASSESSMENT: EXPLORATIONS

1. Check to see how well your search procedure works by using arbitrary values for the starting energy. For example, because no bound-state energies can lie below the bottom of the well, try $E = -V_0$ as well as some arbitrary fractions of V_0. In every case examine the resulting wave function and check that it is both symmetric and continuous.

2. Increase the depth of your potential progressively until you have three or four bound states. Look at the wave function in each case and determine the relation between the number of nodes in the wave function and the position of the bound state in the well.

3. Explore how a bound-state energy changes as you change the depth V_0 of the well. In particular, as you keep decreasing the depth, watch the eigenenergy move closer to $E = 0$, and see if you can find the depth at which the bound state just exists.

4. For a fixed well depth V_0, explore how the energy of a bound state changes as the well radius a is varied.

5. Conduct some explorations in which you discover different values of (V_0, a) that give the same ground-state energies. The existence of several different combinations means that knowledge of a ground-state energy is not enough to determine a unique depth of the well.

6. Modify the procedures to solve for the eigenvalue and eigenfunction for odd wave functions.

7. Solve for the wave function of a linear potential:

$$V(x) = -V_0 \begin{cases} |x|, & \text{for } |x| < a, \\ 0, & \text{for } |x| > a. \end{cases} \tag{10.37}$$

   There is less potential than for a square well, so you may expect lower binding energies and a less-confined wave function. (There are no analytic results with which to compare.)

10.9 EXTENSION: NEWTON'S RULE FOR FINDING ROOTS

Our bisection method of finding the root of $f(\xi) = 0$ always works, but it is slow. A more sophisticated method known as the *Newton–Raphson* method provides a solution for the local tangent to the curve $y = f(\xi)$ in the region of the sign change, and then determines the exact zero of this tangent. When all goes well, this provides a smaller range $[\xi_-, \xi_+]$, which means that the procedure converges rapidly. The scheme corresponds to guesses at ξ_0, ξ_1, ...:

$$\xi_0 = \xi_- - \frac{f(\xi_-)}{f'(\xi_-)}, \tag{10.38}$$

$$\xi_1 = \xi_0 - \frac{f(\xi_0)}{f'(\xi_0)}, \tag{10.39}$$

$$\xi_2 = \xi_1 - \frac{f(\xi_1)}{f'(\xi_1)}, \tag{10.40}$$

$$\vdots$$

Extend the eigenvalue search of this section by using the Newton–Raphson method, and determine how much of a difference it makes.

11

Anharmonic Oscillations

In Chapter 9, *Differential Equations and Oscillations*, we studied the oscillations that result when a mass is attached to a nonlinear spring. Now that you have confidence in being able to solve nonlinear differential equations, we will look at some realistic physical systems that are sometimes harmonic and sometimes anharmonic.

On the Web, you will find tutorials that give you answers to check with, that plot up the solutions to the differential equations, and that let you hear the sounds generated by these oscillatory systems.

11.1 PROBLEM 1: NONLINEARLY PERTURBED HARMONIC OSCILLATOR

Consider the motion of a mass m connected to a spring that is linear for small displacements, but that becomes somewhat nonlinear for large displacements:

$$V(x) \simeq \tfrac{1}{2}kx^2\left(1 - \tfrac{2}{3}\alpha x\right). \tag{11.1}$$

Here α is a measure of the nonlinear perturbation, and for nearly harmonic systems we expect $\alpha x \ll 1$.

Your **problem** is to solve for the position of the mass as a function of time $x(t)$, and to determine how the nature of the motion changes as the nonlinear term becomes more important.

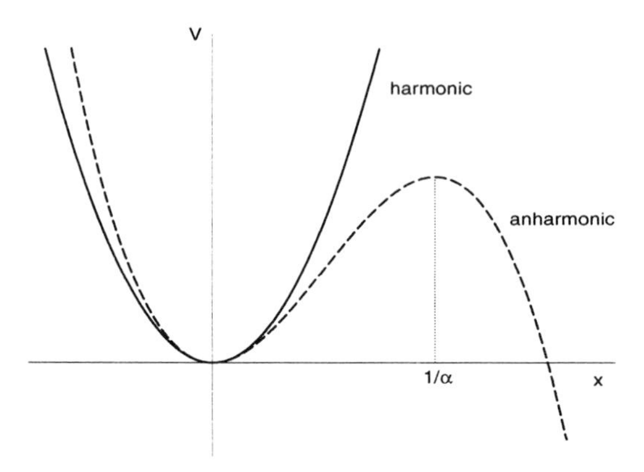

Fig. 11.1 The potential of the traditional harmonic oscillator (solid curve) and that of an oscillator with an anharmonic correction (dashed curve).

11.2 THEORY: NEWTON II

We can understand the basic physics of this problem by looking at Fig. 11.1. As long as $x < 1/\alpha$, there will be a *restoring force* and the motion will be periodic, though not necessarily harmonic (in fact, we can see that it is not even symmetric about the equilibrium position).

To describe the system mathematically, we apply Newton's second law of motion:

$$F(x) = -\frac{dV(x)}{dx} = -kx(1 - \alpha x) = m\frac{d^2x}{dt^2}, \qquad (11.2)$$

where we leave off from the LHS any external, time-dependent driving force. Equation (11.2) is a nonlinear equation that must be solved numerically. However, as long as α is small, we know from first-year physics that the approximate solution is simple harmonic motion

$$x(t) \simeq A\cos(\omega_0 t + \phi), \qquad \omega_0 = \sqrt{\frac{k}{m}}, \qquad (11.3)$$

where A is the amplitude, ω_0 the natural frequency, and ϕ the phase.

11.3 IMPLEMENTATION: ODE SOLVER, RK4.F (.C)

1. Write or revise the program from Chapter 9 that solves the equation of motion (11.2) using a fourth-order Runge–Kutta algorithm.

2. Develop an algorithm that determines the period T of the solution by

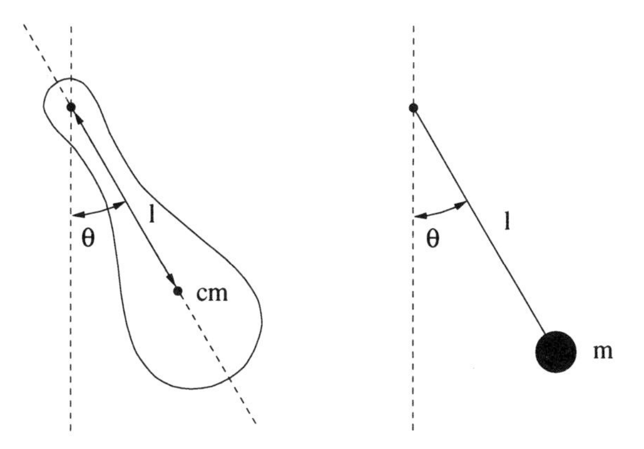

Fig. 11.2 A physical and a simple pendulum. The position of the center of mass in both cases is denoted by l. The theory is the same for both, but with different expression for the natural frequency ω_0.

recording the times at which the mass passes through its equilibrium point. Be careful, the motion is not symmetric and so several times will be needed.

11.4 ASSESSMENT: AMPLITUDE DEPENDENCE OF FREQUENCY

1. Test your program for the linear case, $\alpha = 0$, by verifying that
   (a) Your solution is harmonic with frequency $\omega_0 = \sqrt{k/m}$.
   (b) The frequency ω_0 is independent of the amplitude A.
2. Progressively increase α to add in anharmonicity.
   (a) For a fairly large value of α, plot the actual frequency of the system as a function of A.
   (b) Increase the initial energy of the system until the solution becomes unstable. Can you get rid of the instability by improving the numerics?

11.5 PROBLEM 2: REALISTIC PENDULUM

A *physical pendulum* is an object of arbitrary shape that can swing freely about a horizontal axis, as indicated on the left in Fig. 11.2. A simple pendulum, indicated on the right in Fig. 11.2, has all its mass concentrated at its tip.

Your **problem** is describe the behavior of either pendulum as its maximum displacement is made larger and larger, and the oscillations become less and less harmonic.

11.6 THEORY: NEWTON II FOR ROTATIONS

Let I be the moment of inertia about the axis of rotation, θ the angular displacement from equilibrium, and x the distance to the center of mass from the axis. The equation of motion follows from setting the gravitational torque about the axis of rotation equal to the moment of inertia I times the angular acceleration:

$$-lmg\sin\theta = I\frac{d^2\theta}{dt^2} \tag{11.4}$$

$$\Rightarrow \quad \frac{d^2\theta}{dt^2} = -\omega_0^2\sin\theta. \tag{11.5}$$

Here ω_0 is the natural frequency of the system, and is related to the natural period T_0 and to the parameters by

$$\boxed{\omega_0 \equiv \frac{2\pi}{T_0} = \begin{cases} lmg/I, & \text{for physical pendulum,} \\ g/l, & \text{for simple pendulum.} \end{cases}} \tag{11.6}$$

Equation (11.5) is the *nonlinear equation* of motion for the pendulum. If the displacement is small, we can approximate $\sin\theta$ by θ and obtain simple harmonic motion

$$\frac{d^2\theta}{dt^2} \simeq -\omega_0^2\theta, \tag{11.7}$$

$$\Rightarrow \quad \theta_{\text{shm}}(t) = \theta_0\sin(\omega_0 t + \phi), \tag{11.8}$$

where θ_0 is the amplitude and ϕ the phase.

In analogy with the anharmonic oscillator of §11.1, we expect the solution of (11.5) for the realistic pendulum to be periodic with frequency ω, but with ω equal to the natural frequency ω_0 only for small oscillations. Furthermore, because the actual restoring torque is proportional to $\sin\theta$, which is less than θ ($\sin\theta \simeq \theta - \frac{1}{2}\theta^2$), the actual restoring torque is *less than* that assumed in a harmonic oscillator. This means that that real pendula run slower and slower (their periods get longer and longer) as their angular displacements are made larger and larger.

11.7 METHOD, ANALYTIC: ELLIPTIC INTEGRALS

The closed-form solution to the large amplitude pendulum is a classic text book problem [L&L 69, M&T 88, Schk 94]. The solution is not the result of some ingenious scheme for solving nonlinear equations, but rather a straightforward integration of the energy integral for motion with a maximum displacement angle θ_m:

$$E = \mathrm{KE} + \mathrm{PE} \tag{11.9}$$

$$2mgl\sin^2\tfrac{\theta_m}{2} = \tfrac{1}{2}I\left(\frac{d\theta}{dt}\right)^2 + 2mgl\sin^2\tfrac{\theta}{2}, \tag{11.10}$$

$$\Rightarrow \frac{d\theta}{dt} = 2\omega_0\left[\sin^2\tfrac{\theta_m}{2} - \sin^2\tfrac{\theta}{2}\right]^{1/2}, \tag{11.11}$$

$$\Rightarrow T = \frac{T_0}{\pi}\int_0^{\theta_m}\frac{d\theta}{\left[\sin^2\frac{\theta_m}{2} - \sin^2\frac{\theta}{2}\right]^{1/2}}. \tag{11.12}$$

Here the equation for the period (11.12) uses the symmetry argument that it takes one-quarter of a period for the pendulum to travel from $\theta = 0$ to θ_m.

The integral in (11.12) is an *elliptic integral of the first kind.* If you think of an elliptic integral as a generalization of a trigonometric function, then this is a closed-form solution; otherwise it's an integral needing computation. [We will determine the period by solving the equation of motion for $\theta(t)$.] The denominator can be expanded and the series integrated term by term to yield a series for the amplitude dependence of the period

$$T \simeq T_0\left[1 + \tfrac{1}{4}\sin^2\tfrac{\theta_m}{2} + \tfrac{9}{64}\sin^4\tfrac{\theta_m}{2} + \cdots\right]. \tag{11.13}$$

In this way, an amplitude of 23° leads to a 1% slowdown, while an amplitude of 80° leads to a 10% slowdown.

11.8 IMPLEMENTATION, RK4 FOR PENDULUM

Specialize your differential equation program to solve the nondriven ($F_{\text{ext}} = 0$) realistic pendulum equation (11.5). Solve for the angular displacement θ and the angular velocity $d\theta/dt$ as functions of time for arbitrary (input) initial values of θ_0 and ω_0. We suggest the fourth-order Runge–Kutta integration algorithm. Set $\theta(t = 0) = 0$; that is, start off with the total energy equal to kinetic as the pendulum passes through the equilibrium position.

1. Verify that if the initial energy is less than $2mgl$, the motion is periodic but not harmonic.

2. For $E = 2mgl$, the motion changes from vibrational to rotational. See how close you can get to this *separatrix* and try to verify that at $E = 2mgl$ it takes an infinite time for the pendulum to reach the top.

3. Devise an algorithm to determine the period of the oscillation (or use your previous one that counted times when passing through the origin).

4. Construct a graph of the period deduced from your $\theta(t)$ as a function of θ_m. Compare your answer to the power series (11.13).

11.9 EXPLORATION: RESONANCE AND BEATS

Include a sinusoidal, external driving force

$$F_{\text{ext}}(t) = F_0 \sin \omega t \tag{11.14}$$

into the equation of motion. Study how *resonance* and *beating* occur for both harmonic and anharmonic oscillations. You can use any of the three anharmonic systems we have studied so far: the x^p oscillator, the perturbed harmonic oscillator, or the realistic pendulum.

You may recall that when a *harmonic* oscillator is driven by an $F_{\text{ext}}(t)$ with ω equal to the natural frequency ω_0, the system resonates. In that case, the amplitude of oscillation increases without bound if there is no friction. You may also recall that if the external driving frequency ω is close to, but not equal to, ω_0, *beating* occurs. In beating, the natural response at ω_0 and the driven response at ω add:

$$x \simeq x_0 \sin \omega t + x_0 \sin \omega_0 t = 2x_0 \cos \tfrac{\omega-\omega_0}{2} t \sin \tfrac{\omega+\omega_0}{2} t. \tag{11.15}$$

When the $\cos \frac{\omega-\omega_0}{2} t$ term varies slowly, it appears that the mass is oscillating at the average frequency $\frac{\omega+\omega_0}{2}$ with an amplitude varying at the slow frequency $\frac{\omega-\omega_0}{2}$.

1. Add the time-dependent external force (11.14) to the space-dependent restoring force in your program.

2. First use a very large value for the magnitude of the force F_0 in (11.14). This should lead to *mode locking*, where the system is overwhelmed by the driving force and after a short time oscillates in phase with the driver regardless of the frequency.

3. Now lower F_0 till it is close to the magnitude of the natural restoring force of the system. You need to have this for beating to occur.

4. For a harmonic system, verify that the "beat frequency," that is, the number of oscillations per unit time, equals the frequency difference in

cycles per second. You will be able to do this only if ω and ω_0 are close.

5. Explore what happens when you make a linear and nonlinear system resonate. If the nonlinear system is close to being harmonic, you should get beating. (This is because the natural frequency changes as the amplitude increases.)

6. (Optional) Make a plot of the amplitude of oscillation versus driving frequency, and try to identify hysteresis (that is, when two amplitudes are possible for one frequency) [Abar 93].

7. Notice how the character of the resonance changes as the exponent p in the potential

$$V(x) = \frac{1}{p}k|x|^p \tag{11.16}$$

is made larger and larger. At some point the mass effectively "hits" the wall and falls out of phase with the driver.

11.10 EXPLORATION: PHASE-SPACE PLOT

The conventional solution to an equation of motion is given by the position $x(t)$ and the velocity $v(t)$ as functions of time. It is illuminating to make a *phase-space* plot of the velocity $v(t)$ (as dependent variable) versus the position $x(t)$ (as independent variable), this is called a *phase-space* plot.

1. Take your solution to the *harmonic* oscillator and plot the paths in phase space for

   (a) $x_0 = 0, v_0 \neq 0$,

   (b) $x_0 \neq 0, v_0 = 0$,

   (c) $x_0 \neq 0, v_0 \neq 0$.

2. Your solutions should form elliptical orbits with all initial conditions of the same energy tracing out the same ellipse. Determine, by watching the points being plotted, whether the motion is clockwise or counterclockwise.

3. Take your solution to an *anharmonic* oscillator and plot the path in phase space for a variety of initial conditions.

4. Deduce the differences and similarities in the orbit *shapes* for the harmonic and anharmonic oscillations.

11.11 EXPLORATION: DAMPED OSCILLATOR

Our lives are full of friction, and it is rather academic for us to be ignoring it. The simplest models for friction are *static*, *kinetic* (or sliding), and *viscous* friction:

$$F_f \leq -\mu_s N, \qquad \text{(static)}, \tag{11.17}$$

$$F_f = -\mu_k N \frac{v}{|v|}, \qquad \text{(kinetic)}, \tag{11.18}$$

$$F_f = -bv, \qquad \text{(viscous)}. \tag{11.19}$$

Here N is the *normal force*, μ and b are parameters, and v is the velocity. Include static plus kinetic, or viscous, friction into your equation of motion and explore the effects. If you include viscous friction into a harmonic, or nearly harmonic, oscillator, you may be able to identify three regimes:

Underdamped: $\frac{b}{2m} < \omega_0$ The solution oscillates within an exponentially decaying envelope.

Critically damped: $\frac{b}{2m} = \omega_0$ The solution goes directly to equilibrium without oscillating.

Over damped: $\frac{b}{2m} > \omega_0$ The solution decays slowly, reaching equilibrium only after infinite time.

12

Fourier Analysis of Nonlinear Oscillations

In this chapter we examine *Fourier series* and *Fourier integrals (or transforms)*. This is the traditional tool for decomposing both periodic and nonperiodic motions, respectively, into an infinite number of harmonic functions. Whereas a Fourier series is just one example of an expansion of a function as a series of orthogonal polynomials, it has the distinguishing characteristic of always generating a periodic approximation. This clearly is desirable if the function we wish to approximate is periodic, in which case we construct the series to have the correct period. A nonperiodic function can frequently be well approximated by a Fourier series *over some limited time*, but the Fourier series will eventually show its periodicity for values outside this range. In this latter case of limited range, the Fourier integral is more appropriate.

12.1 PROBLEM 1: THE HARMONICS OF NONLINEAR OSCILLATIONS

Consider a particle oscillating in a nonharmonic potential. This could be the nonharmonic oscillator (9.2),

$$V(x) = \frac{1}{p} k |x|^p, \tag{12.1}$$

for $p \neq 2$, the perturbed harmonic oscillator (11.1),

$$V(x) = \tfrac{1}{2} k x^2 \left(1 - \tfrac{2}{3} \alpha x\right), \tag{12.2}$$

or the realistic pendulum of §11.6. While free oscillations in these potentials are always periodic, they are not truly sinusoidal. Your **problem** is to take the solution of one of these nonlinear oscillators and relate it to the solution

$$x(t) = A_0 \sin(\omega t + \phi_0) \tag{12.3}$$

of the linear, harmonic oscillator. (In your future study of chaos, you will want to extend this analysis to the response of a damped, oscillating system driven by an external force.) If your oscillator is sufficiently nonlinear to behave like the sawtooth function shown on the left of Fig. 12.1, then the Fourier spectrum you obtain should be similar to that shown on the right of Fig. 12.1.

In general, when we want to undertake a spectral analysis, we want to analyze the steady-state behavior of a system. This means that the initial transient behavior has had a chance to die off. Just what is the initial transient is easy to identify for linear systems, but may be less so for nonlinear systems in which the "steady state" jumps among a number of configurations.

12.2 THEORY: FOURIER ANALYSIS

Nonlinear oscillations are interesting in part because they hardly ever are studied in traditional courses. This is true even though the linear term is just a first approximation to a naturally oscillating system. If the force on a particle is always toward its equilibrium position (a restoring force), then the resulting motion will be *periodic*, but not necessarily *harmonic*. A good example is the motion in a highly anharmonic well $p \approx 10$, which produces an $x(t)$ looking like a series of pyramids; this is periodic but not harmonic.

On a computer, the distinction between a Fourier integral and a Fourier series is less clear because the integral is approximated as a finite sum. We will illustrate both methods by analyzing anharmonic oscillations with the series and by analyzing the charge density of elementary systems with the integral.

In a sense, our approach is the inverse of the traditional one in which the *fundamental* oscillation is determined analytically, and the higher-frequency *overtones* are determined in perturbation theory [L&L 69]. We start with the full (numerical) periodic solution and then decompose it into what may be called "harmonics." When we speak of fundamentals, overtones, and harmonics, we speak of solutions to the *boundary-value problem*, for example, of waves on a plucked violin string. In this latter case, and when given the correct conditions (enough musical skill), it is possible to excite individual harmonics or sums of them within the series

$$y(t) = b_0 \sin \omega_0 t + b_1 \sin \frac{n}{m}\omega_0 t + \cdots \tag{12.4}$$

The anharmonic oscillator vibrates with a single frequency (which may change with changing amplitude) but not a sinusoidal waveform. Expanding the anharmonic vibration as a Fourier series does not imply that the individual harmonics can be "played."

You may recall from classical mechanics that the most general solution for some vibrating physical system can be expressed as the sum of the *normal modes* of that system. These expansions are possible because we have *linear operators* and, subsequently, the *principle of superposition: If* $x_1(t)$ *and* $x_2(t)$ *are solutions of some linear equation, then* $\alpha_1 x_1(t)+\alpha_2 x_2(t)$ *is also a solution.*

The principle of linear superposition does not hold when we solve nonlinear problems. Nevertheless, it is always possible to expand a *periodic* solution of a *nonlinear* problem in terms of trigonometric functions that have frequencies that are integer multiples of the true frequency of the nonlinear oscillator. This is a consequence of *Fourier's theorem* being applicable to any single-valued, periodic function with only a finite number of discontinuities. We assume we know the period T; that is, that

$$y(t+T) = y(t). \tag{12.5}$$

This tells us the "true" frequency ω:[1]

$$\omega \equiv \omega_1 = \frac{2\pi}{T}. \tag{12.6}$$

Any such periodic function can be expanded as a series of harmonic functions with frequencies that are multiples of the true frequency:

$$\boxed{y(t) = \frac{a_0}{2} + \sum_{n=1}^{\infty} \left(a_n \cos n\omega t + b_n \sin n\omega t\right).} \tag{12.7}$$

The Fourier series (12.7) is a "best fit" in the least-squares sense of Chapter 5, *Data Fitting*, because it minimizes $\sum_i [y(t_i) - y_i]^2$. This means that the series converges to the average behavior of the function, but misses the function at discontinuities (at which points it converges to the mean) or at sharp corners (where it overshoots). The coefficients a_n and b_n in (12.7) measure the amount of $\cos n\omega t$ and $\sin n\omega t$ present in the $y(t)$. A general function $y(t)$ may contain an infinite number of Fourier components, although a good approximation is usually possible with $\sim$10 harmonics.

The coefficients a_n and b_n are determined by the standard techniques for orthogonal function expansion. To find them, you multiply both sides of (12.7) by $\cos n\omega t$ or $\sin n\omega t$, integrate over one period, and project out a single a_n

[1] We remind the reader that every periodic system by definition has a period T and consequently a "true" frequency ω. Nonetheless, this does not imply that the system behaves like $\sin \omega t$. Only harmonic oscillators do that.

or b_n:

$$\begin{pmatrix} a_n \\ b_n \end{pmatrix} = \frac{2}{T} \int_0^T dt \begin{pmatrix} \cos n\omega t \\ \sin n\omega t \end{pmatrix} y(t), \qquad \omega \stackrel{\text{def}}{=} \frac{2\pi}{T}. \tag{12.8}$$

Awareness of the *symmetry* of the function $y(t)$ may eliminate the need to evaluate all the expansion coefficients. For example

- a_0 is twice the average value of y.

$$a_0 = 2 \langle y(t) \rangle . \tag{12.9}$$

- For an *odd function*; that is, one for which $y(-t) = -y(t)$, the coefficient $a_n \equiv 0$ and only half the integration range is needed to determine b_n:

$$b_n = \frac{4}{T} \int_0^{T/2} dt \; y(t) \sin n\omega t. \tag{12.10}$$

- For an *even function*; that is, one for which $y(-t) = y(t)$, the coefficient $b_n \equiv 0$ and only half the integration range is needed to determine a_n:

$$a_n = \frac{4}{T} \int_0^{T/2} dt \; y(t) \cos n\omega t. \tag{12.11}$$

12.2.1 Example 1: Sawtooth Function

A sawtooth function is shown in the Fig. 12.1. It clearly is periodic, nonharmonic, and discontinuous. While it is relatively easy to reproduce the general shape of this function with only a few terms, many components are needed to reproduce the sharp corners. Because the function is odd, the Fourier series is a sine series, and (12.8) determines the values

$$\begin{aligned} b_n &= \frac{\omega^2 A}{2\pi^2} \int_{-\pi/\omega}^{+\pi/\omega} dt t \sin n\omega t = \frac{A}{n\pi}(-1)^{n+1}, & (12.12) \\ \Rightarrow \quad y(t) &= \frac{A}{\pi} \left[\sin \omega t - \tfrac{1}{2} \sin 2\omega t + \tfrac{1}{3} \sin 3\omega t - \cdots \right]. & (12.13) \end{aligned}$$

12.2.2 Example 2: Half-Wave Function

The half-wave function is

$$y(t) = \begin{cases} \sin \omega t, & \text{for } 0 < t < T/2 \,, \\ 0, & \text{for } T/2 < t < T. \end{cases} \tag{12.14}$$

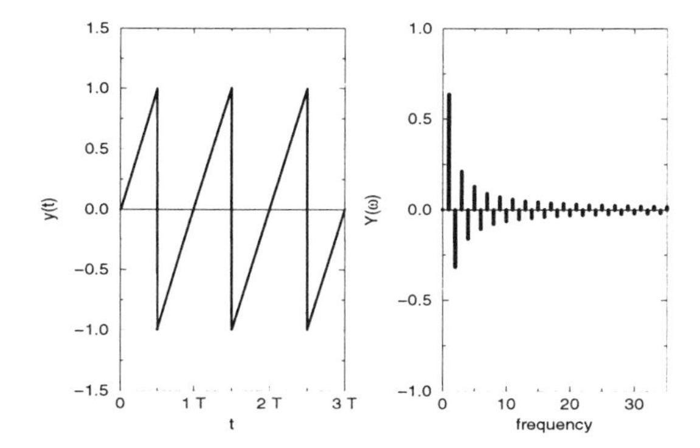

Fig. 12.1 (*Left*) A sawtooth function with period T; (*right*) the Fourier spectrum of frequencies contained in this sawtooth function, with frequencies measured in units of natural frequency ω.

It is periodic, nonharmonic (actually the upper half of a sine wave), continuous, but with discontinuous derivatives. Because it lacks the sharp "corners" of the sawtooth, it is easier to reproduce with a finite Fourier series. Equations (12.8) determine

$$a_n = \begin{cases} \frac{-2}{\pi(n^2-1)}, & n \text{ even or } 0, \\ 0, & n \text{ odd}, \end{cases} \tag{12.15}$$

$$b_n = \begin{cases} \frac{1}{2}, & n = 1, \\ 0, & n \neq 1, \end{cases} \tag{12.16}$$

$$\Rightarrow \quad y(t) = \tfrac{1}{2}\sin\omega t + \frac{1}{\pi} - \frac{2}{3\pi}\cos 2\omega t - \frac{2}{15\pi}\cos 4\omega t + \cdots. \tag{12.17}$$

12.3 ASSESSMENT: SUMMATION OF FOURIER SERIES

1. **Sawtooth function:** Sum the Fourier series for the *sawtooth function* up to order $n = 2, 4, 10, 20$, and plot the results over two periods.

   (a) Check that in each case the series gives the mean value of the function *at* the points of discontinuity.

   (b) Check that in each case the series *overshoots* by about 9% the value of the function on either side of the discontinuity (the *Gibbs phenomenon*).

2. **Half-wave function:** Sum the Fourier series for the *half-wave function* up to order $n = 2, 4, 10$, and plot the results over two periods. (The

series converges quite well, doesn't it?)

12.4 THEORY: FOURIER TRANSFORMS

While a Fourier *series* is the right tool for approximating or analyzing periodic functions, the Fourier *transform* or *integral* is the right tool for nonperiodic functions. We transform the series formalism to the integral formalism by imagining a system described by a continuum of "fundamental" frequencies. We therefore deal with *wave packets* containing continuous rather than discrete frequencies.[2] While the difference between series and transform methods may appear clear, if we transform a function known only over a finite number of points, or approximate the Fourier integral as a finite sum, then the Fourier transform and the Fourier series become quite similar.

As an analogy to (12.7), we will now imagine our function or signal $y(t)$ expressed in terms of a continuous series of harmonics

$$\boxed{y(t) = \frac{1}{\sqrt{2\pi}} \int_{-\infty}^{+\infty} d\omega Y(\omega) e^{-i\omega t},} \qquad (12.18)$$

where for compactness we use a complex exponential function.[3] Here the expansion amplitude $Y(\omega)$ is analogous to (a_n, b_n) and is called the *Fourier transform* of $y(t)$.

A plot of the squared modulus of Y, $|Y(\omega)|^2$ versus ω is called the *power spectrum*. Actually, (12.18) is more properly called the *inverse transform* because it converts $Y(\omega)$ to $y(t)$. The conversion of $y(t)$ to $Y(\omega)$,

$$\boxed{Y(\omega) = \frac{1}{\sqrt{2\pi}} \int_{-\infty}^{+\infty} dt\, e^{i\omega t} y(t),} \qquad (12.19)$$

is the *Fourier transform*.

You will note in (12.18) and (12.19) that only the relative sign in the exponential matters. In addition, you should note that we have chosen a symmetric $1/\sqrt{2\pi}$ normalization factor, which is common in quantum mechanics [L 96] but differs from that used in engineering.

If $y(t)$ represents the response of some system as a function of time, $Y(\omega)$ is a *spectral function* that measures the amount of frequency ω making up

[2] We follow convention and consider time t as the function's variable and frequency ω as the transform's variable. Nonetheless, these can be reversed or other variables such as position r and momentum p may also be used.

[3] Recall the principle of linear superposition and that $\exp(-i\omega t) = \cos\omega t + i \sin\omega t$. This means that the real part of y gives the cosine series and the imaginary part the sine series.

this response. While usually we think of measuring $y(t)$ in the laboratory and numerically transforming it to obtain $Y(\omega)$, some experiments may well measure $Y(\omega)$ directly [in which case a transform is needed to obtain $y(t)$]. Clearly, the mathematics is symmetric even if the real world is not.

If the Fourier transform and its inverse are consistent with each other, we should be able to substitute (12.18) into (12.19) and obtain an identity:

$$Y(\omega) = \frac{1}{\sqrt{2\pi}} \int_{-\infty}^{+\infty} dt\; e^{i\omega t} \frac{1}{\sqrt{2\pi}} \int_{-\infty}^{+\infty} d\omega' e^{-i\omega' t} Y(\omega') \tag{12.20}$$

$$= \frac{1}{2\pi} \int_{-\infty}^{+\infty} d\omega' \left\{ \int_{-\infty}^{+\infty} dt\; e^{i(\omega-\omega')t} \right\} Y(\omega'). \tag{12.21}$$

For (12.21) to be an identity, the term in braces must be the *Dirac delta function*:

$$\int_{-\infty}^{+\infty} dt\; e^{i(\omega-\omega')t} = 2\pi\delta(\omega-\omega'), \tag{12.22}$$

$$\text{where} \quad \int_{-\infty}^{+\infty} d\omega \delta(\omega-\omega') f(\omega) \stackrel{\text{def}}{=} f(\omega'). \tag{12.23}$$

While the delta function is one of the most common and useful functions in theoretical physics, it is not well behaved in a mathematical sense and is terribly misbehaved in a computational sense. While it is possible to create numerical approximations to $\delta(\omega-\omega')$, they may well be borderline pathological. It is probably better for you to do the delta function part of an integration analytically and give the nonsingular leftovers to the computer.

12.5 METHOD: DISCRETE FOURIER TRANSFORM

If $y(t)$ or $Y(\omega)$ is known analytically, the integral (12.18) or (12.19) can be evaluated analytically or numerically using the integration techniques studied earlier (particularly Gaussian quadrature). Likewise, if a table of N values for $y(t)$ is known, interpolations within the table can be used to evaluate the integral.

Here we will consider a technique for directly Fourier transforming functions that are known only for a finite number N of times t (for example, as sampled in an experiment). This *discrete Fourier transform* (DFT) is an "approximate" procedure because the integrals are evaluated numerically.[4] By sampling a function at N times, we can determine N values of the Fourier transform of this function [N independent $y(t)$ values can produce N inde-

[4]More discussion can be found in the book [B&H 95] devoted to just this topic.

pendent $Y(\omega)$ values]. We can then use those values of the transform to approximate the original function at any value of time. In this way the DFT can also be thought of as a technique for interpolating and extrapolating data.

Assume that the function $y(t)$ we wish to transform is measured or sampled at a discrete number $N+1$ of times (N time intervals)

$$y_k \equiv y(t_k), \quad k = 0, 1, 2, \ldots, N. \tag{12.24}$$

Assume that these times are evenly spaced with a time step h:

$$t_k = kh, \quad h = \Delta t. \tag{12.25}$$

In other words, we measure the signal $y(t)$ once every h seconds during a total time interval of

$$T \stackrel{\text{def}}{=} Nh. \tag{12.26}$$

While the time T is determined by the period we choose for our sampling, the mathematics produces a function $y(t)$ that is periodic with this period. To make this consistent and ensure that there are only N independent function values being used, we require the first and last y values to be the same:

$$y(t+T) = y(t) \quad \Rightarrow y_0 = y_N. \tag{12.27}$$

If we are, in fact, analyzing a truly periodic function, then the first N points should all be within one period to guarantee their independence. Unless we make further assumptions, these N independent input data $y(t_k)$ can determine no more than N independent output Fourier components $Y(\omega_k)$.

The time interval T (which should be made to be the period for periodic functions) is the largest time over which we consider variation of $y(t)$. Consequently, it determines the lowest frequency,

$$\omega_1 = \frac{2\pi}{T}, \tag{12.28}$$

contained in our Fourier representation of $y(t)$ (unless you want to be picky and argue that there may also be an $\omega = 0$ or "dc" component).

While we will be able to compute N independent values $Y(\omega_n)$ for $n = 1, N$, the values for the frequencies ω_n are determined by the number of samples taken and the total sampling time T. To make the connection with Fourier series, we choose the frequencies as

$$\omega_n = n\omega_1 = n\frac{2\pi}{Nh}, \quad n = 0, 1, \ldots, N. \tag{12.29}$$

Here the $n = 0$ value corresponds to the zero-frequency or dc component, $\omega_0 = 0$. We now see clearly that by limiting the time interval over which we sample the input function, we are making an approximation that limits the maximum frequency of the Fourier components we can compute.

A consequence of our approximations is that Fourier components higher in frequency than those we include in our analysis will contribute, erroneously, to those lower in frequency that we include in our analysis. This effect is called *aliasing* and makes the computed higher-frequency components less accurate. If, for some reason, accurate values for the high frequencies are required, then we may increase the number N of samples taken. We keep T constant, so this makes h smaller and in this way picks out higher frequencies. By increasing the number of frequencies you compute, you move the higher-frequency components you are interested in farther from the error-prone ends.

The discrete Fourier transform follows from evaluating (12.19) with a trapezoid rule:[5]

$$Y(\omega_n) \stackrel{\text{def}}{=} \frac{1}{\sqrt{2\pi}} \int_{-\infty}^{+\infty} e^{i\omega_n t} y(t) dt \tag{12.30}$$

$$\approx \frac{1}{\sqrt{2\pi}} \sum_{k=1}^{N} y(t_k) e^{i\omega_n t_k} h = \frac{h}{\sqrt{2\pi}} \sum_{k=1}^{N} y_k e^{2\pi ikn/N}. \tag{12.31}$$

To keep the final notation more symmetric, the step size h is factored from the transform Y and a discrete value Y_n is defined:

$$Y_n \stackrel{\text{def}}{=} \frac{1}{h} Y(\omega_n) \tag{12.32}$$

$$= \frac{1}{\sqrt{2\pi}} \sum_{k=1}^{N} y_k e^{2\pi ikn/N}. \tag{12.33}$$

With this same care in accounting, we invert the Y_n's into y_n's;

$$y(t) \stackrel{\text{def}}{=} \frac{1}{\sqrt{2\pi}} \int_{-\infty}^{+\infty} e^{-i\omega t} Y(\omega) d\omega \tag{12.34}$$

$$\approx \frac{\sqrt{2\pi}}{Nh} \sum_{n=1}^{N} e^{-i\omega_n t_k} Y(\omega_n) = \frac{\sqrt{2\pi}}{N} \sum_{n=1}^{N} Y_n e^{-2\pi itn/Nh}. \tag{12.35}$$

Once we know the N values of the transform, (12.35) is clearly useful for evaluating $y(t)$ for any value of the time t.

There is nothing illegal about evaluating the DFT expressions for Y_n and y_k for arbitrary values of n and k. There is no such thing as a free lunch, as the case may be; because the trigonometric functions are periodic, we just get

[5]The alert reader may be wondering what has happened to the $h/2$ with which the trapezoid rule weights the initial and final points. Actually, they are there, but because we have set $y_0 \equiv y_N$, two $h/2$'s have been added to produce one h.

the old answer back:

$$y(t_{k+N}) = y(t_k), \quad Y(\omega_{n+N}) = Y(\omega_n). \tag{12.36}$$

Another way of stating this is that none of the equations change if we replace $\omega_n t$ by $\omega_n t + 2\pi n$. There are still just N independent output numbers for N independent inputs.

While periodicity is expected for Fourier *series*, it is somewhat surprising for Fourier *integrals*, which have been touted as the right tool for nonperiodic functions. The periodicity arises from approximating the integral as a sum over a finite number of points. Clearly, if we input values of the function for longer lengths of time, it will take longer before the function repeats. If the repeat period is very long, it may be of little consequence for times short compared to the period.

If $y(t)$ is actually periodic with period Nh, then the integration formulas converge very rapidly and the DFT is an excellent way of obtaining Fourier series. If the input function is not periodic, then the DFT can be a bad approximation near the endpoints of the interval or once the transform starts repeating.

The discrete Fourier transform and its inverse can be written in a concise way and evaluated efficiently by introducing a complex variable Z for the exponential:

$$\boxed{\begin{aligned} Y_n &= \tfrac{1}{\sqrt{2\pi}} \textstyle\sum_{k=1}^{N} Z^{nk} y_k = \tfrac{1}{h} Y(\omega_n), \\ y_k &= \tfrac{\sqrt{2\pi}}{N} \textstyle\sum_{n=1}^{N} Z^{-nk} Y_n, \\ Z &= e^{2\pi i/N}. \end{aligned}} \tag{12.37}$$

Now the computer needs to compute only powers.

If your preference is to avoid complex numbers, or if your programming language does not support them, we can rewrite (12.37) in terms of separate real and imaginary parts. We start by applying Euler's theorem:

$$Z = e^{i\theta}, \tag{12.38}$$

$$\Rightarrow Z^{nk} = e^{ink\theta} = \cos nk\theta + i \sin nk\theta, \tag{12.39}$$

$$Z^{-nk} = e^{-ink\theta} = \cos nk\theta - i \sin nk\theta, \tag{12.40}$$

$$\text{where} \quad \theta \stackrel{\text{def}}{=} \frac{2\pi}{N}. \tag{12.41}$$

If we now make explicit that y_k and Y_n have real and imaginary parts, we obtain

$$Y_n = \frac{1}{\sqrt{2\pi}} \sum_{k=1}^{N} [(\cos nk\theta \mathrm{Re} y_k - \sin nk\theta \mathrm{Im} y_k)$$

$$+i(\cos nk\theta \mathrm{Im} y_k + \sin nk\theta \mathrm{Re} y_k)] , \qquad (12.42)$$

$$y_k = \frac{\sqrt{2\pi}}{N} \sum_{n=1}^{N} [(\cos nk\theta \mathrm{Re} Y_n + \sin nk\theta \mathrm{Im} Y_n)$$
$$+i(\cos nk\theta \mathrm{Im} Y_n - \sin nk\theta \mathrm{Re} Y_n)] . \qquad (12.43)$$

Equation (12.42) is interesting in that it shows that a real function produces a real Fourier transform only if all the $\sin nk\theta$ terms cancel out; that is, only if $y(t)$ is an even function of t.

The actual computation time for a discrete Fourier transform can be reduced even further by use of the *fast Fourier transform (FFT)* algorithm. An examination of (12.37) shows that the DFT is evaluated as a matrix multiplication of a vector of length N of Z values times a vector of length N of y value. The time for this DFT scales like N^2. With the FFT algorithm, the time would scale like $N \log_2 N$. While this may not seem like much at first, for $N = 10^{2-3}$, the difference of a factor 10^{3-5} is the difference between a minute and a week. This is the reason FFT is often used for the on-line analysis of data. We will not discuss FFT techniques and suggest those interested in them consult the references.

12.6 METHOD: DFT FOR FOURIER SERIES

For simplicity let us consider the Fourier cosine series:

$$y(t) = \sum_{n=0}^{\infty} a_n \cos(n\omega t), \quad (\omega = \frac{2\pi}{T}), \qquad (12.44)$$

$$a_k = \frac{2}{T} \int_0^T dt \cos(k\omega t) y(t). \qquad (12.45)$$

Here T is the actual period of the system (not necessarily the period of the simple harmonic motion occurring for small amplitude). We assume that the function $y(t)$ is sampled for a discrete set of times

$$y(t = t_k) \equiv y_k, \quad k = 0, 1, \ldots, N. \qquad (12.46)$$

Because we are analyzing a periodic function, we will retain the conventions used in the DFT and require the function to repeat itself with period $T = Nh$; that is, we assume that the amplitude is the same at the first and last points:

$$y_0 = y_N. \qquad (12.47)$$

This means that there are only N independent values of y being used as input. For these N independent y_k values, we can determine uniquely only N expansion coefficients a_k. If we use the trapezoid rule to approximate the integration in (12.45), we determine the N independent Fourier components as

$$a_n \simeq \frac{2h}{T} \sum_{k=1}^{N} \cos\left(n\omega t_k\right) y(t_k) = \frac{2}{N} \sum_{k=1}^{N} \cos\left(\frac{2\pi nk}{N}\right) y_k, \quad n = 0, \ldots, N. \tag{12.48}$$

Because for N independent $y(t)$ values we can determine only N Fourier components, our Fourier series for the function $y(t)$ must be in terms of only these components:

$$y(t) \simeq \sum_{n=0}^{N} a_n \cos(n\omega t) = \sum_{n=0}^{N} a_n \cos\left(\frac{2\pi nt}{Nh}\right). \tag{12.49}$$

In summary, we sample the function $y(t)$ at N times, t_1, ..., t_N. Because $y(t)$ is periodic, if we sample within one period, we are ensured of independent input data. You see that all N values of y sampled contribute to each a_k. Consequently, if we increase N in order to determine more coefficients, we must recompute all the a_n values. In the model-independent approach discussed in §12.15, the theory is reformulated so that additional samplings determine higher Fourier components without affecting lower ones.

12.7 IMPLEMENTATION: DFT, FOURIER.F (.C), INVFOUR.C

Understand the programs on the diskette before using packaged routines.

12.8 ASSESSMENT: SIMPLE ANALYTIC INPUT

The simple checks here are generally good to do before examining more complex problems. If your system has some Fourier analysis packages (such as the graphing package *Ace/gr*), you may want to compare your results with those from the packages. Once you understand how the packages work, it makes sense to use them.

1. Sample the even signal,

$$y(t) = \cos(\omega t) + 2\cos(3\omega t) + 3\cos(5\omega t). \tag{12.50}$$

Decompose this into its components and check that they are real and in the ratio 1:2:3 (or 1:4:9 if a power spectrum is plotted).

2. Sample the odd signal,

$$y(t) = \sin(\omega t) + 2\sin(3\omega t) + 3\sin(5\omega t). \quad (12.51)$$

Decompose this into its components and check that they are imaginary and in the ratio 1:2:3 (or 1:4:9 if a power spectrum is plotted).

3. Sample the signal

$$y(t) = \sin(\omega t) + 2\cos(3\omega t) + \sin(5\omega t). \quad (12.52)$$

Decompose this into its components and see if there are three of them in the ratio 1:2:1 (or 1:4:1 if a power spectrum is plotted). Then check that your Y_n values can be resummed to reproduce this input.

12.9 ASSESSMENT: HIGHLY NONLINEAR OSCILLATOR

Recall the numerical solution for oscillations of the spring with power $p = 11$, (12.1). Decompose the solution into a Fourier series and determine the number of higher harmonics that contribute at least 10% (e.g., determine the n for which $|b_n/b_1| < 0.1$). Also check that summing your series reproduces your original solution. (*Warning:* The ω you use in your series must correspond to the actual frequency of the system, not just that in the small oscillation limit.)

12.10 ASSESSMENT: NONLINEARLY PERTURBED OSCILLATOR

Recall the harmonic oscillator with a nonlinear perturbation (11.1):

$$V(x) = \tfrac{1}{2}kx^2\left(1 - \tfrac{2}{3}\alpha x\right), \quad F(x) = -kx(1 - \alpha x). \quad (12.53)$$

For very small amplitudes of oscillation ($x \ll 1/\alpha$), the solution $x(t)$ will essentially be only the first term of a Fourier series.

1. Fix your value of α to about 10%.

2. Decompose your numerical solution into a Fourier series.

3. Plot a graph of the percentage importance of the first *two* Fourier components as a function of the initial displacement for $0 < x_0 < \frac{1}{2\alpha}$. (You may find that higher harmonics are more important as the amplitude increases.)

4. As always, make sure to check by resuming the series for $y(t)$ and seeing if the input is reproduced.

(*Warning:* The ω you use in your series must correspond to the *true* frequency of the system, not just ω in the small oscillation limit.)

12.11 EXPLORATION: DFT OF NONPERIODIC FUNCTIONS

Consider a simple model of a "localized" electron that moves through space and time. We assume that the electron is described by a wave packet $\psi(x)$ that is a function of the spatial coordinate x. A good model for an electron initially localized around $x = 5$ is a Gaussian multiplying a plane wave:

$$\psi(x, t = 0) = \exp\left[-\frac{1}{2}\left(\frac{x - 5.0}{\sigma_0}\right)^2\right] e^{ik_0 x}. \tag{12.54}$$

This wave packet is not an eigenstate of the momentum operator[6] $p = id/dx$, and in fact contains a spread of momenta. The **problem** is evaluate the Fourier transform

$$\psi(p) = \frac{1}{\sqrt{2\pi}} \int_{-\infty}^{+\infty} dx e^{ipx} \psi(x) \tag{12.55}$$

as a way of determining the momenta spectrum in (12.54).

12.12 EXPLORATION: PROCESSING NOISY SIGNALS

You have measured the function of time $y(t)$ with a noisy detector. The **problem** is to remove the noise from $y(t)$. To do that we assume that the measured signal is a sum of the true signal $s(t)$, in which you have an interest, and some *noise* $n(t)$, in which you have no interest:

$$y(t) = s(t) + n(t). \tag{12.56}$$

Your problem is to deduce $s(t)$.

12.13 MODEL: AUTOCORRELATION FUNCTION

To remove noise from $y(t)$, we examine y's *autocorrelation function:*

$$A(\tau) \stackrel{\text{def}}{=} \int_{-\infty}^{+\infty} dt y(t) y(t + \tau). \tag{12.57}$$

[6]We use natural units in which $\hbar = 1$.

Here the time τ is called the *lag time*, and the integral can be evaluated for all values of τ. To understand how this integration removes noise, we express $y(t)$ in terms of its Fourier transform:

$$y(t) = \frac{1}{\sqrt{2\pi}} \int_{-\infty}^{+\infty} d\omega \; Y(\omega) e^{-i\omega t}. \tag{12.58}$$

We substitute this representation for $y(t)$ and $y(t+\tau)$ into the definition (12.57) and assume that the integrals converge well enough to be rearranged:

$$\begin{aligned} A(\tau) &= \frac{1}{2\pi} \int_{-\infty}^{+\infty} d\omega \int_{-\infty}^{+\infty} d\omega' Y(\omega) Y(\omega') e^{-i\omega'\tau} \int_{-\infty}^{+\infty} dt' e^{-i(\omega+\omega')t} \\ &= \frac{1}{2\pi} \int_{-\infty}^{+\infty} d\omega \int_{-\infty}^{+\infty} d\omega' Y(\omega) Y(\omega') e^{-i\omega'\tau} 2\pi\delta(\omega+\omega') \\ &= \int_{-\infty}^{+\infty} d\omega Y(\omega) Y(-\omega) e^{i\omega\tau}, \end{aligned} \tag{12.59}$$

where we have used the Dirac delta function (12.22). The autocorrelation function is seen to be the inverse transform of the product $Y(\omega)Y(-\omega)$. If $y(t)$ is real, then $Y(-\omega) = Y^*(\omega)$ and $A(t)$ is the inverse transform of the *power spectrum* $|Y(\omega)|^2$:

$$A(\tau) = \int_{-\infty}^{+\infty} d\omega |Y(\omega)|^2 e^{i\omega\tau}. \tag{12.60}$$

According to (12.60), we can think of the autocorrelation function as related to the intensity of our input signal at time τ, $|y(\tau)|^2$. But much of the noise present in $|y(t)|^2$ is absent in $A(\tau)$. If we take the Fourier transform of (12.56), we obtain the simple sum of transforms:

$$Y(\omega) = S(\omega) + N(\omega), \tag{12.61}$$

$$\text{where} \quad S(\omega) = \frac{1}{\sqrt{2\pi}} \int_{-\infty}^{+\infty} dt \; s(t) e^{i\omega t}, \tag{12.62}$$

$$N(\omega) = \frac{1}{\sqrt{2\pi}} \int_{-\infty}^{+\infty} dt \; n(t) e^{i\omega t}. \tag{12.63}$$

The autocorrelation function (12.57), involving the second power of y, is not a linear function; that is, $A_y \neq A_s + A_n$, but rather

$$A_y(\tau) = \int_{-\infty}^{+\infty} dt \left[s(t)s(t+\tau) + s(t)n(t+\tau) + n(t)n(t+\tau) \right]. \tag{12.64}$$

If we assume that the noise $n(t)$ is random, it should have an average of zero for long times, with its value at times t and $t+\tau$ uncorrelated. In these cases both integrals involving the noise vanish and we obtain

$$A_y(\tau) \approx \int_{-\infty}^{+\infty} dt s(t) s(t+\tau) \ = A_s(\tau). \tag{12.65}$$

The autocorrelation function of y in this way provides a picture of the power spectrum of y with the random noise removed.

12.14 ASSESSMENT: DFT AND AUTOCORRELATION FUNCTION

Write a discrete Fourier transform routine to transform

$$f(t) = \frac{1}{1 + 0.9 \sin t}. \tag{12.66}$$

Observe that while there is only one frequency in the denominator, the expansion of the function as a power series in $\sin t$ contains all frequencies:

$$f(t) \simeq 1 - 0.9 \sin t + \frac{(0.9 \sin t)^3}{3!} - \frac{(0.9 \sin t)^5}{5!} + \cdots. \tag{12.67}$$

1. Represent this function on a numerical mesh at times $t_i = 2\pi i p/n$, for $i = 0, 1, \ldots, N$, with p and n arbitrary integers. Increasing n makes the sampled time steps finer and finer. Increasing p makes the total time sampled longer and longer.

2. Plot the Fourier transform and the *autocorrelation function* of $f(t)$ for different values of n and p.

3. Observe what happens as you increase p and n. You may see that using more than one period to sample a periodic function does not improve the quality of your transform, but using more independent time steps (finer scales) does improve the quality of the high-frequency components.

4. Take values of p and n for which you obtained good results and add some random noise by sampling the function

$$y(t_i) = f(t_i) + \alpha(2r_i - 1), \quad 0 < r_i < 1, \tag{12.68}$$

where α is an adjustable parameter and r_i is a random number in the interval 0 to 1.

5. For different values of α, plot your noisy data, their Fourier transform, and their autocorrelation function. Notice how the noise appears to decrease.

6. Look for a relation between $A(t)$ and $|f(t)|^2$.

7. For which value of α do you essentially lose all information?

12.15 PROBLEM 2: MODEL DEPENDENCE OF DATA ANALYSIS ⊙

The scattering of electrons and x-rays from solids, atoms, molecules, and nuclei provides a means of determining the charge density $\rho(r)$ of the target. The experiments actually measure the *form factor* or *structure function* of the target,[7] that is, the Fourier transform of the charge density:

$$F(q) = \int d^3 r e^{i\mathbf{q}\cdot\mathbf{r}} \rho(r) = 4\pi \int_0^\infty r^2 dr \frac{\sin qr}{qr} \rho(r). \tag{12.69}$$

Here $\mathbf{q}$ is the momentum transferred during scattering and the 1-D integral obtains for spherically symmetric $\rho(r)$. The **problem** is to determine $\rho(r)$ from experimental measurements of $F(q)$; that is, to *invert* the transform in (12.69) to obtain ρ as a function of r. The real problem is that a laboratory beam of particles has a finite momentum, which means that there is a finite limit to the largest q value $q_{\max}$ at which $F(q)$ can be measured, and that experiments must be finished in a finite period of time, which means that only a finite number of q values can be measured. The $q < q_{\max}$ limitation leads to uncertainties in those parts of $\rho(r)$ that oscillate with high frequencies. The discrete number of measurements means that there are uncertainties in lower-frequency components as well.

12.16 METHOD: MODEL-INDEPENDENT DATA ANALYSIS

The traditional solution to this problem has been to *assume* that some specific functional form for the density $\rho(r)$ containing a number of adjustable parameters, and then find a best fit to the data by varying the parameters. A shortcoming with this approach is that the values deduced for the density depend somewhat on the functional form assumed for $\rho(r)$.

A more general approach is called *model-independent analysis*. In it, $\rho(r)$ is expanded in a complete set of functions

$$\rho(r) = \sum_{n=1}^{\infty} \rho_n(r), \tag{12.70}$$

with, typically, one parameter in each $\rho_n(r)$. If the ρ_n values form a complete set and if the sum actually goes out to $n = \infty$, then this is an exact

[7]While the form factor can be deduced directly from experiment only in first Born approximation, the method described here is more general.

representation of any $\rho(r)$. In practice, the discrete and finite nature of the measurements limit the number of ρ_n values that can be determined to a maximum number N, and so we have an approximate representation:

$$\rho(r) \approx \sum_{n=1}^{N} \rho_n(r). \tag{12.71}$$

To make a clear separation of how the deduced $\rho_n(r)$ values are affected by the measurements of $F(q)$ for $q > q_{\max}$, we impose the constraint that the ρ_n values determined for measurement with $q < q_{\max}$ be orthogonal to those determined with $q > q_{\max}$. While the equations will appear similar to those in DFT, in DFT *all* N values of y_k are used to determine each Y_n, while here we require each ρ_n to be determined by only *one* measured $F(q)$. It is then manifest that each new experimental measurement determines one additional expansion coefficient.

Now for a specific example. The density $\rho(r)$ within a specific atomic nucleus is known to vanish rapidly beyond some radius $r \simeq R$ and to be approximately constant for $r \simeq 0$. We therefore model $\rho(r)$ as the sine series:

$$r\rho(r) = \begin{cases} \sum_{n=1}^{N} \rho_n(q_n r) = \sum_{n=1}^{N} b_n \sin(q_n r), & \text{for } r \leq R, \\ 0, & \text{for } r > R\,. \end{cases} \tag{12.72}$$

For simplicity, we use a simple model for the measured momentum-transfer values:

$$q_n = \frac{n\pi}{R}. \tag{12.73}$$

This makes it clear that each additional measurement is made at a higher q value. The coefficients b_n are determined from the inversion formula (12.8):

$$b_n = \frac{2}{R}\int_0^R \sin(q_n r) r\rho(r) dr \;=\; \frac{q_n F(q_n)}{2\pi R}. \tag{12.74}$$

The corresponding expression for the form factor as determined by the measured values $F(q_n)$ is

$$F(q) = \frac{2\pi R}{q}\sum_{n=1}^{N} b_n \left\{ \frac{\sin\left[(q-q_n)R\right]}{(q-q_n)R} - \frac{\sin\left[(q+q_n)R\right]}{(q+q_n)R} \right\}. \tag{12.75}$$

These relations show exactly how measurements at larger and larger momentum transfers, q_n values, determine the higher and higher Fourier components b_n values, which, in turn, are related to higher- and higher-frequency ripples in the charge density. (This is sometimes stated somewhat loosely as "large q are needed to measure small r.") Regardless of the model assumed for the density ρ, any experiment with a definite $q_{\max}$ has a limit on the components of the density it can deduce. As higher and higher q measurements are made, more components of ρ are determined.

12.17 ASSESSMENT

1. As a simple exercise, verify the sensitive relation between each measurement at q_n and the Fourier component b_n by evaluating the term in braces in (12.75) for $n = 10$ and plotting it as a function of qR. You should find a peaking that indicates the region of sensitivity; that is, the region to which a single experimental measurement is most sensitive.

2. Now try a computer experiment that simulates a model-independent data analysis. Assume at first that you are mother nature and so you know that the actual charge distribution $\rho(r)$ and form factor for some nucleus is

$$\rho(r) = \rho(0)\left[1 + \alpha(r/a)^2\right] e^{-(r/a)^2}, \qquad (12.76)$$

$$F(q) = \left[1 - \frac{\alpha(qa)^2}{2(2+3\alpha)}\right] e^{-(qa)^2/4}. \qquad (12.77)$$

   Use this analytic expression for $F(q)$ to determine the values of the experimental $F(q_m)$ as needed. In a real-world situation you would measure data and then fit them to determine the Fourier coefficients b_m.

   (a) Consider the nucleus ^{16}O which has $\alpha = \frac{4}{3}$. We will measure distances in fermis, fm $= 10^{-13}$ cm and momentum transfers q in inverse fermis. In these units, mother nature knows that $a \simeq 1.66$, and you as an outsider would have to try values of $R \simeq 5$ and $R \simeq 6$ as the radius beyond which the nuclear density vanishes.

   (b) Generate from (12.77) a table of $F(q_n)$ values for both values of R and for q values starting at zero and increasing to the point where $F(q) \approx 10^{-11}$. Plot up both (they should fall on the same curve).

   (c) Use (12.72) to calculate and plot the contribution to $\rho(r)$ coming from progressively larger and larger values of q_n.

   (d) For 10 terms, examine how the sum for $\rho(r)$ differs for the two R values used. This is a good measure of the *model dependence* in this "model independent" analysis.

   (e) Again, for the sum of 10 terms, examine how the Fourier series for $\rho(r)$ differs from the actual functional form (12.76).

   (f) Examine the series expansion for $F(q)$ and $\rho(r)$ and note any unphysical oscillations.

   (g) Do a computer experiment in which you assume that a different form for the large q behavior of $F(q)$ [e.g., $1/q^4, \exp(-bq)$], and then see how this affects the deduced $\rho(r)$. You should find that the small r ripples are most sensitive to the assumed large q behavior.

13 Unusual Dynamics of Nonlinear Systems

13.1 PROBLEM: VARIABILITY OF BUG POPULATIONS

The population of insects and the patterns of weather do not appear to follow any simple laws.[1] At times they appear stable, at other times they vary periodically, and at other times they appear chaotic, only to settle down to something simple again. Your **problem** is to deduce if a simple law might be producing such complicated behavior.

13.2 THEORY: NONLINEAR DYNAMICS

In many ways nonlinear dynamics is the glory of computational science. The computer helps us solve equations that are otherwise inaccessible, and because it is rather painless, it encourages exploration. And it works! The computed solutions have led to the discovery of new phenomena such as *solitons*, *chaos*, and *fractals*.

While volumes have been written on nonlinear dynamics, we will spend only two short chapters studying some simple systems. Nevertheless, you will uncover unusual properties on your own, and in the process cannot help but getting convinced that simple systems can have very complicated behaviors.

In some cases these complicated behaviors will be *chaotic*, but unless you

[1]Other than in Oregon, where storm clouds come to spend their weekends.

have a bug in your program, they will not be random.[2] We define *chaos* as the deterministic behavior of a system displaying no discernible regularity. This may seem contradictory; if a system is deterministic, it must have step-to-step correlations (which when added up means long-range correlations), but if it is chaotic, this means that it is too complicated to understand.

In an operational sense, a chaotic system is one with an extremely high sensitivity to parameters or initial conditions. This sensitivity to even minuscule changes is so high that, in practice, it is impossible to predict the long-range behavior unless the parameters are known to infinite precision (which they never are, in practice).

13.3 MODEL: NONLINEAR GROWTH, THE LOGISTIC MAP

Imagine a bunch of insects reproducing, generation after generation. We start with N_0 bugs, then in the next generation we have to live with N_1 of them, and after i generations there are N_i bugs to bug us. We want to model the time dependence of N_i, with time measured discretely in steps or generations.

For guidance, we look to the radioactive decay simulation in Chapter 7, *Monte Carlo Applications*, where the discrete decay law, $\Delta N/\Delta t = -\lambda N$, led to exponential decay when the numbers were large and the time steps short. So, if we assume that the bug breeding rate is proportional to the number of bugs:

$$\frac{\Delta N_i}{\Delta t} = \lambda' \, N_i, \tag{13.1}$$

we know that this will lead to unbounded exponential growth with a rate parameter λ'.

We can improve the model by realizing that bugs cannot live on love alone. They must also eat. But bugs, not being farmers, do not produce more food if they need it to sustain their growing numbers, and so the competition for a finite food supply tends to limit their number to a maximum N_*. We build this into our model by defining a new birth-rate parameter λ' that is proportional to the difference of the present and maximum populations:

$$\lambda' = \lambda(N_* - N_i), \tag{13.2}$$

$$\Rightarrow \quad \frac{\Delta N_i}{\Delta t} = \lambda'(N_* - N_i)N_i. \tag{13.3}$$

Physically, we expect that when their numbers are small, the bugs will grow exponentially, and as their numbers increase, we will see some type of modulation in the growth.

[2]You may recall from Chapter 6, *Deterministic Randomness*, that a random sequence of events does not even have step-by-step correlations.

13.3.1 The Logistic Map

Equation (13.3) is known mathematically as the *logistic map*. It is usually written in dimensionless form as an equation for the number of bugs in generation $i+1$:

$$\begin{aligned} N_{i+1} &= N_i + \lambda' \Delta t (N_* - N_i) N_i, && (13.4) \\ &= N_i \left(1 + \lambda' \Delta t N_*\right) \left[1 - \frac{\lambda' \Delta t}{1 + \lambda' \Delta t N_*} N_i\right]. && (13.5) \end{aligned}$$

We define a dimensionless growth parameter μ and a dimensionless population variable x_i:

$$\begin{aligned} \mu &\stackrel{\text{def}}{=} 1 + \lambda' \Delta t N_*, && (13.6) \\ x_i &\stackrel{\text{def}}{=} \frac{\lambda' \Delta t}{\mu} N_i \simeq \frac{N_i}{N_*}. && (13.7) \end{aligned}$$

Observe that the *growth rate* μ equals 1 when the breeding rate $\lambda' = 0$, and is otherwise expected to be larger than 1. If the number of bugs born per generation $\lambda' \Delta t$ is large, then $\mu \approx \lambda' \Delta t N_*$ and $x_i \approx N_i / N_*$. That is, x_i is essentially the fraction of the maximum population N_*. Consequently, we consider x values in the range

$$0 \leq x_i \leq 1, \qquad (13.8)$$

where the value $x = 0$ corresponds to no bugs and $x = 1$ to the maximum population.

Dressed in these natural variables, the difference equation (13.5) assumes the standard form for the *logistic map*:

$$\boxed{x_{i+1} = \mu x_i (1 - x_i).} \qquad (13.9)$$

In general, a map may use any function $f(x)$ to map one number in a sequence to the next:

$$x_{i+1} = f(x_i). \qquad (13.10)$$

For the logistic map, $f(x) = \mu x(1-x)$. The quadratic dependence of f on x makes this a nonlinear map. The dependence of (13.9) on only the one variable x makes it a *one-dimensional* map.

We have developed a discrete model for the bug population and have expressed it as a *difference equation.* We will study this equation and see that its nonlinear dependence on x leads to some unusual behaviors. Similar behaviors are found in nonlinear differential equations, as we will see in Chapter 14, *Differential Chaos in Phase Space*; Chapter 28, *Solitons, the KdeV Equation*; and Chapter 29, *Sine–Gordon Solitons*.

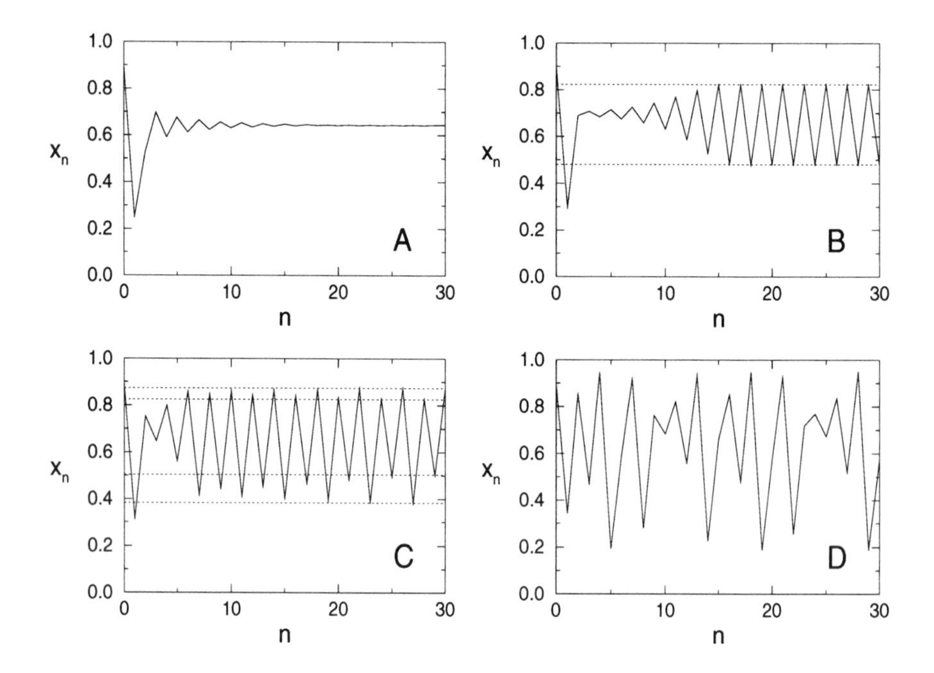

Fig. 13.1 The insect population x_n versus generation number n for various growth rates: (A) $\mu = 2.8$, a period-one cycle; (B) $\mu = 3.3$, a period-two cycle; (C) $\mu = 3.5$, a period-four cycle; (D) $\mu = 3.8$, a chaotic regime.

13.4 THEORY: PROPERTIES OF NONLINEAR MAPS

Some typical nonlinear behavior is shown in Fig. 13.1. The initial population x_0 is known as the *seed*, and as long as it is not equal to zero, its exact value generally has little effect on the population dynamics (similar to what we found when generating pseudo-random numbers). In contrast, the dynamics are controlled by the value of the growth parameter μ. For those values of μ at which the dynamics are chaotic, there is an extreme sensitivity to the initial condition, x_0, as well as on the exact value of μ.

13.4.1 Fixed Points

An important property of the map (13.9) is the possibility of a sequence of x_i values reaching a *fixed* point at which x_i remains or keeps returning to. We denote fixed points by x_*. At a one-cycle fixed point there is no change in the population from generation i to generation $i+1$, and so it must satisfy the equation

$$x_{i+1} = x_i = x_*. \tag{13.11}$$

If we now use the logistic map (13.9) to relate x_{i+1} to x_i, we obtain an algebraic equation to solve

$$\mu x_*(1 - x_*) = x_* \tag{13.12}$$

$$\Rightarrow \quad x_* = 0, \text{ or } \quad x_* = \frac{\mu - 1}{\mu}. \tag{13.13}$$

The nonzero fixed point $x_* = (\mu - 1)/\mu$ corresponds to a stable population with an equilibrium between birth and death that is reached regardless of the initial population. In contrast, the $x_* = 0$ point is unstable and the population remains static only as long as no bugs exist; if even a few bugs are introduced, exponential growth occurs.

Further analysis [Rash 90] tells us that the stability of a population is determined by the magnitude of the derivative of the mapping function $f(x_i)$ at the fixed point:

$$\left|\frac{df}{dx}\right|_{x_*} < 1 \quad \text{(stable)}. \tag{13.14}$$

For the logistics map (13.9) this means that

$$\left.\frac{df}{dx}\right|_{x_*} = \mu - 2\mu x_* = \begin{cases} \mu, & \text{stable at } x_* = 0 \text{ if } \mu < 1, \\ 2 - \mu, & \text{stable at } x_* = \frac{\mu-1}{\mu} \text{ if } \mu < 3. \end{cases} \tag{13.15}$$

13.4.2 Period Doubling, Attractors

Equation (13.15) tells us that while the equation for fixed points (13.13) may be satisfied for all values of μ, the points will not be stable if $\mu > 3$. For $\mu \geq 3$, the system's long-term population *bifurcates* into two populations (*period doubling*). Because the system then acts as if it were attracted now to two populations, these populations are called *attractors* or cycle points. We can easily predict the x values for these two-cycle attractors by requiring that generation $i + 2$ have the same population as generation i:

$$x_* = x_i = x_{i+2} = \mu x_{i+1}(1 - x_{i+1}), \tag{13.16}$$

$$\Rightarrow \quad x_* = \frac{1 + \mu \pm \sqrt{\mu^2 - 2\mu - 3}}{2\mu}. \tag{13.17}$$

We see that as long as $\mu > 3$, the square root produces a real number and thus physical solutions (complex or negative x_* values are unphysical).

We leave it for your computer explorations to discover how the system continues to double periods as the value of μ gets larger. In all cases the pattern is the same: one population bifurcates into two.

13.5 IMPLEMENTATION AND ASSESSMENT: EXPLICIT MAPPING

Program the logistic map to produce a sequence of population values x_i as a function of the generation number i. These are called *map orbits.* The assessment consists of confirmation of Feigenbaum's observations [Feig 79] of the different behavior patterns shown in parts A, B, C, and D of Fig. 13.1. These occur for growth parameter $\mu = (0.4,\ 2.4,\ 3.2,\ 3.6,\ 3.8304)$ and seed population $x_0 = 0.75$. Notice the following on your graphs of x_i versus i:

1. **Transients:** Each map undergoes a *transient behavior* before reaching a steady state. These transients differ for different seeds.

2. **Asymptotes:** In some cases the steady state is reached after only 20 generations, while for larger μ values, hundreds of generations may be needed. These steady-state populations are independent of the seed.

3. **Extinction:** As shown in Fig. 13.1A, if the growth rate is too low, the population dies off.

4. **Stable states:** The single-population stable states attained for $\mu < 3$ agree with the prediction (13.13).

5. **Period doubling:** Examine the map orbits for a growth parameter μ increasing continuously through 3. Try to discover how the system continues to double periods as μ increases. For example, in Fig. 13.1C with $\mu = 3.5$, we notice a steady state in which the population alternates among four attractors (a four-cycle point).

6. **Intermittency:** Try to find solutions for $3.8264 < \mu < 3.8304$. Here the system appears stable for a while but then jumps all around, only to become stable again.

7. **Chaos:** The chaotic region is critically dependent on the value of μ, with the x_i values obtained critically dependent on x_0.

   (a) Verify this by running the logistics map with $\mu = 4$ and what should essentially be two identical seeds:

   $$x_0 = 0.75, \text{ and } x_0' = 0.75(1+\epsilon). \tag{13.18}$$

   Here ϵ should be a number very close to machine precision, for example; $\epsilon \simeq 2 \times 10^{-14}$ for double precision.

   (b) Now repeat the experiment with $x_0 = 0.75$ but with what should essentially be two identical growth parameters:

   $$\mu = 4.0, \text{ and } \mu' = 4.0(1-\epsilon). \tag{13.19}$$

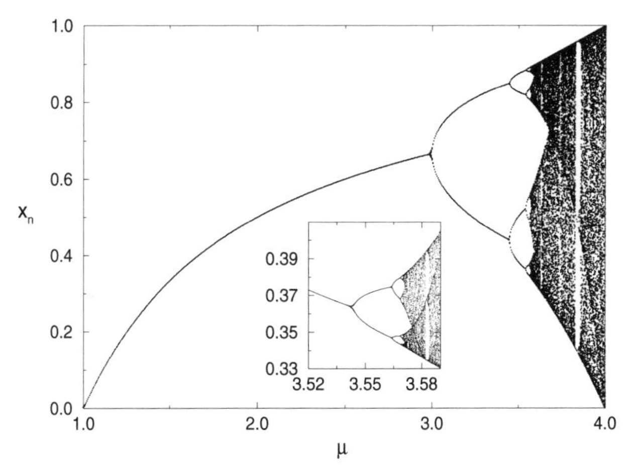

Fig. 13.2 The bifurcation plot, attractor populations versus growth rate, for the logistics map.

13.6 ASSESSMENT: BIFURCATION DIAGRAM

Computing and watching the population change with each generation gives a good idea of the basic phenomena, at least until the time dependence gets so complicated that it is hard to discern patterns. In particular, as the number of bifurcations keeps increasing and the system becomes chaotic, it is hard for our minds to see a simple underlying structure within the complicated behavior.

One way to visualize what is going on is to concentrate on the attractors; that is, those populations that appear to attract the solutions and to which the solutions continuously return. A plot of these attractors (long-term iterates) of the logistics map as a function of the growth parameter μ is an elegant way to summarize the results of extensive computer simulations.

One such *bifurcation diagram* is given in Fig. 13.2 (another is given in Fig. 13.3). For each value of μ, hundreds of iterations were made to make sure that all transients died out. Then the points (μ, x_*) were written to a file for hundreds of iterations after that. To be sure to reach all attractors, various values for the initial populations x_0 are stepped throughout.

13.7 IMPLEMENTATION: BIFURCATION DIAGRAM, BUGS.F (.C)

You should reproduce Fig. 13.2.[3] While the best way to make a visualization of this sort would be with a visualization program that permits you to vary the intensity of each individual point on the screen, we outline a method in which you control the density at each point in your plot by varying the number of pixels plotted.

Notice that your screen resolution may be $\sim$100 dots per inch and your laser printer resolution $\sim$300 dots per inch. This means that you need to plot $\sim 3000 \times 3000 \simeq 10$ million elements. But *beware* for this can take some time to print and enough memory space on your hard disk and printer to choke them.

1. Break up the range $1 \leq \mu \leq 4$ into a 1000 steps and loop through them.
2. To be complete, loop through a range of x_0 values as well.
3. Wait at least 200 generations for the transients to die and then print out the next several hundred values of (μ, x_*) to a file.
4. Print out your x_* values to no more than 3–4 decimal places (you will not be able to resolve more places on your plot).
5. Sort[4] your (μ, x_*) file to remove any duplicate points (this may be a slow process, but it takes much less time than waiting for a choked printer).
6. Plot up your file of x_* versus μ. Use small symbols for the points and do not connect them.
7. Enlarge sections of your plot and notice that a similar bifurcation diagram tends to be contained within portions of the original (*self similarity*).
8. Notice the series of bifurcations at

   $$\mu_k \simeq 3, 3.449, 3.544, 3.5644, 3.5688, 3.569692, 3.56989, \ldots. \qquad (13.20)$$

   The end of this series is chaotic behavior.
9. Notice that after the sequence ends, others begin, only to end in chaos again. (The changes are rather quick, and plots with an enlarged μ scale are illuminating.)
10. Close examination of Fig. 13.2 shows regions containing very few populations (which are not artifacts of the video display). These are *windows*

[3] You can listen to a sonification of it on the Web.
[4] For example, with the Unix command *sort -u*.

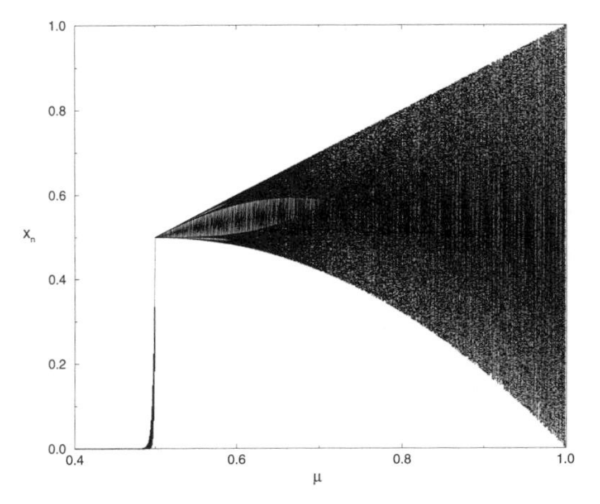

Fig. 13.3 A bifurcation plot for the tent map showing fixed points versus growth rate.

in which a slight increase in the growth rate μ changes a chaotic population into one with a finite number of fixed points. Check that at $\mu = 3.828427$ chaos turns into a three-cycle population.

13.8 EXPLORATION: RANDOM NUMBERS FROM LOGISTIC MAP?

There are claims [P&R 95] that the logistics map in the chaotic region

$$x_{i+1} = 4x_i(1 - x_i), \tag{13.21}$$

can be used to generate random numbers. While we know that successive bug populations are correlated, if the population for every $\sim$ 6th generation is examined, the correlations die out and random numbers result. To help make the sequence uniform, a trigonometric transformation is used

$$y_i = \frac{1}{\pi} \cos^{-1}(1 - 2x_i). \tag{13.22}$$

Use the random-number tests discussed in Chapter 6, *Deterministic Randomness*, to confirm this claim.

13.9 EXPLORATION: FEIGENBAUM CONSTANTS

It was discovered by [Feig 79] that the sequence of μ_k values (13.20) converges geometrically when expressed in terms of the variable δ, which is the distance

Table 13.1 Several nonlinear maps to explore

Name	$f(x)$
Tent	$\mu(1-2\left\|x-1/2\right\|)$
Ecology	$xe^{\mu(1-x)}$
Quartic	$\mu[1-(2x-1)^4]$

between bifurcations:

$$\mu_k \rightarrow \mu_\infty - \frac{c}{\delta^k}, \tag{13.23}$$

$$\delta = \lim_{k\rightarrow\infty} \frac{\mu_k - \mu_{k-1}}{\mu_{k+1} - \mu_k}. \tag{13.24}$$

Use your sequence of μ_k values to determine the constants in (13.23) and compare to those found by Feigenbaum:

$$\mu_\infty \simeq 3.56995, \quad c \simeq 2.637, \quad \delta \simeq 4.6692. \tag{13.25}$$

Amazingly, the value of the *Feigenbaum constant* δ is a universal constant for all second order maps.

13.10 EXPLORATION: OTHER MAPS

Only nonlinear systems exhibit unusual behavior like chaos. Yet systems can be nonlinear in any number of ways. Table 13.1 lists three maps that you may use to generate x_i sequences and bifurcation plots. The tent map is illustrated in Fig. 13.3, which makes clear the origin of the name.

14

Differential Chaos in Phase Space

14.1 PROBLEM: A PENDULUM BECOMES CHAOTIC

In Fig. 14.1 we see a realistic pendulum *driven* through viscous air by an external, sinusoidal force. When the driving force is turned off, the system is observed to oscillate at the pendulum's natural frequency ω_0. When the driving force is strong, the system is observed to oscillate at the driver's frequency ω (*mode locking*). When ω_0 and ω are equal, or nearly equal, and the driving force is not too strong, a slow and periodic variation of the pendulum's amplitude is observed (*beating*). As the initial conditions are changed, the motion gets very complicated and appears chaotic. Your **problem** is to describe this behavior with a simple equation of motion, and to determine whether even the most complicated motion has some simple structure underlying it all.

In Chapter 13, *Unusual Dynamics of Nonlinear Systems*, we discovered that a simple nonlinear *difference* equation yields solutions that may be simple, complicated, or chaotic. In this chapter we search for similar behavior in *differential* equations. We also reveal the beauty and simplicity underlying chaotic systems by observing their flow in phase space.

Our study is based on the description given by [Rash 90]. An excellent, analytic discussion of the related *parametric oscillator* is given in [L&L 69, §25–30]. A similar system [G&T 96] has a vibrating pivot—in contrast to our periodic driving torque.

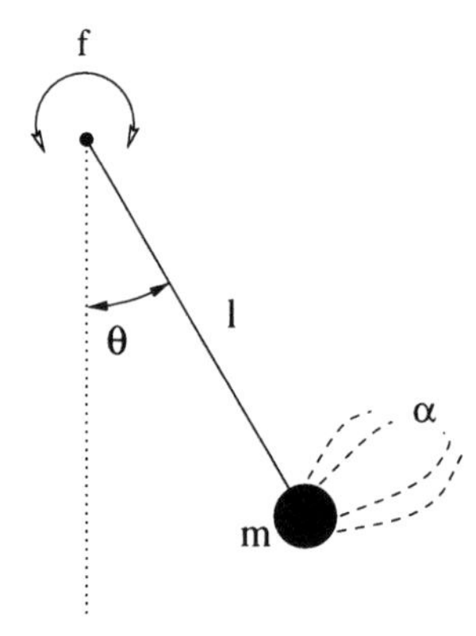

Fig. 14.1 A pendulum of length l driven through air by an external, sinusoidal torque. The strength of the torque is given by f and that of air resistance by α.

14.2 THEORY AND MODEL: THE CHAOTIC PENDULUM

The basic theory for the chaotic pendulum is the same as given in our discussion of the realistic pendulum in Chapter 11, *Anharmonic Oscillators.* Newton's laws of rotational dynamics tell us that the sum of the gravitational torque τ_g, the frictional torque τ_f, and the external torque τ_{ext} equals the moment of inertia of the pendulum times its angular acceleration:

$$\tau_g + \tau_f + \tau_{\text{ext}} = I \frac{d^2\theta}{dt^2}. \tag{14.1}$$

If we assume that the external torque is sinusoidal and that the viscous friction is proportional to the velocity of the pendulum's bob, we obtain the nonlinear differential equation

$$-\frac{mgl}{I}\sin\theta - \frac{\beta}{I}\frac{d\theta}{dt} + \frac{\tau_0}{I}\cos\omega t = \frac{d^2\theta}{dt^2}, \tag{14.2}$$

where l is the distance to the center of mass and the nonlinearity arises from the $\sin\theta$, as opposed to θ, dependence of the gravitational torque. In a more standard form, (14.2) is

$$\boxed{\frac{d^2\theta}{dt^2} = -\omega_0^2 \sin\theta - \alpha\frac{d\theta}{dt} + f\cos\omega t,} \tag{14.3}$$

where

$$\omega_0 \stackrel{\text{def}}{=} \frac{mgl}{I}, \quad \alpha \stackrel{\text{def}}{=} \frac{\beta}{I}, \quad f \stackrel{\text{def}}{=} \frac{\tau_0}{I}. \tag{14.4}$$

In review, ω_0 is the natural frequency of the system arising from the gravitational restoring force, the α term arises from friction, and the f term arises from the driving force.

Equation (14.3) is a second order, time-dependent, nonlinear differential equation. In the standard form we use to solve differential equations,[1] it is the two simultaneous first-order equations:

$$y^{(1)} \stackrel{\text{def}}{=} \theta, \quad y^{(2)} \stackrel{\text{def}}{=} \frac{d\theta}{dt}, \tag{14.5}$$

$$\frac{dy^{(1)}}{dt}(t) = y^{(2)}(t), \tag{14.6}$$

$$\frac{dy^{(2)}}{dt}(t) = -\omega_0^2 \sin y^{(1)}(t) - \alpha y^{(2)}(t) + f \cos \omega t. \tag{14.7}$$

14.3 THEORY: LIMIT CYCLES AND MODE LOCKING

It is easy to solve (14.6)–(14.7) on the computer. The difficulty is that the types of solutions that occur are so rich that it is not easy to figure out what's going on. Accordingly, we will examine some simple limiting cases where the motion is easy to understand.

When the chaotic pendulum is driven by a weak driving force (small value for f) with frequency ω equal to the natural frequency ω_0, it is possible to have a periodic steady-state motion with frequency of ω_0. In this case it is possible to pick the magnitude for the external torque such that after the initial transients die off, the average energy put into the system during one period exactly balances the average energy dissipated by friction during that period:

$$\langle f \cos \omega t \rangle = \langle \alpha \frac{d\theta}{dt} \rangle = \langle \alpha \frac{d\theta}{dt}(0) \cos \omega t \rangle, \tag{14.8}$$

$$\Rightarrow \quad f = \alpha \frac{d\theta}{dt}(0). \tag{14.9}$$

This leads to a *limit cycle* in which the motion is stable even in the presence of friction.

A somewhat opposite extreme occurs when the magnitude f of the driving torque is made much larger than that in (14.9). In this case the driving torque overpowers the natural oscillations of the pendulum and the steady-state motion is at the frequency of the driver. This is an example of *mode locking.* While mode locking can occur for linear or nonlinear systems, something unusual occurs for nonlinear systems. We have already seen how the oscillations

[1]See Chapter 9, *Differential Equations and Oscillations.*

of a nonlinear system may contain higher Fourier components (overtones). This means that that under the right conditions, the driving force may lock onto the system by exciting its overtones. In this case the driving frequency and natural frequency are rationally related:

$$\frac{\omega}{\omega_0} = \frac{n}{m}, \tag{14.10}$$

where n and m are integers.

14.4 IMPLEMENTATION 1: SOLVE ODE, RK4.F (.C)

Take your solution to the harmonic or anharmonic oscillator and extend it to (14.6)–(14.7). This means adding velocity and time-dependent terms to the derivative function subroutine. Make α and f input parameters. Run some checks before you attack the full problem; namely, verify that

1. If $\alpha = f = 0$, you get the realistic pendulum studied previously.

2. If $f = 0$ but $\alpha \neq 0$, you get a uniformly decaying solution.

14.5 ASSESSMENT AND VISUALIZATION: PHASE-SPACE ORBITS

The conventional solution to an equation of motion is the position $x(t)$ and the velocity $v(t)$ as functions of time. In contrast, it is illuminating to go to an abstract space, *phase space*, where the ordinate is the velocity $v(t)$ and the abscissa is the position $x(t)$. As we can see in Figs. 14.2–14.6, the solutions of the equations of motion of classical dynamics form geometric objects in phase space. In this way, periodic motion, which is rather complicated to describe in terms of the time dependence of a position and a velocity, becomes recognizable geometric objects. Likewise, chaos, which seems beyond our descriptive abilities in ordinary space, becomes easily recognized structures in phase space.

Phase-space plots are useful in visualizing the solutions to the equations of motion. To understand why, we look at the nondriven one-dimensional harmonic oscillator with no friction. The position and velocity of the oscillating mass as functions of time are

$$x(t) = A\sin(\omega t), \tag{14.11}$$

$$v(t) = \frac{dx}{dt} = \omega A\cos(\omega t). \tag{14.12}$$

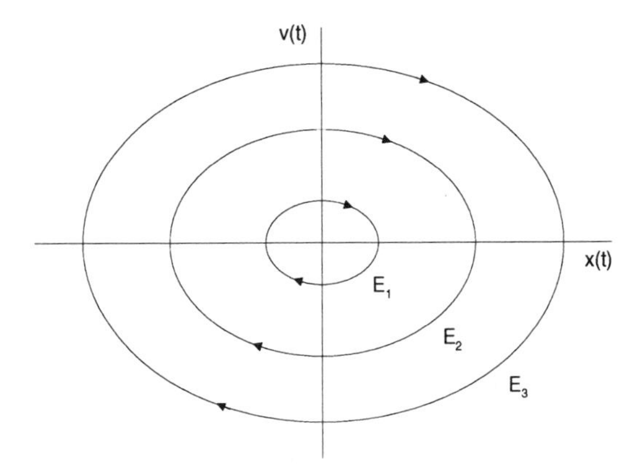

Fig. 14.2 A phase-space plot of velocity versus position for a harmonic oscillator. Because the ellipses close, the system must be periodic. The different ellipses correspond to different energies.

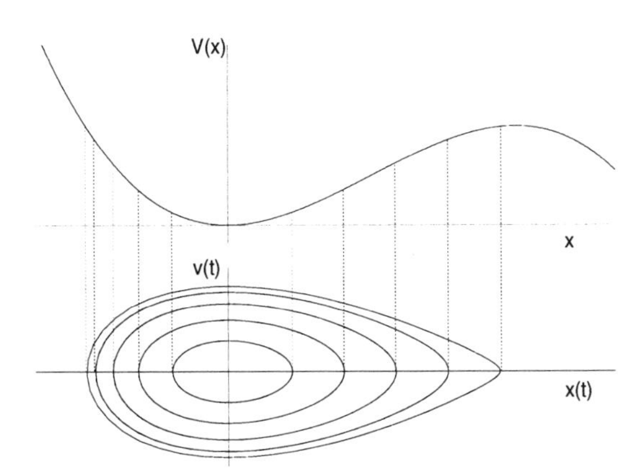

Fig. 14.3 (*Upper*) A potential-energy plot for a nonharmonic oscillator; (*lower*) a phase-space plot for the same nonharmonic oscillator. The ellipse-like figures are neither ellipses nor symmetric with respect to the v axis. The different orbits in phase space correspond to different energies.

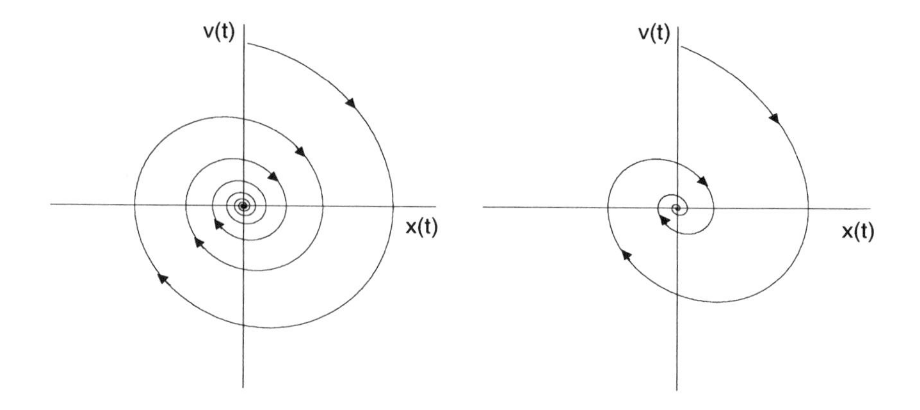

Fig. 14.4 Two phase-space orbits for oscillators with friction. The system eventually comes to rest at the origin. The damping is greater for the oscillator on the right.

When we use these solutions to form the total energy, we obtain a constant:

$$E = K(v) + V(x) = \tfrac{1}{2}mv^2 + \tfrac{1}{2}\omega^2 m^2 x^2 \tag{14.13}$$

$$= \frac{\omega^2 m^2 A^2}{2m}\cos^2(\omega t) + \tfrac{1}{2}\omega^2 m^2 A^2 \sin^2(\omega t) \tag{14.14}$$

$$= \tfrac{1}{2}m\omega^2 A^2. \tag{14.15}$$

Equation (14.13) implies that the harmonic oscillator follows closed elliptical orbits in phase space, with the size of the ellipse increasing with the system's energy. Different initial conditions produce the same ellipse if they have the same energy.

In Figs. 14.2–14.6 we show some typical structures encountered in phase space. Some generalities to note are

- For nonharmonic oscillators, the orbits will still be ellipse-like but with "corners" that become more distinct with increasing nonlinearity.
- Closed trajectories describe periodic oscillations [the same (x, v) occur again and again].
- The closed figures in the plots are stable *predictable attractors* or *stable attractors*. A nearby orbit in phase space is attracted to these (which means that they are stable).
- Phase-space orbits move clockwise for restoring forces (negative v after maximum x).
- Open orbits correspond to nonperiodic or "running" motion (a pendulum rotating like a propeller).

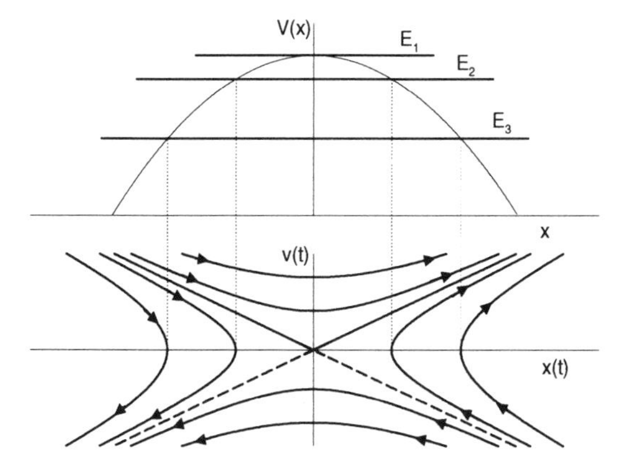

Fig. 14.5 Phase-space trajectories for a particle in a repulsive potential. Notice the absence of trajectories in the regions forbidden by energy conservation.

- Friction may cause the energy to decrease with time and the phase-space orbit to spiral into the a *fixed point.* However, for certain parameters, the energy pumped in by the external force exactly balances that lost by friction and a closed orbit results.

- For given initial conditions, different orbits do not cross because solutions are unique. Nonetheless, open orbits do come together at the points of unstable equilibrium (*hyperbolic points*).

14.6 IMPLEMENTATION 2: FREE OSCILLATIONS IN PHASE SPACE

You may recall the assessment in §11.10 in which you investigated phase-space orbits for anharmonic oscillations. That is the best place to start visualizing the chaotic pendulum.

1. Take your solution to the harmonic pendulum (θ, not $\sin\theta$, restoring torque) and plot the phase-space orbits for a variety of initial conditions:

   (a) $\theta(0) = 0$, $d\theta/dt(0) \neq 0$,

   (b) $\theta(0) \neq 0$, $d\theta/dt(0) = 0$,

   (c) $\theta(0) \neq 0$, $d\theta/dt(0) \neq 0$.

   Make sure that some of the initial conditions correspond to energies large enough to send the bob over the top ($E > 2ml$).

2. The phase-space flow should follow closed elliptic orbits. Determine whether the motion is clockwise or counterclockwise.

3. Plot the phase-space orbits for the same initial conditions but this time for the *realistic* pendulum ($\sin\theta$ not θ).

4. Deduce how the *shapes* of the orbits differ as the amplitude of oscillation becomes larger and larger.

5. Include friction into your model and describe the changes.

14.7 THEORY: CHAOTIC AND RANDOM MOTION IN PHASE SPACE

Random motion, when viewed in phase space, appears as a diffuse cloud filling all accessible regions of phase space. Periodic motion appears as smooth closed curves, ellipse-like in shape. As shown in Fig. 14.7, chaotic motion falls someplace in between. In particular, if viewed for long times and many initial conditions, the *flow* in phase space may contains dark *bands* rather than lines. The continuity within a band means that there is a continuum of solutions possible, and this is what causes the coordinate space solutions to appear chaotic. So even though the motion may be chaotic, the definite shape of the band means that there is a well-defined and simple structure within the chaos.

14.8 IMPLEMENTATION 3: CHAOTIC PENDULUM

The challenge with the computer study of the chaotic pendulum (14.6)–(14.7) is that the 4-D parameter space is immense. For normal behavior, sweeping through ω should show us resonances and beating; sweeping through α should show us underdamping, critical damping, and overdamping; sweeping through f should show us mode locking (at least for certain values of ω). These behaviors can be found here, yet they get mixed together.

Worse yet, when in the chaotic region, a minuscule change of a parameter or an initial condition may drastically change the solution. Accordingly, the exact locations of the characteristic regions in phase space are highly sensitive. For that reason, we are specific as to what parameters to use in the implementation to follow. You should not be surprised to require slight variations to obtain results similar to ours.

1. In this project you will reproduce the phase-space diagrams in Fig. 14.7. The different behaviors in this figure correspond to the different initial conditions (from top to bottom): $x(0) = -0.0885$, $v(0) = 0.8$; $x(0) = -0.0883$, $v(0) = 0.8$; $x(0) = -0.0888$, $v(0) = 0.8$. (The Web tutorials give animations showing an actual pendulum and sonifcations of these and other motions.)

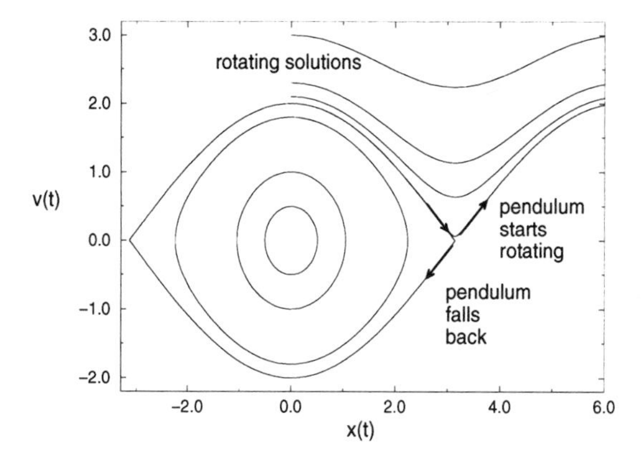

Fig. 14.6 Phase-space trajectories for a plane pendulum including "over the top" or rotating solutions. (Although not shown completely, the trajectories are symmetric with respect to vertical and horizontal reflections through the origin.)

(a) For this first step, set $\alpha = 0.2$ and $\omega_0 = 1$.

(b) To save time and storage, you may use a larger time step for plotting the orbits than used to solve the differential equations (a step size h equal to approximately $T/100$ is usually good). Plot $[\theta(t), d\theta/dt(t)]$ for ever-increasing time steps and see where your plots start losing detail.

(c) Indicate which part of the orbits are transients.

(d) Correlate phase-space structures with the behavior of $\theta(t)$ by plotting θ versus t on the same page as $d\theta/dt$ versus θ.

(e) Gain some physical intuition about the flow in phase space by watching how it builds up with time.

2. For the second part of the study, use the same parameters as in first part, but now sweep through a range of ω values.

(a) Use initial conditions: $d\theta/dt(0) = 0.8$, and $\theta(0) = -0.0888$.

(b) For $\omega \simeq 0.6873$, you should find a period-three limit cycle where the pendulum jumps between three major orbits in phase space. (More precisely, there are three dominant Fourier components.)

(c) For $\omega \simeq 0.694 - 0.695$, you should find running solutions where the pendulum keeps going over the top. Try to determine how many vibrations are made before the pendulum rotates.

(d) For $\omega \simeq 0.686$, you should find chaotic motion in which the paths in phase space become bands of motion and the Fourier spectrum becomes broad (if you let the solution run long enough). Try to

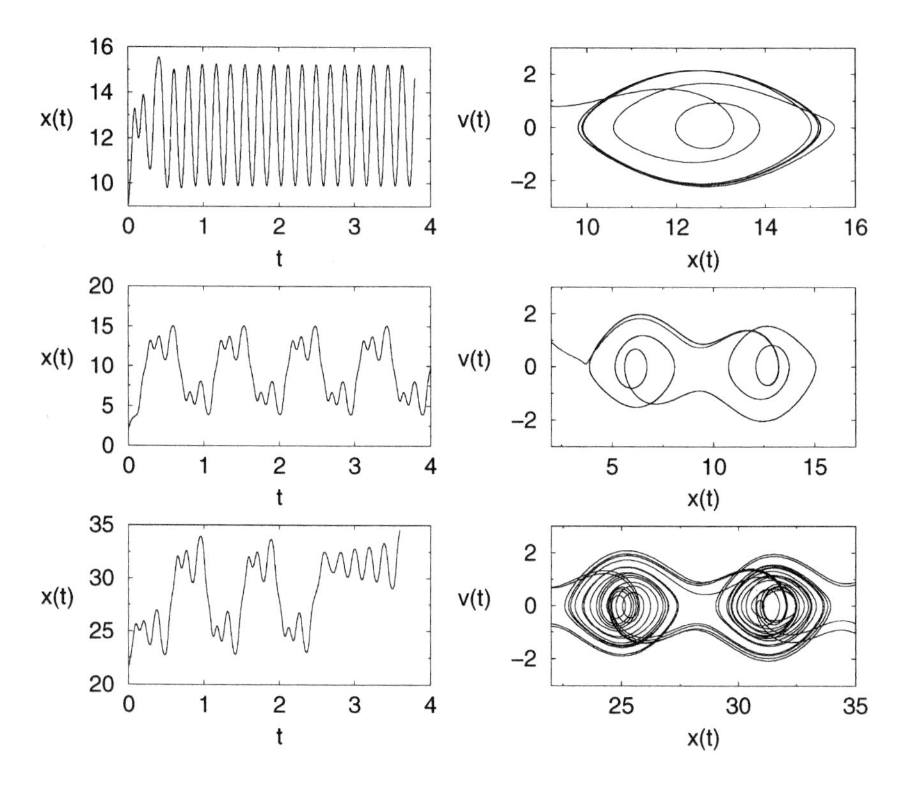

Fig. 14.7 A phase-space plot for the chaotic pendulum with $\omega_0 = 1$, $\alpha = 0.2$, $f = 0.52$, and $\omega = 0.666$. The angular position is $x(t)$ and the angular velocity is $v(t)$. The chaotic regions are the dark bands in the bottom figure.

determine just how small a difference in ω values separates the regular and the chaotic behaviors.

(e) Decrease your time step and try to determine how the bands get filled. Try to distinguish short-term and long-term behaviors in phase space.

14.9 ASSESSMENT: CHAOTIC STRUCTURE IN PHASE SPACE

Look at the plots you have produced and identify the following characteristic structures:

1. **Limit cycles:** ellipse-like figures with frequencies greater than ω_0. There may be sporadic changes among the limit cycles.
2. **Strange attractors:** well-defined, yet complicated semiperiodic behaviors that appear to be uncorrelated to the motion at an earlier time. These are highly sensitive to initial conditions. Even after millions of observations, the motion remains *attracted* to these paths.
3. **Predictable attractors:** well-defined, yet fairly simple periodic behaviors that are not particularly sensitive to initial conditions. These are orbits, such as fixed points and limit cycles, into which the system settles. If your location in phase space is near a predictable attractor, ensuing times will bring you to it.
4. **Chaotic paths:** regions of phase space that appear as filled-in bands rather than lines. The continuity within the bands implies some very complicated $\theta(t)$ behaviors, yet the general motion still has some underlying structure.

14.10 ASSESSMENT: FOURIER ANALYSIS OF CHAOTIC PENDULUM

We have seen that a realistic pendulum contains nonlinear terms in the resorting torque that lead to overtones; that is, frequencies other than just the fundamental. In addition, when a realistic pendulum is driven by an external sinusoidal force, the pendulum and driver may mode lock, and this leads to the pendulum moving at a frequency that is rationally related to the driver's. Consequently, the behavior of our chaotic pendulum is expected to be a combination of various periodic behaviors, sometimes occurring simultaneously and sometimes occurring sequentially.

In this assessment you determine the Fourier components present in the pendulum's complicated and chaotic behaviors. This should show that a "three-cycle structure," for example, contains three major Fourier components. You should also notice that when the pendulum goes over the top, its spectrum contains a steady-state ("dc") component.

1. Dust off your program which analyzes a $y(t)$ into Fourier components. (Alternatively, you may use a Fourier analysis tool contained in your graphics program or system library.)

2. Apply your analyzer to the solution of the forced, damped pendulum for the cases where there are one-, three-, and five-cycle structures in phase space. Deduce the major frequencies contained in these structures. (Try *not* to analyze the transient behavior.)

3. Try to deduce a relation between the Fourier components, the natural frequency ω_0, and the driving frequency ω.

4. A classic signal of chaos is a broadband, although not necessarily flat, Fourier spectrum. Examine your system for parameters that give chaotic behavior and verify this statement.

14.11 EXPLORATION: PENDULUM WITH VIBRATING PIVOT

A pendulum with a vibrating pivot point is an example of a *parametric resonance.* It is similar to our chaotic pendulum (14.3), but with the driving force depending on $\sin\theta$:

$$\frac{d^2\theta}{dt^2} = -\alpha\frac{d\theta}{dt} - \left(\omega_0^2 + f\cos\omega t\right)\sin\theta. \tag{14.16}$$

One way of understanding the physics of this equation is to go to the rest frame of the pivot (an accelerating reference frame) where you would say that there is a fictitious force that effectively leads to a sinusoidal variation of g or ω_0^2.

Analytic [L&L 69, §25-30] as well as numeric [DeJ 92, G&T 96] studies of this system exist. A fascinating aspect of this system is that the excitation of its modes of vibration (overtones) occurs through a series of *bifurcations.* In fact, when the instantaneous angular velocity $d\theta/dt$ is plotted as a function of the strength of the driving force, the bifurcation diagram in Fig. 14.8 results. Although the physics is very different, this behavior is manifestly similar to the bifurcation diagram for bug populations studied in Chapter 13, *Unusual Dynamics of Nonlinear Systems.* This behavior is, apparently, the result of mode locking and beating with the overtones.

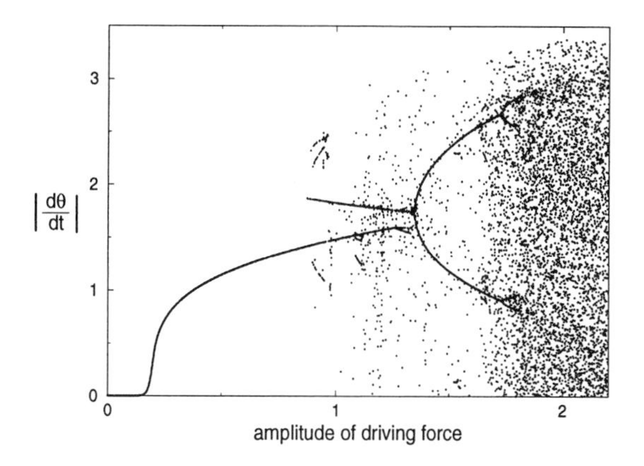

Fig. 14.8 Bifurcation diagram for the damped pendulum with a vibrating pivot. The ordinate is $|d\theta/dt|$, the absolute value of the instantaneous angular velocity at the beginning of the period of the driver, and the abscissa is the magnitude of the driving force d. The heavy line results from the overlapping of points, not from connecting the points. (Produced with the assistance of Melanie Johnson and Hans Kowallik.)

14.11.1 Implementation: Bifurcation Diagram of Pendulum

We obtained the bifurcation diagram of Fig. 14.8 by following these steps (a modification of those in [DeJ 92]):

1. Set $\alpha = 0.1, \omega_0 = 1, \omega = 2$, and let f vary through the range in Fig. 14.8.

2. Use the initial conditions: $\theta(0) = 1$ and $\frac{d\theta}{dt}(0) = 1$.

3. Sample (record) the instantaneous angular velocity $\frac{d\theta}{dt}$ whenever the driving force passes through zero.

4. Wait 150 periods before sampling to permit transients to die off.

5. Sample $\frac{d\theta}{dt}$ for 150 times and plot the results.

6. Plot $\left|\frac{d\theta}{dt}\right|$ versus f.

7. Repeat the procedure for each new value of f.

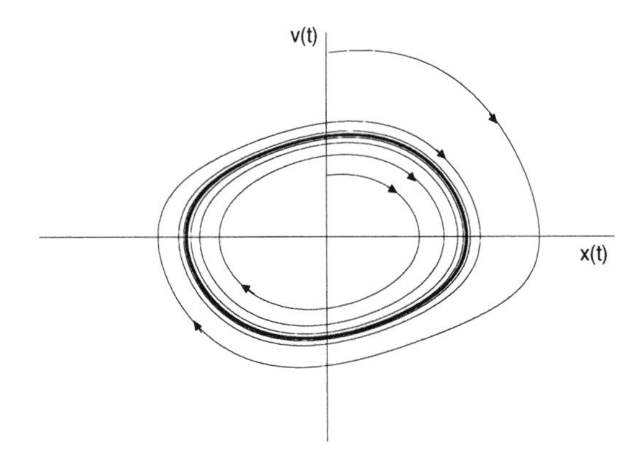

Fig. 14.9 Two phase-space trajectories corresponding to solutions of the van der Pol equation. One trajectory approaches the limit cycle (the dark curve) from the inside, while the other approaches it from the outside.

14.12 FURTHER EXPLORATIONS

1. The nonlinear behavior in once-common objects such as vacuum tubes and metronomes are described by the **van der Pool equation**

$$\frac{d^2x}{dt^2} + \mu(x^2 - x_0^2)\frac{dx}{dt} + \omega_0^2 x = 0. \tag{14.17}$$

   The behavior predicted for these systems is *self-limiting* because the equation contains a limit cycle that is also a predictable attractor. You can think of (14.17) as describing an oscillator with x-dependent damping (the μ term). If $x > x_0$, friction slows the system down; if $x < x_0$, friction speeds the system up. Some phase-space orbits are shown in Fig. 14.9. The heavy curve is the *limit cycle*. Orbits internal to the limit cycle spiral out until they reach the limit cycle; orbit external to it spiral in.

2. **Duffing oscillator:** Here we have a nonlinear oscillator that is damped and driven,

$$\frac{d^2\theta}{dt^2} - \tfrac{1}{2}\theta(1 - \theta^2) = -\alpha\frac{d\theta}{dt} + f\cos\omega t. \tag{14.18}$$

   This is similar to the chaotic pendulum we studied, but has some advantage in the ease with which the multiple attractor sets can be found. It has been studied by [M&L 85].

3. **Lorenz attractor:** In 1962 Lorenz was looking for a simple model for weather prediction and simplified the heat-transport equations to the

three equations [Tab 89]:

$$\frac{dx}{dt} = 10(y - x), \tag{14.19}$$

$$\frac{dy}{dt} = -xz + 28x - y, \tag{14.20}$$

$$\frac{dz}{dt} = xy - \tfrac{8}{3}z. \tag{14.21}$$

The solution of these simple nonlinear equations gave the complicated behavior that has led to the modern interest in chaos (after considerable doubt regarding the numerical solutions).

4. **A 3-D computer fly:** Plot, in 3-D space, the equations

$$x = \sin ay - z\cos bx, \tag{14.22}$$

$$y = z\sin cx - \cos dy, \tag{14.23}$$

$$z = e\sin x. \tag{14.24}$$

Here the parameter e controls the degree of randomness.

5. **Hénon–Heiles potential:** The potential and Hamiltonian

$$V(x,y) = \tfrac{1}{2}x^2 + \tfrac{1}{2}y^2 + x^2 y - \tfrac{1}{3}y^3, \tag{14.25}$$

$$H = \tfrac{1}{2}p_x^2 + \tfrac{1}{2}p_y^2 + V(x,y), \tag{14.26}$$

are used to describe three astronomical objects interacting. They bind the objects near the origin, but releases them if they move far out. The equations of motion for this problem follow from the Hamiltonian equations

$$\frac{dp_x}{dt} = -x - 2xy, \qquad \frac{dp_y}{dt} = -y - x^2 + y^2, \tag{14.27}$$

$$\text{where} \quad \frac{dx}{dt} = p_x, \qquad \frac{dy}{dt} = p_y. \tag{14.28}$$

(a) Numerically solve for the position $[x(t), y(t)]$ for a particle in the Hénon–Heiles potential.

(b) Plot $[x(t), y(t)]$ for a number of initial conditions. Check that the initial condition $E < \frac{1}{6}$ leads to a bounded orbit.

(c) Produce a Poincare section in the (y, p_y) plane. Plot (y, p_y) each time an orbit passes through $x = 0$.

15

Matrix Computing and Subroutine Libraries

In this chapter you face a number of **problems** that require you to extract, call, and run programs from scientific subroutine libraries. In the process of solving these problems, we review the terminology for various general classes of problems. For information on the numerical methods used in the subroutines, we refer you to the References (at the end of this book) and to the information in the subroutine libraries themselves. An easy way to become familiar with and obtain these libraries is through the Web tutorials on scientific libraries.

15.1 PROBLEM 1: MANY SIMULTANEOUS LINEAR EQUATIONS

You are investigating a physical system that you model as the $N = 100$ coupled, linear equations in N unknowns:

$$a_{11}x_1 + a_{12}x_2 + \cdots + a_{1N}x_N = b_1, \tag{15.1}$$

$$a_{21}x_1 + a_{22}x_2 + \cdots + a_{2N}x_N = b_2, \tag{15.2}$$

$$\vdots = \vdots \tag{15.3}$$

$$a_{N1}x_1 + a_{N2}x_2 + \cdots + a_{NN}x_N = b_N. \tag{15.4}$$

In many cases, the a and b values are known, so your **problem** is to solve for all the x values.

To be more specific, we take a as the *Hilbert* matrix

$$[a_{ij}] = A = \left[\frac{1}{i+j-1}\right] = \begin{pmatrix} 1 & \frac{1}{2} & \frac{1}{3} & \frac{1}{4} & \cdots & \frac{1}{100} \\ \frac{1}{2} & \frac{1}{3} & \frac{1}{4} & \frac{1}{5} & \cdots & \frac{1}{101} \\ \vdots & & & & & \\ \frac{1}{100} & \frac{1}{101} & \cdots & & \cdots & \frac{1}{199} \end{pmatrix}, \tag{15.5}$$

and b as

$$[b_i] = B = \left[\frac{1}{i}\right] = \begin{pmatrix} 1 \\ \frac{1}{2} \\ \frac{1}{3} \\ \vdots \\ \frac{1}{100} \end{pmatrix}. \tag{15.6}$$

Your **problem** is to solve for all the x values numerically, and then compare to the analytic solution

$$\begin{pmatrix} x_1 \\ x_2 \\ \vdots \\ x_N \end{pmatrix} = \begin{pmatrix} 1 \\ 0 \\ \vdots \\ 0 \end{pmatrix}. \tag{15.7}$$

15.2 PROBLEM 1, FORMULATION: LINEAR INTO MATRIX EQUATIONS

For computational and conceptual purposes, it is helpful to cast our linear equations into matrix form:

$$\begin{pmatrix} a_{11} & a_{12} & \cdots & a_{1N} \\ a_{21} & a_{22} & \cdots & a_{2N} \\ \vdots & & & \\ a_{N1} & a_{N2} & \cdots & a_{NN} \end{pmatrix} \begin{pmatrix} x_1 \\ x_2 \\ \vdots \\ x_N \end{pmatrix} = \begin{pmatrix} b_1 \\ b_2 \\ \vdots \\ b_N \end{pmatrix}, \tag{15.8}$$

$$AX = B. \tag{15.9}$$

So now we can say that our **problem** is to solve the simple equation (15.9).

15.3 PROBLEM 2: SIMPLE BUT UNSOLVABLE STATICS

Three known masses are connected by four pieces of string of known length as shown in the left of Fig. 15.1. Your **problem** is to determine the four tensions

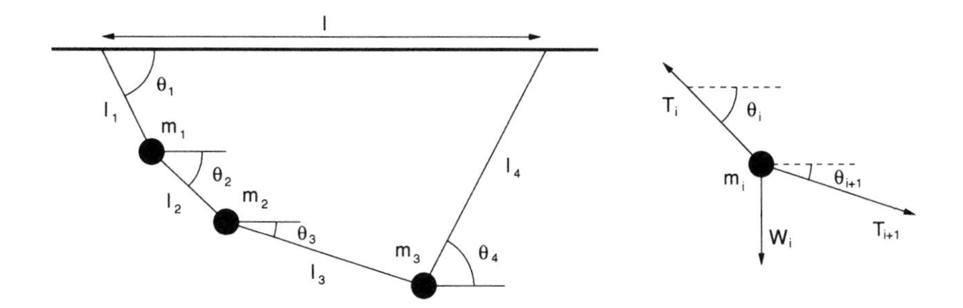

Fig. 15.1 Three masses connected by four pieces of string, and suspended from a horizontal bar of length l. The four angles and the tensions in the strings are unknown.

in the strings *and* the four unknown angles θ_1–θ_4.[1]

15.4 PROBLEM 2: THEORY, STATICS

This is a statics problem because there is no motion. There are the geometric constraints that the horizontal length of the structure is l and that the strings begin and end at the same height:

$$l_1 \cos\theta_1 + l_2 \cos\theta_2 + l_3 \cos\theta_3 + l_4 \cos\theta_4 = l, \tag{15.10}$$

$$l_1 \sin\theta_1 + l_2 \sin\theta_2 + l_3 \sin\theta_3 - l_4 \sin\theta_4 = 0. \tag{15.11}$$

The basics physics says that since there are no accelerations, the sum of the forces on each mass must equal zero. A typical free-body diagram is shown on the right of Fig. 15.1. The equations resulting from balancing the horizontal and vertical components of the forces on each mass are

$$T_1 \sin\theta_1 - T_2 \sin\theta_2 - W_1 = 0, \tag{15.12}$$

$$T_1 \cos\theta_1 - T_2 \cos\theta_2 = 0, \tag{15.13}$$

$$T_2 \sin\theta_2 - T_3 \sin\theta_3 - W_2 = 0, \tag{15.14}$$

$$T_2 \cos\theta_2 - T_3 \cos\theta_3 = 0, \tag{15.15}$$

$$T_3 \sin\theta_3 + T_4 \sin\theta_4 - W_3 = 0, \tag{15.16}$$

$$T_3 \cos\theta_3 - T_4 \cos\theta_4 = 0, \tag{15.17}$$

where $W_i = m_i g$ is the weight of mass i, and T_i is the tension in string i.

[1] We thank Pat Canan for pointing this problem out to us. Hans Kowallik and Paul Hillard have created a Web tutorial for solving it.

15.5 PROBLEM 2, FORMULATION: NONLINEAR SIMULTANEOUS EQUATIONS

Equations (15.10)–(15.17) are eight simultaneous equations in the four unknown angles and the four unknown tensions. Because the tensions and angles are multiplied together, these are simultaneous nonlinear equations, *not* linear equations. If one were clever and patient enough, one might be able to reduce these equations analytically. As a general approach applicable to a wide class of (even more complicated) problems, we will let the computer search for a solution. To do that, we write (15.10)–(15.17) in the standard form:

$$\mathbf{f}(\mathbf{x}) = 0, \tag{15.18}$$

where $\mathbf{x}$ is a vector of the eight unknowns

$$\mathbf{x} = \begin{pmatrix} x_1 \stackrel{\text{def}}{=} \theta_1 \\ x_2 \stackrel{\text{def}}{=} \theta_2 \\ x_3 \stackrel{\text{def}}{=} \theta_3 \\ x_4 \stackrel{\text{def}}{=} \theta_4 \\ x_5 \stackrel{\text{def}}{=} T_1 \\ x_6 \stackrel{\text{def}}{=} T_2 \\ x_7 \stackrel{\text{def}}{=} T_3 \\ x_8 \stackrel{\text{def}}{=} T_4 \end{pmatrix}, \tag{15.19}$$

and $\mathbf{f}$ is a vector of eight functions:

$$\mathbf{f} = \begin{pmatrix} f_1 \\ f_2 \\ f_3 \\ f_4 \\ f_5 \\ f_6 \\ f_7 \\ f_8 \end{pmatrix} \stackrel{\text{def}}{=} \begin{pmatrix} l_1 \cos x_1 + l_2 \cos x_2 + l_3 \cos x_3 + l_4 \cos x_4 - l \\ l_1 \sin x_1 + l_2 \sin x_2 + l_3 \sin x_3 - l_4 \sin x_4 \\ T_1 \sin\theta_1 - T_2 \sin\theta_2 - W_1 \\ T_1 \cos\theta_1 - T_2 \cos\theta_2 \\ T_2 \sin\theta_2 - T_3 \sin\theta_3 - W_2 \\ T_2 \cos\theta_2 - T_3 \cos\theta_3 \\ T_2 \sin\theta_3 + T_4 \sin\theta_4 - W_3 \\ T_3 \cos\theta_3 - T_4 \cos\theta_4 \end{pmatrix}. \tag{15.20}$$

Now that we have them, we must tell you that there is no truly reliable way to solve simultaneous nonlinear equations. Furthermore, there is no guarantee that a solution exists [Pres 94, Rhei 74]. Yet for our statics problem, it is clear from intuition and the physics that a solution exists. So if you can make a fair guess as to what that solution is, the *Newton–Raphson* method is pretty good at searching for it with a trial-and-error procedure. (Essentially, the nonlinear equations are expanded about the guess, the resulting linear equations are solved to obtain a better guess, and the process is continued until convergence or exhaustion.)

15.6 THEORY: MATRIX PROBLEMS

As models of physical systems are made more realistic, the matrices in the models become larger and larger. This is not all bad because the computer is an excellent tool for the matrix manipulations that intrinsically involve continued repetition of a small number of simple instructions. Because the steps are so predictable and straightforward, the codes can be made to run extraordinarily quickly by using clever techniques and by *tuning* them to a particular machine's architecture.[2]

For the reasons just discussed, it becomes more and more important for the computational scientist to use industrial-strength matrix subroutines from a well-established scientific library. These library subroutines are usually an order of magnitude or more faster than an elementary approaches, are designed to minimize roundoff error, and are "robust," that is, they have a high chance of being successful for a broad class of problems. For these reasons we recommend that you *do not* write your own matrix subroutines. Instead, get them from a library.[3] An additional value of the library routine is that you can run the same program on a workstation and a supercomputer and automatically have the numerically most intensive parts of it adapted to the RISC, vector, or parallel architecture of the individual computer.

The thoughtful and pensive reader may be wondering when a matrix is "large" enough to be worth the effort of using a library routine. Basically, if the summed sizes of all your matrices is a good fraction of your computer's RAM, if virtual memory is needed to run your program, or if you have to wait minutes or hours for your job to finish, then it's a good bet your matrices are "large."

Now that you have heard the sales pitch, you may be asking "What's the cost?" In this chapter we pay the costs of having to find what libraries are on your computer, of having to find the name of the routine in that library, of having to find the names of the subroutines your routine calls, and then of having to figure out how to call all these routines. If you are a C programmer, you will also be taxed by having to call a Fortran routine from your C program!

15.6.1 Classes of Matrix Problems

It helps to remember that the rules of mathematics apply even to the world's most powerful computers. For example, you *should* have problems solving equations if you have more unknowns than equations, or if your equations

[2]See Chapter 19, *High-Performance Computing: Profiling and Tuning*, for a discussion of computer architectures.

[3]Although we prize the book [Pres 94] and what it has accomplished, we cannot recommend taking subroutines from it. They are neither optimized nor documented for easy and stand-alone use. The subroutine libraries recommended in this chapter are.

are not linearly independent. But do not fret, while you cannot obtain a unique solution when there are not enough equations, you may still be able to map out a space of allowable solutions. At the other extreme, if you have more equations than unknowns, you have an *overdetermined* problem, which may not have a unique solution. This overdetermined problem is sometimes treated like data fitting in which a solution to a sufficient set of equations is found, tested on the unused equations, and then improved if needed. Not surprisingly, this is known as the *linear least-squares method*, and finds the best solution "on the average."

The most basic matrix problem is the system of linear equations you have to solve for **problem 1**:

$$AX = B. \tag{15.21}$$

Here A is a known $N \times N$ matrix, X is an unknown vector of length N, and B is a known vector of length N:

$$[A]_{N\times N}[X]_{N\times 1} = [B]_{N\times 1}. \tag{15.22}$$

The best ways to solve this equation is by Gaussian elimination or LU (lower-upper) decomposition. They provide you with the vector X without explicitly calculating A^{-1}. Another, albeit slower, method is to determine the inverse of A, and then form the solution by multiplying both sides of (15.21) by A^{-1}:

$$X = A^{-1}B. \tag{15.23}$$

Inversion is usually performed with some type of Gauss–Jordan elimination.

If you have to solve the matrix equation,

$$AX = \lambda X, \tag{15.24}$$

with X an unknown vector and λ an unknown constant (scalar), then the direct solution (15.23) will not be of much help because the matrix $B = \lambda X$ contains the unknown. Equation (15.24) is, of course, *the eigenvalue problem.* It is harder to solve than (15.21) because solutions exist only for certain values of the constant λ (or possibly none depending on A).

If we write (15.24) as

$$[A - \lambda I]X = 0, \tag{15.25}$$

we see that multiplication by $[A - \lambda I]^{-1}$ yields the *trivial solution*:

$$X = 0, \qquad \text{(trivial solution)}. \tag{15.26}$$

While the trivial solution is a bona fide solution, it is trivial. For a more interesting solution to exist ,there must be something that forbids us from multiplying both sides of (15.25) by $[A - \lambda I]^{-1}$. That something is the nonexistence of the inverse. If you recall that Cramer's rule for the inverse includes division by $\det[A - \lambda I]$, it becomes clear that the inverse fails to exist (and in this way eigenvalues *do* exist) when

$$\det[A - \lambda I] = 0. \tag{15.27}$$

Those values of λ that satisfy (15.27) are the eigenvalues of the original equation (15.24).

If you were interested only in the eigenvalues, you would have the computer solve (15.27). To do that, you need a subroutine to calculate the determinant of a matrix, and then a search routine to help you zero in on the solution of (15.27). Such routines are available in the libraries.

The traditional way to solve the eigenvalue problem (15.24) is by *diagonalization.* This is equivalent to successive changes of basis vectors, each change leaving the eigenvalues unchanged while continually decreasing the values of the off-diagonal elements of A. The sequence of transformations is equivalent to continually operating on the original equation with a matrix U:

$$UA(U^{-1}U)X = \lambda UX, \tag{15.28}$$

$$(UAU^{-1})(UX) = \lambda(UX), \tag{15.29}$$

until UAU^{-1} is diagonal:

$$UAU^{-1} = \begin{pmatrix} a'_{11} & & \cdots & 0 \\ 0 & a'_{22} & \cdots & 0 \\ 0 & \cdots & a'_{33} & \cdots \\ 0 & \cdots & & a'_{NN} \end{pmatrix}. \tag{15.30}$$

The values of UAU^{-1} along the diagonal are then the eigenvalues of the original problem. In addition, we note from (15.29) that we can choose the vectors UX to be the basis vectors:

$$UX_i = e_i. \tag{15.31}$$

In this case the eigenvectors are

$$X_i = U^{-1}e_i, \tag{15.32}$$

that is, the eigenvectors are the columns of the matrix U^{-1}. A number of routines of this type are found in the subroutine libraries.

15.7 METHOD: MATRIX COMPUTING

Many scientific programming problems arise from the improper use of arrays on computers.[4] This may be due to the extensive use of matrices in scientific computing or the complexity of keeping track of indices and dimensions. In any case, here are some rules of thumb to observe:

Computers are finite: Unless you are careful, you can run out of memory or run very slowly when dealing with large matrices. For example, let's say that you store data in a four-dimensional array with each index having a *physical dimension* of 100: `A(100,100,100,100)` or `A[100] [100] [100] [100]`. This array occupies $(100)^4$ words $\approx$ 1GB (gigabyte) of memory (if you are lucky enough to have that much RAM).

Complex, double precision, double size: Making a single-precision matrix double doubles the size of the matrix. Making that matrix complex doubles the size yet again. Doubling the dimensions of our 4-D matrix A leads to a 16-fold increase in size. This helps explain why it is so easy to run out of memory.

Processing time: Matrix operations such as inversion require on the order of N^3 steps for a square matrix of dimension N. Therefore, doubling the dimensions of a 2-D square matrix (as happens when the number of integration steps are doubled) leads to an *eightfold* increase in processing time.

Paging: Many computer systems have *virtual memory* in which disk space or some other *slow* memory is used when a program runs out of RAM (see Chapter 18, *Computing Hardware Basics: Memory and CPU*, for a discussion of how computers do memory). This is a slow process that requires writing a full *page* of words to either slow memory or to the disk. If your program is near the memory limit at which paging occurs, even a slight increase in the physical dimensions of a matrix may lead to an order-of-magnitude increase in running time.

Matrix storage: While we may think of matrices as multidimensional blocks of stored numbers, the computer stores them sequentially as linear strings of numbers. For example, a matrix `a(3,3)` in Fortran is stored in *column-major order* as

$$\mathsf{a}(1,1)\mathsf{a}(2,1)\mathsf{a}(3,1)\mathsf{a}(1,2)\mathsf{a}(2,2)\mathsf{a}(3,2)\mathsf{a}(1,3)\mathsf{a}(2,3)\mathsf{a}(3,3), \tag{15.33}$$

[4]Even a vector $V(N)$ is an "array," albeit a 1-D one.

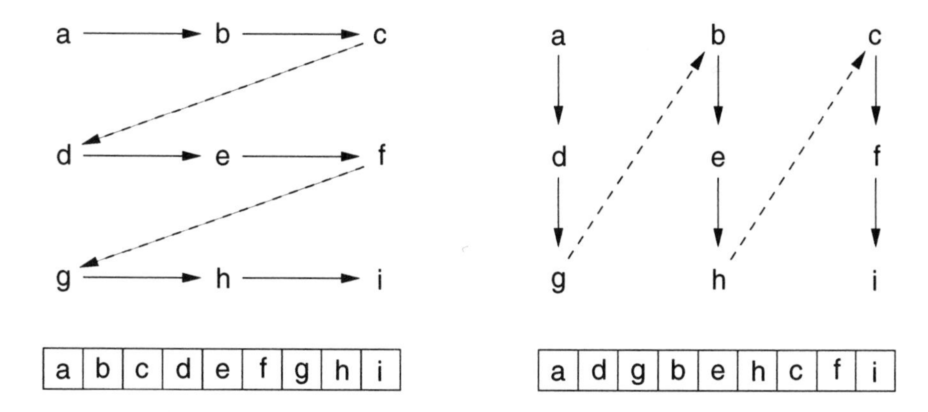

Fig. 15.2 (*Left*) Row-major order used for matrix storage in C and Pascal; (*right*) column-major order used for matrix storage in Fortran. On the bottom is shown how successive matrix elements are stored in a linear fashion in memory.

as shown on the right of Fig. 15.2. In the C language, `a[3][3]` is stored in *row-major order* as shown on the left of Fig.15.2. It is important to keep this linear storage scheme in mind in order to write proper code and to permit mixing Fortran and C programs. In addition, Fortran subscripts traditionally start at 1 while C subscripts usually start at 0. On that account, the same location can have data with different meanings in different languages:

Location	C Element	Fortran Element
Lowest	`a(0,0)`	`a(1,1)`
	`a(0,1)`	`a(2,1)`
	`a(1,0)`	`a(3,1)`
	`a(1,1)`	`a(1,2)`
	`a(2,0)`	`a(2,2)`
Highest	`a(2,1)`	`a(3,2)`

Physical and logical dimensions: When you run a program, you issue commands such as `double a[3][3]` or `Dimension a(3,3)` that tell the computer how much memory it needs to set aside for arrays. This is called *physical memory* and is usually made large enough to handle all foreseeable cases. Often you run programs without the full complement of values declared in the declaration statements (perhaps because you are running small test cases or perhaps because you like to declare all dimensions to be 1000 to impress the hacker at the next keyboard). The amount of memory you actually use to store numbers is the matrix's *logical size.* For example, if you declared `Dimension a(3,3)` but only use a logical dimension of `(2,2)`, then the a in storage would look like

$$\texttt{a}(1,1)'\texttt{a}(2,1)'\texttt{a}(3,1)\texttt{a}(1,2)'\texttt{a}(2,2)'\texttt{a}(3,2)\texttt{a}(1,3)\texttt{a}(2,3)\texttt{a}(3,3), \tag{15.34}$$

where only the prime elements have values assigned to them. Clearly, the defined a values do not occupy sequential locations in memory, and so an algorithm processing this matrix cannot assume that the next element in memory is the next element in your array. This is the reason why subroutines from a library usually need to know *both* the physical and logical sizes of your arrays.

Passing sizes to subprograms: not only must you pass the logical and physical dimensions separately to subprograms, you must also watch that the sizes of your matrices do not exceed the bounds which have been declared in the subprograms. This usually occurs *without* an error message, and most likely will give you the wrong answers.[5] You should always check the declaration statements in subprograms for their limits and storage schemes, and build warning messages into your program to notify you if the bounds are exceeded. In addition, if you are a C programmer calling Fortran, you need to pass pointers to variables and not the actual values of the variables to the Fortran subprograms (Fortran deals with pointers only). Here we have a program possibly ruining some data stored nearby:

```
Program Main                                        In main program.
Dimension a(100), b(400)
Subroutine Sample(a)                                  In subroutine.
Dimension a(10)                                   Smaller dimension.
a(300) = 12                      Way out of bounds, but no message.
```

One way to ensure size compatibility among main programs and subroutines is to declare array sizes only in your main program and then to pass those sizes along to your subprograms as arguments.

Equivalence: Using an *Equivalence* statement to overlay several arrays in the same memory location saves memory. It also can ruin some of the matrices already stored in memory (especially if the equivalenced matrices are of different sizes). Equivalenced variables also make vectorizing difficult. Use it only if you must or if the savings are too great to pass up (and do include comments).

Say what's happening: You decrease errors and problems by using self-explanatory labels for your indices and providing frequent comments explaining your particularly clever storage schemes before you forget it.

Clarity versus efficiency: When dealing with matrices, you have to bal-

[5]There will be error messages if you are running with a version of your program compiled with the *debugging option*, which watches array bounds.

ance the clarity of the operations being performed against the efficiency with which the computer does them. For example, having one matrix with many indices such as `V(L,Nre,Nspin,k,kp,Z,A)` may be neat packaging, but it may require the computer to jump through large blocks of memory to get to the particular values needed (large *strides*). As a compromise, you may want to get a clear but slow version of your programming running first, and then modify it to produce the speedier version.

Subscript 0: It is standard in C to have array indices begin with the value 0. Although not the default, Fortran compilers now permit statements such as `Dimension a(0:99)` which start indexing the array at zero (and end at 99). The default, and the old standard in Fortran, is to start all indexing at 1. Care is needed when programming up equations in the two schemes because the codes must differ. For example, here are two equivalent recursion relations for the different indexing schemes ($P_i \equiv P_{l+1}$):

$$(l+1)P_{l+1} - (2l+1)xP_l + lP_{l-1} = 0, \quad (l = 0, 1, \ldots), \tag{15.35}$$

$$iP_i - (2i-1)xP_{i-1} + (i-1)P_{i-2} = 0, \quad (i = 1, 2, \ldots). \tag{15.36}$$

Tests: Always test a library routine on a small problem whose answer you know. Then you'll know if you are supplying it with the right arguments and if you have all the links working. We give some exercises in §15.11.

15.8 IMPLEMENTATION: USING SCIENTIFIC LIBRARIES, WWW

Some major scientific and mathematical libraries available include:

Some Standard Libraries

NETLIB	A WWW metalibrary of free math libraries
IMSL	International Mathematical and Statistical Libraries
ESSL	Engineering and Scientific Subroutine Library (IBM)
DXML	Advanced Mathematical Library (DEC)
NAG	Numerical Algorithms Group (UK Labs)
SLATEC	Comprehensive Mathematical and Statistical Package
LAPACK	Linear Algebra Package
ScaLAPACK	LAPACK for distributed memory
CERN	European Center for Nuclear Research
BLAS	Basic Linear Algebra Subprograms (building blocks)

The LAPACK, SLATEC, CERN, and Netlib libraries are in the public domain, while ESSL, IMSL, and NAG are proprietary ($$$), but they are frequently available on a mainframe or via an institution-wide site license. General subroutine libraries are treasures to possess because they typically contain routines for almost everything you might want to do. For example

Linear algebra manipulations	Matrix operations
Solution of linear equations	Eigensystem analysis
Signal processing	Sorting and searching
Interpolation, fitting	Differential equations
Roots, zeros, and extrema	Random-number operations
Statistical functions	Numerical quadrature

What is known as LAPACK [LAP 95] is a transportable library of Fortran 77 linear equation and eigenproblem solvers.[6] LAPACK, which may come with a computer's operating system, supersedes and extends Linpack and Eispack. An advantage of LAPACK is that it has recently been restructured to achieve high efficiency on vector processors, high-performance RISC workstations, and shared-memory multiprocessors. The high efficiency is usually accomplished through the use of optimized (hardware-tuned) elementary BLAS subprograms.

In contrast to LAPACK, the SLATEC library (also called CML) contains general purpose mathematical and statistical Fortran routines, and is consequently more general. Nonetheless, it is not tuned to the architecture of a particular machine the way LAPACK is, and usually is not included with a computer's operating system.

Often a subroutine library will supply only Fortran routines, and this requires the C programmer to call a Fortran routine (we describe how to do that in §15.9.4). In some cases, C-language routines may also be available, but they may not be optimized for a particular machine.

As is evident in the examples to follow, you must do some work to use a library. You will need to call the routine from your source code, feed it exactly the type of arguments it requires, and possibly link in all the subroutines called by this routine. If you can give compiler options like `-llibename` to have the entire subroutine library loaded, then you will not need to concern yourself with the subroutines which your routine calls—but you will have to know what to substitute for "`libname`." For example, to load DEC's (Digital Equipment Corporation's) extended math library while compiling:

```
> f77 -ldmxl prog.f                    Compile prog.f with DMXL libe.
```

[6]There is some inevitable confusion regarding names of functions and variables in Fortran. In general, Fortran compilers converts all words to uppercase, so case does not matter within a Fortran program (unless you evoke a -U compiler option to make it matter). However, operating systems are case-sensitive, and so to load some library, or to look up some the man pages, you may have to say `man lapack` and not `man LAPACK`—only to find that the man pages use yet a different convention. Our attitude is that this is something you just can't worry about. If one convention fails, try another!

One of the choices you may have to make in using a library is to select a routine from some general categories and to choose the level of precision. In our experience, serious scientific calculations, and especially those dealing with large matrices, require words greater than 32 bits in length to avoid roundoff error accumulation.

This is what the general categories mean:

General Categories	
Single precision	4 B = 32-bit word length (64 on "big" machines)
Double precision	8 B= 64-bit word length (128 on "big" machines)
Complex	Complex numbers (single or double precision)
Vector	Tuned for special vector processor
Scalar	Not vector
Superscalar	Tuned for a RISC processor
Parallel	Tuned for parallel processors
Matrix type	Real, complex, symmetric, and so on

15.9 IMPLEMENTATION: DETERMINING AVAILABLE LIBRARIES

Before you worry about calling a subroutine, you need to know what subroutine libraries are on your system. Alternatively, or if your system does not have a scientific subroutine library that meets your needs, you can use the internet or the Web to get what you need directly (we show how in §15.10 and our Web tutorials). An excellent way to find out what math libraries are on your computer is to ask your system administrator. Another is to ask your computer to search through its documentation (manuals) for anything containing the word "library." On Unix systems this is done with the `apropos` or `man -k` command:

```
> apropos library          Search manual pages for ''library.''
> man -k library           Search manual pages for ''library.''
```

Be prepared, the response to a `man` request is sometimes a computer's stream of consciousness. A typical IBM response (along with lots of other junk) is

```
Lapack is a transportable library of Fortran 77 subroutines ...
fdxtrn(1)       - Resolves cross-database links within a library
hdf (1)         - Hierarchical Data Format library
```

A typical DEC response (along with lots of other junk) is

```
ar (1)               - Archive and library maintainer
array-math (3DXML)   - A library of linear algebra routines
intro (3hpf)         - Introduction to the High-Performance Fortran library
lapack (3DXML)       - A library of linear algebra routines
```

In both cases we see the library name LAPACK mentioned, and so it's a good bet (but not a sure one) that this library is on these computers somewhere.

15.9.1 Determining Contents of a Library

To get information about a known library, issue the `man` command:

```
> man lapack                                    Show man pages for  LAPACK
```

If your computer has anything to tell you about LAPACK, it will probably be a lot, and we give a sample output in §15.12. Here we summarize some of LAPACK's computational routines used to perform distinct algorithmic tasks, such as matrix inversion, as well as driver routines used to solve complete problems, such as solving a system of linear equations. There are both *simple* as well as *expert* drivers, the latter for power users who may want more options or better error estimates. The expert routines are obtained from the simple ones by adding an x to the name, for example; $sgesv \rightarrow sgesvx$.

The name of a LAPACK routine may not appear too revealing at first, but it may be once you have broken the code. All driver and computational routines have names of the form $xyyzzz$, where for some driver routines the 6th character is blank.

Data type indicated by first letter x	
s	Real, single precision
d	Double precision
c	Complex
z	Complex*16, double complex

The next two letters yy indicate the type of matrix. The last three letters zzz indicate the computation performed. For example, *sgebrd* is a single-precision (s) routine for a general (ge) matrix, performing a bidiagonal reduction (brd).

15.9.2 Determining the Needed Routine

To get an idea how all this works, we will first review the steps you follow. You have a problem to solve, for example, the linear equation (15.9). First you determine the name of the supplementary library on your computer, say, by issuing the `man -k library` command. From that you discover that LAPACK is on your machine (or you can get it from Netlib as described in §15.10). Then you examine what's in LAPACK by issuing the `man LAPACK` command. You get an output like that in §15.12, and from that conclude that the routine *dgesv* will provide you with a double-precision (d) solution of the linear equations $AX = B$, with A a general (ge) matrix. To see how to call *dgesv*, you then look within the subroutine for self-documentation or ask for documentation with the `man` command:

```
> man dgesv                                    Get manual pages for dgesv
```

dgesv(l): LAPACK driver routine

dgesv computes solution to real linear equations A * X = B

Subroutine dgesv(n, nrhs, A, lda, ipiv, B, ldb, info)
Integer info, lda, ldb, n, nrhs, ipiv(*)
Double Precision A(lda, *), B(ldb, *)

Purpose: dgesv computes the solution to a real linear equations A * X = B, where A is an n-by-n matrix and X and B are n-by-nrhs matrices. The LU decomposition with partial pivoting and row interchanges is used to factor A as A = P * L * U, where P is a permutation matrix, L is unit lower triangular, and U is upper triangular. The factored form of A is then used to solve the equations A * X = B.

Arguments:

n (input) Integer The number of linear equations, i.e., the order of the matrix A. $n \geq 0$.

nrhs (input) Integer The number of right-hand sides, i.e., the number of columns of the matrix B. nrhs $\geq$ 0.

A (input/output) Double Precision array, Dimension (lda,n) On entry, the n-by-n coefficient matrix A. On exit, the factors L and U from the factorization A = P*L*U; the unit diagonal elements of L are not stored.

lda (input) Integer The leading dimension of the array A. lda $\geq$ max(1,n).

Ipiv (output) Integer array, dimension (n) The pivot indices that define the permutation matrix P; row i of the matrix was interchanged with row Ipiv(i).

B (input/output) double-precision array, Dimension (ldb,nrhs) On entry, the n-by-nrhs matrix of right-hand side matrix B. On exit, if info = 0, the n-by-nrhs solution matrix X.

ldb (input) Integer The leading dimension of the array B, ldb $\geq$ max(1,n).

info (output) Integer = 0: successful exit
$<$ 0: if info = -i, the i-th argument had an illegal value
$>$ 0: if info = i, U(i,i) is exactly zero. The factorization has been completed, but the factor U is exactly singular, so the solution could not be computed.

Now that you know what to do, you need to modify your source code to call *dgesv* properly. Here's an example of that

15.9.3 Implementation: Calling LAPACK from DEC Fortran, lineq.c

```
Program LAPACK
c Call a LAPACK routine to solve Ax=B with A = Hilbert matrix.
c compile with DEC extended math lib: f77 call.f -ldxml
    Implicit none
c       Declarations
    Real*8 matrix(100,100), result(100)
    Integer i, j, pivot(100), ok
c create Hilbert matrix
    do i=1, 100
       do j=1, 100
          matrix(i, j)=1.0/(j+i-1)
       end do
c create solution vector
    result(i)=1.0/i
    end do
```

Table 15.1 Matching data types in C and Fortran

C	Fortran
char	Character
signed char	Integer*1
unsigned char	Logical*1
short signed int	Integer*2
short unsigned int	Logical*2
signed int (long int)	Integer*4
unsigned int	Logical*4
float	Real (Real*4)
structure of 2 floats	Complex
double	Real*8
structure of 2 doubles	Complex*16
char[n]	Character*n

```
call dgesv(100, 1, matrix, 100, pivot, result, 100, ok)
do i=1, 100
   write(*,*) result(i)
end do
End
```

15.9.4 Calling LAPACK from C

The good news is that Unix permits intermixing subroutines compiled from C and Fortran sources [L&F 93]. The bad news is that it won't work unless you are careful to account for the somewhat different ways the compilers store subroutine names, for the quite different ways they store arrays with more than one subscript, and for the different data types available in the two languages.

The first thing you must do is ensure that the data types of the variables in the two languages are matched. The matching data types are given in Table 15.1. Note, if the data are stored in arrays, your C calling program must convert to the storage scheme used in the Fortran subroutine before you can call the Fortran subroutine (and then convert back to the C scheme after the subroutine does its job if you have overwritten the original array and intend to use it again).

When a function is called in the C language, usually the actual value of the argument is passed to the function. In contrast, Fortran passes the address in memory where the value of the argument is to be found (a *reference pass*). If you do not ensure that both languages have the same passing protocol, your program will process the numerical value of the address of a variable as if it were the actual value of the variable (we are willing to place a bet on the correctness of the result). Here are some procedures for **calling Fortran from C**:

1. Use pointers for all arguments in your C programs. Generally this is done with the address operator &.

2. Do not have your program make calls such as `sub(1, N)` where the actual value of the constant "1" is fed to the subroutine. Instead, assign the value one to a variable, and feed that variable (actually a pointer to it) to the subroutine. For example

```
one = 1.                                  Assign value to variable
sub(one)          In Fortran, where all calls are reference calls
sub(&one)                    In C, where we make pointer explicit
```

   This is important because the value "1" in the subroutine call, in Fortran, anyway, is actually the address where the value "1" is stored. If the subroutine modified that variable, it will modify the value of "1" every place in your program! In C, it would change the first value in memory.

3. Depending on the particular operating system you are using, you may have to append an underscore _ to the called Fortran subprogram names. For example, `sub(one, N)` → `sub_(one, N)`. Generally, the Fortran compiler appends an underscore automatically to the names of its subprograms, while the C compiler does not (but you'll need to experiment).

4. Use lowercase letters for the names of external functions. The exception is when the Fortran subprogram being called was compiled with a `-U` option, or the equivalent, for retaining uppercase letters.

15.9.5 Calling LAPACK Fortran from C

```
/* Calling LAPACK from C */
  /* Solves AX=B,  A = Hilbert matrix; compiled as:  cc call.c -ldxml  */
  #include <stdio.h>hwit
  #define size 100                          dimension of Hilbert matrix
  main()
  {
  int i, j , c1, c2, pivot[size], ok;
  double matrix[size][size], help[size*size], result[size];
  c1=size;                         arguments (pointers) for function call
  c2=1;                                     numbers stored as variables
  for (i = 0; i < c1; i++)                        create Hilbert matrix
  {
```

```
for (j = 0; j < c1; j++) matrix[i][j]  = 1.0/(i+j+1);
result[i] = 1.  / (i+1);                          create solution vector
}
for (i=0; i<size; i++)                                  transform matrix
{                                     storage block, reordered elements
for(j=0; j<size; j++) help[j+size*i]=matrix[j][i];
}
dgesv_(&c1, &c2, help, &c1, pivot, result, &c1, &ok);   call + pointers
for (j=0; j<size; j++) printf("%e\n", result[j]);
}
```

You will notice here that the call to the Fortran subroutine `dgesv` is made as

```
dgesv_(&c1, &c2, help, &c1, pivot, result, &c1, &ok)
```

That is, lowercase letters are used and an underscore is added to the subroutine name. In addition, we convert the matrix A,

```
for (i=0; i<size; i++)                            matrix transformation
  for (j=0; j<size; j++)
    help[j+size*i]=matrix[j][i];
```

This changes C's row-major order to Fortran's column-major order using a scratch vector.

15.9.6 C Compiling Calling Fortran

Multilanguage programs actually get created when the compiler links the object files together. The tricky part is that while Fortran automatically includes its math library, if your final linking is done with the C compiler, you may have to explicitly include the Fortran library as well as others:

```
> cc -O call.c f77sub.o -lm -ldxml                  DEC extended mathlibe
> cc -O call.c f77sub.o -L/usr/lang/SC0.0 -lF77            Link, SunOS
> cc -O c_fort c_fort.o area_f.o -lxlf                       Link, AIX
```

15.10 EXTENSION: MORE NETLIB LIBRARIES

Our example of using LAPACK has assumed that someone has been nice and placed the library on the computer. Life can be rough, however, and you may have to get the routines yourself. Probably the best place to start looking for them is Netlib, a repository of free software, documents, and databases of interest to computational scientists. We give an abridged and edited table of Netlib's contents in §15.13. More information, as well as subroutines, are available over the internet via the following routes:

Netlib Availability

World Wide Web	*http://www.netlib.org/*
Anonymous ftp	*ftp.netlib.org*
Search for routine	http://www.netlib.org /utk/misc/netlib_query.html
Browse libes	http://www.netlib.org/liblist.html
GAMS classes	http://www.netlib.org/bib/gams.html
Documentation	Self-documenting via code comments
Check Dependency	Auto retrieve of dependent routines
Technical Support	Direct to the author of subroutine
NIST Guide	http://gams.nist.gov

15.10.1 SLATEC's Common Math Library

SLATEC (Sandia, Los Alamos, Air Force Weapons Laboratory Technical Exchange Committee) is a portable, nonproprietary, mathematical subroutine library available from Netlib. We recommend it highly and also recommend that you get more information about it from Netlib. SLATEC (CML) contains over 1400 general purpose mathematical and statistical Fortran routines. It is more general than LAPACK, which is devoted to linear algebra, yet not necessarily tuned to the particular architecture of a machine the way LAPACK is. The full library contains a guide, table of contents, and documentation via comments in the source code. The subroutines are classified by the Guide to Available Mathematical Software (GAMS) system. We give a listing of the available SLATEC routines in §15.12.

Let's now return to our **problem 2**, the statics problem. After some exploration, we have found the needed routines:

From SLATEC

snsq-s, dnsq-d	Find zero of n-variable, nonlinear function
snsqe-s, dnsqe-d	Easy-to-use snsq

If you extract these routines, you will find that they need the following:

enorm.f	j4save.f	r1mach.f	xerprn.f	fdjac1.f	r1mpyq.f
xercnt.f	xersve.f	fdump.f	qform.f	r1updt.f	xerhlt.f
xgetua.f	dogleg.f	i1mach.f	qrfac.f	snsq.f	xermsg.f

Of particular interest in these "helper" routines, are *i1mach.f, r1mach.f*, and *d1mach.f*. They tell LAPACK the characteristic of your particular machine when the library is first installed. Without that knowledge, LAPACK does not know when convergence is obtained or what step sizes to use.

15.11 EXERCISES: TESTING MATRIX CALLS

Before you direct the computer to go off crunching numbers on the million elements of some matrix, it's a good idea for you to try out your procedures on a small matrix, especially one for which you know the right answer. In this way it will only take you a short time to realize how hard it is to get the calling procedure perfectly right! Here are some exercises:

1. Find the inverse of

$$A = \begin{pmatrix} 4 & -2 & 1 \\ 3 & 6 & -4 \\ 2 & 1 & 8 \end{pmatrix}. \tag{15.37}$$

   (a) As a general procedure, applicable even if you do not know the analytic answer, check your inverse in both directions; that is, check that

$$AA^{-1} = A^{-1}A = 1. \tag{15.38}$$

   (b) Verify that

$$A^{-1} = \frac{1}{263} \begin{pmatrix} 52 & 17 & 2 \\ -32 & 30 & 19 \\ -9 & -8 & 30 \end{pmatrix}. \tag{15.39}$$

2. Consider the same matrix A as in (15.37), now used to describe three simultaneous linear equations,

$$AX = B, \tag{15.40}$$

$$\begin{pmatrix} a_{11} & a_{12} & a_{13} \\ a_{21} & a_{22} & a_{23} \\ a_{31} & a_{32} & a_{33} \end{pmatrix} \begin{pmatrix} x_1 \\ x_2 \\ x_3 \end{pmatrix} = \begin{pmatrix} b_1 \\ b_2 \\ b_3 \end{pmatrix}. \tag{15.41}$$

   Here the vector B on the RHS is assumed known, and the problem is to solve for the vector X. Use an appropriate subroutine to solve these equations for the three different X vectors appropriate to these three different B values on the RHS:

$$B_1 = \begin{pmatrix} +12 \\ -25 \\ +32 \end{pmatrix}, \quad B_2 = \begin{pmatrix} +4 \\ -10 \\ +22 \end{pmatrix}, \quad B_3 = \begin{pmatrix} +20 \\ -30 \\ +40 \end{pmatrix}. \tag{15.42}$$

   The solutions should be

$$X_1 = \begin{pmatrix} +1 \\ -2 \\ +4 \end{pmatrix}, \quad X_2 = \begin{pmatrix} +0.312 \\ -0.038 \\ +2.677 \end{pmatrix}, \quad X_3 = \begin{pmatrix} +2.319 \\ -2.965 \\ +4.790 \end{pmatrix}. \tag{15.43}$$

3. Consider the matrix

$$A = \begin{pmatrix} \alpha & \beta \\ -\beta & \alpha \end{pmatrix}, \tag{15.44}$$

where you are free to use any values you want for α and β. Use a numerical eigenproblem solver to show that the eigenvalues and eigenvectors are the complex conjugates:

$$X_1 = \begin{pmatrix} +1 \\ -i \end{pmatrix}, \quad \lambda_1 = \alpha - i\beta, \tag{15.45}$$

$$X_2 = \begin{pmatrix} +1 \\ +i \end{pmatrix}, \quad \lambda_2 = \alpha + i\beta. \tag{15.46}$$

4. Use your eigenproblem solver to find the eigenvalues of the matrix

$$A = \begin{pmatrix} -2 & +2 & -3 \\ +2 & +1 & -6 \\ -1 & -2 & +0 \end{pmatrix}. \tag{15.47}$$

(a) Verify that you obtain the eigenvalues

$$\lambda_1 = 5, \quad \lambda_2 = \lambda_3 = -3. \tag{15.48}$$

Notice that double roots can cause problems. In particular, there is a uniqueness problem with their eigenvectors because any combinations of these eigenvectors would also be an eigenvector.

(b) Verify that the eigenvector for $\lambda_1 = 5$ is proportional to

$$X_1 = \begin{pmatrix} -1 \\ -2 \\ +1 \end{pmatrix}. \tag{15.49}$$

(c) The eigenvalue 5 corresponds to a double root. This means that the corresponding eigenvectors are degenerate, which, in turn, means that they are not unique. Two linearly independent ones are:

$$X_2 = \begin{pmatrix} -2 \\ +1 \\ +0 \end{pmatrix}, \quad X_3 = \begin{pmatrix} 3 \\ 0 \\ 1 \end{pmatrix}. \tag{15.50}$$

In this case it's not clear what your eigenproblem solver will give for the eigenvectors. Try to find a relationship between your computed eigenvectors with the eigenvalue -3 to these two linearly independent ones.

15.12 IMPLEMENTATION: LAPACK SHORT CONTENTS

The general categories in LAPACK are:

Data type indicated by first letter of routine	
s	Real, single precision
d	Double precision
c	Complex
z	Complex*16, double complex

Expert routines are obtained from the simple drivers below by adding an x to the name, e.g., *sgesv* → *sgesvx*:

Simple LAPACK Drivers	
sgesv, dgesv, cgesv, zgesv	General linear equations AX=B.
dgesv, sgbsv, cgbsv, zgbsv	General banded linear equations AX=B.
sgtsv, dgtsv, cgtsv, zgtsv	General tridiagonal linear equations AX=B.
sposv, dposv, cposv, zposv	Symmetric/Hermitian + definite linear equations.
sppsv, dppsv, cppsv,zppsv	Symmetric/Hermitian + definite linear equations, packed
spbsv,dpbsv,cpbsv,zpbsv	Symmetric/Hermitian + definite banded linear equations.
spbsv,dpbsv,cpbsv,zpbsv	Symmetric/Hermitian + definite tridiagonal linear equatio
ssysv, dsysv, csysv, zsysv	Real, complex, symmetric indefinite linear equations.
ssysv, dsysv, csysv, zsysv, chesv, zhesv	Hermitian indefinite linear equations.
sspsv, dspsv, cspsv, zspsv	Real, complex, symmetric indefinite linear equations, pack
chpsv, zhpsv	Hermitian indefinite linear equations, packed A.
sgels, dgels, cgels, zgels	Least-squares linear equations.
sgelss, dgelss, cgelss, zgelss	Least-squares linear equations, singular value.
sgglse, dgglse, cgglse, zgglse	Constrained linear least-squares problem, GRQ.
sggglm, dggglm, cggglm, zggglm	Generalized linear regression.
ssyev, dsyev, cheev, zheev	Eigenvalues/vectors symmetric/Hermitian matrix.
sspev, dspev, chpev, zhpev	Eigenvalues/vectors symmetric/Hermitian, packed matrix.
ssbev, dsbev, chbev, zhbev	Eigenvalues/vectors symmetric/Hermitian, banded matrix
sstev, dstev	Eigenvalues/vectors real symmetric tridiagonal matrix.
sgeev, dgeev, cgeev, zgeev	Eigenvalues, Schur factorization, general matrix.
sgesvd, dgesvd, cgesvd, zgesvd	Singular value decomposition general rectangular matrix.
ssygv, dsygv, chegv, zhegv	General eigenproblem, A X= λ B X, ABX= λ X, BAX=
sspgv, dspgv, chpgv, zhpgv	General, packed eigenproblem.
sgegs, dgegs, cgegs, zgegs	General eigenvalues, Schur form, nonsymmetric.
sgegv, dgegv, cgegv, zgegv	General eigenproblem, nonsymmetric matrices.
sggsvd, dggsvd, cggsvd, zggsvd	General singular value decomposition.

LAPACK Computational Routines (edited)

sbdsqr, dbdsqr, cbdsqr, zbdsqr	Singular value decomposition, real bidiagonal.
sgbcon, dgbcon, cgbcon, zgbcon	Reciprocal condition number, general banded.
sgbequ, dgbequ, cgbequ, zgbequ	Equilibrate general banded matrix.
sgbrfs, dgbrfs, cgbrfs, zgbrfs	Improve solution, general banded linear equations.
sgbtrf, dgbtrf, cgbtrf, zgbtrf	LU factorization,general banded matrix.
sgbtrs, dgbtrs, cgbtrs, zgbtrs	General banded linear equations.
sgebak, dgebak, cgebak, zgebak	Transforms eigenvectors, balanced matrix.
sgebal, dgebal, cgebal, zgebal	Balances general matrix.
sgebrd, dgebrd, dgebrd, zgebrd	Reduces rectangular matrix to bidiagonal form.
sgecon, dgecon, cgecon, zgecon	Reciprocal condition number, general matrix.
sgeequ, dgeequ, cgeequ, zgeequ	Equilibrate general rectangular matrix.
sgehrd, dgehrd, cgehrd, zgehrd	Reduce general matrix to upper Hessenberg form.
sgelqf, dgelqf, cgelqf, zgelqf	LQ factorization, general rectangular matrix.
sgeqlf, dgeqlf, cgeqlf, zgeqlf	QL factorization, general rectangular matrix.
sgeqpf, dgeqpf, cgeqpf, zgeqpf	QR factorization, general rectangular matrix.
sgeqrf, dgeqrf, cgeqrf, zgeqrf	QR factorization, general rectangular matrix.
sgerfs, dgerfs, cgerfs, zgerfs	Improves solution, general linear equations.
sgerqf, dgerqf, cgerqf, zgerqf	RQ factorization, general rectangular matrix.
sgetrf, dgetrf, cgetrf, zgetrf	LU factorization, general matrix.
sgetri, dgetri, cgetri	Inverse, general matrix.
sgetrs, dgetrs, cgetrs, zgetrs	General linear equations $AX=B$, $A\hat{T}\ X=B$ or $A\hat{H}\ X=B$.
sggbak, dggbak, cggbak, zggbak	R/L eigenvectors, general eigenvalue problem.

LAPACK Computational Routines (continued)

sggbal, dggbal, cggbal, zggbal	Balances pair of matrices.
sgghrd, dgghrd, cgghrd, zgghrd	Reduce matrices to general upper Hessenberg form.
sgtcon, dgtcon, cgtcon, zgtcon	Reciprocal condition number, tridiagonal matrix.
sgtrfs, dgtrfs, cgtrfs, zgtrfs	Improves solution, tridiagonal linear equations.
sgttrf, dgttrf, cgttrf, zgttrf	LU factorization, general tridiagonal matrix.
shgeqz, dhgeqz, chgeqz, zhgeqz	Eigenvalues $\det(A - w(i)B) = 0$.
shsein, dhsein, chsein, zhsein	R/L eigenvectors, upper Hessenberg matrix.
shseqr, dhseqr, chseqr, zhseqr	Eigenvalues, Schur factorization, upper Hessenberg matrix
spbequ, dpbequ, cpbequ, zpbequ	Equilibrate symmetric/Hermitian + definite banded matri
spbrfs, dpbrfs, cpbrfs, zpbrfs	Improve solution, banded linear equations.
spbtrf, dpbtrf, cpbtrf, zpbtrf	Cholesky factorization, banded matrix.
spotrf, dpotrf, cpotrf, zpotrf	Cholesky factorization.
spptrf, dpptrf, cpptrf, zpptrf	Cholesky factorization, packed matrix.
spttrf, dpttrf, cpttrf, zpttrf	LDL^H factorization, tridiagonal matrix.
ssbtrd, dsbtrd, chbtrd, zhbtrd	Reduce symmetric/Hermitian banded matrix.
ssptrd, dsptrd, chptrd, zhptrd	Reduce packed symmetric/Hermitian matrix.
ssptrf, dsptrf, csptrf, zsptrf, zsptrf, zhptrf	Factorization, packed matrix.
sstebz, dstebz	Selected eigenvalues, real symmetric tridiagonal matrix.
sstein, dstein, cstein, zstein	Eigenvectors, real symmetric tridiagonal matrix.
ssteqr, dsteqr, csteqr, zsteqr	Eigenvalues/vectors, real symmetric tridiagonal matrix.
ssterf, dsterf	Eigenvalues, real symmetric tridiagonal matrix,
ssyrfs, dsyrfs, csyrfs, zsyrfs, cherfs, zherfs	Improve solution, linear equations.
ssytrd, dsytrd, chetrd, zhetrd	Reduce symmetric/Hermitian matrix.
ssytrf, dsytrf, csytrf, zsytrf, chetrf, zhetrf	Matrix factorization.
ssytri, dsytri, csytri, zsytri, chetri, zhetri	Matrix inverse.
ssytrs, dsytrs, csytrs, zsytrs, chetrs, zhetrs	Solves linear equations AX=B.
stbcon, dtbcon, ctbcon, ztbcon	Reciprocal condition number, triangular band matrix.
stbrfs, dtbrfs, ctbrfs, ztbrfs	Error bounds, triangular banded linear equations.
stbtrs, dtbtrs, ctbtrs, ztbtrs	Triangular banded linear equations.
stgevc, dtgevc, ctgevc, ztgevc	R/L eigenvectors, pair upper triangular matrices.
stgsja, dtgsja, ctgsja, ztgsja	Singular value decomposition, two upper triangular matric
stpcon, dtpcon, ctpcon, ztpcon	Reciprocal condition number, packed triangular matrix.
stprfs, dtprfs, ctprfs, ztprfs	Error bounds for packed triangular linear equations.
stptri, dtptri, ctptri, ztptri	Inverse, packed triangular matrix.
stptrs, dtptrs, ctptrs, ztptrs	Triangular, packed linear equations.
strcon, dtrcon, ctrcon, ztrcon	Reciprocal condition number, triangular matrix.
strevc, dtrevc, ctrevc, ztrevc	R/L eigenvectors, upper quasitriangular/triangular matrix
strexc, dtrexc, ctrexc, ztrexc	Reorder Schur factorization.
strrfs, dtrrfs, ctrrfs, ztrrfs	Error bounds, triangular linear equations.
strsen, dtrsen, ctrsen, ztrsen	Reorder Schur factorization, etc.
strsna, dtrsna, ctrsna, ztrsna	Reciprocal condition numbers, upper triangular matrix.
strsyl, dtrsyl, ctrsyl, ztrsyl	Sylvester matrix equation A X ± X B=C.
strtri, dtrtri, ctrtri, ztrtri	Inverse, triangular matrix.
strtrs, dtrtrs, ctrtrs, ztrtrs	Triangular linear equations.
stzrqf, dtzrqf, dtzrqf, ztzrqf	RQ factorization, upper trapezoidal matrix.

.13 IMPLEMENTATION: NETLIB SHORT CONTENTS

hort Netlib Library List

a	Approximation algorithms
aicm	Selected material from Advs in Comptnl Math
lliant	Programs, Alliant
mos	Special functions by D. Amos
apollo	Programs, Apollo
enchmark	Benchmarks (timings)
bib	Bibliographies
bihar	Bjorstad's biharmonic solver
blas	Basic linear algebra subs
bmp	Brent's multiple precision package
c	"Misc" C library
chammph	Shallow water equations, spherical geometry
cheney-kincaid	Programs from 1985 text
clapack	C version of LAPACK
confdb	Conference database
conformal	Conformal mapping
contin	Continuation, limit points
c++	Codes in C++
lierckx	Spline fitting, various geometries
lomino	Communication, scheduling, multiple tasks
ispack	Matrix eigenvalues/vectors (see LAPACK)
lefunt	Cody and Waite's tests for elementary functions
rrata	Corrections to numerical books
2c	Fortran to C converter
shpack	Separable elliptic PDEs
loppy	Fortan code syntax and flow control checker
itpack	Cline's splines under tension
ftpack	Swarztrauber's Fourier transforms
mm	Software from Forsythe, Malcolm, & Moler
n	Fullerton's special functions
ortran	Single/double precision converter, static debugger
p	Floating point arithmetic
cv	Generalized cross validation
mat	Multi-processing time line, state graph tools
o	Golden oldies gaussq, zeroin, lowess, ...
raphics	Auto color, ray-tracing benchmark
arwell	MA28 sparse linear system
ence	Heterogeneous network computing environment
ompack	Nonlinear equations, homotopy method
eecss	IEEE Control Systems Society
sa	Intl J of Supercomputer Apps
ntercom	Interprocessor collective comm libe
pack	Iterative linear system
akef	Automatic differentiation of f77 subs
graph	Postscript graphs tool
incaid-cheney	Programs from 1990 text
apack	Libe for common linear algebra probs
anczos	Lanczos programs
anz	Lanczos programs
aso	Lanczos programs
nalg	Collected linear algebra stuff
npack	(See LAPACK)
p	Linear programming

Short Netlib Library List (continued)

machines	Descriptions of computers
microscope	Discontinuity checking
minpack	Nonlinear equations, least-squares
misc	Everything else
mpi	Message passing interface
napack	Numerical algebra programs
news	News column
numeralgo	Algorithms, "Numerical Algorithms"
ode	Ordinary differential equations
odepack	ODEs from Hindmarsh
odrpack	Orthogonal distance regression
opt	Optimization
p4	Portable program, parallel processors
paranoia	Kahan's floating-point test
parmacs	Parallel programmming macros
pascal	Pascal libe
pchip	Hermite cubics
picl	Portable instrumented comm libe
pltmg	Multigrid code
polyhedra	Database, geometric solids
popi	Digital Darkroom, image manipulation
port	Port library
pppack	Subs from Practical Guide to Splines
pvm	Parallel virtual machine
pvm3	Parallel virtual machine version 3
quadpack	Univariate quadrature
research	Miscellanea from AT&T Bell Labs
scalapack	LAPACK routines, MIMD computers
sched	Portable parallel algorithms enverirmomen
sequent	Programs, Sequent
slap	Seager + Greenbaum, iterative methods
slatec	Comprehensive math libe
sparse	C sparse linear algebra
sparse-blas	BLAS by indirection
sparspak	Sparse linear algebra core
specfun	Transportable special functions
spin	Communication simulation & validation
stringsearch	String matching
toeplitz	Linear systems, Toepli form
toms	ACM algorithms
typesetting	Typesetting macros
uncon/data	Optimization test problems
vanhuffel	Total least-squares, partial SVD
vfftpack	Vectorized fftpack
voronoi	Voronoi diagrams, Delaunay triangles
xNetlib	Netlib X window interface
y12	Sparse Linear System

15.14 IMPLEMENTATION: SLATEC SHORT CONTENTS

In much the same way as LAPACK, the SLATEC library contains routines that operate on different types of data but are otherwise equivalent. These are dicated by

Data Types

s	Single precision
d	Double precision
c	Complex
i	Integer
h	Character
l	Logical
a	Pseudotype, not convertible

'e give here a condensed and edited listing of the available SLATEC routines. more complete description of the individual routines is available through etlib on the Web.

User-callable SLATEC Routines

A (gams)	Arithmetic, Error Analysis
xadd-s, dxadd-d, xadj-s, dxadj-d	Floating-point arithmetic, extended range
xc210-s, dxc210-d, xcon-s, dxcon-d	Floating-point arithmetic, extended range
xred-s, dxred-d, xset-s, dxset-d	Floating-point arithmetic, extended range
carg-c	Argument of complex number
r9pak-s, d9pak-d	Pack base 2 exponent into floating-point number
r9upak-s, d9upak-d	Unpack floating-point number x to $x = y2^n$
C (gams)	Elementary and Special Functions (see also L)
fundoc-a	Documentation for fnlib, (elementary & special functions)
binom-s, dbinom-d	Binomial coefficients
fac-s, dfac-d	Factorial function
poch-s, dpoch-d	Generalized Pochhammer's symbol
poch1-s, dpoch1-d	Generalized Pochhammer's symbol, 1st O
cbrt-s, dcbrt-d, ccbrt-c	Cube root
csevl-s, dcsevl-d	Chebyshev series
inits-s, initds-d	# terms, polynomial series, specified accuracy
qmomo-s, dqmomo-d	Modified Chebyshev moments
xlegf-s, dxlegf-d	Normalized Legendre & associated functions
xnrmp-s, dxnrmp-d	Normalized Legendre polynomials
cacos-c, casin-c	Complex arc cosine, sine
catan-c	Complex arc tgent
catan2-c	Complex arc tgent in proper quadrt
cosdg-s, dcosdg-d	Cosine of argument in degrees
cot-s, dcot-d, ccot-c	Cotgent
ctan-c	Complex tangent
sindg-s, dsindg-d	Sine of argument in degrees
alnrel-s, dlnrel-d, clnrel-c	$\ln(1+x)$ accurate in relative error
clog10-c	Principal value of complex base 10 logarithm
exprel-s, dexprl-d, cexprl-c	Relative error exponential $(\exp(x)-1)/x$
acosh-s, dacosh-d, cacosh-c	Arc hyperbolic cosine
asinh-s, dasinh, casinh-c	Arc hyperbolic sine
atanh-s, datanh-d, catanh-c	Arc hyperbolic tangent
ccosh-c	Complex hyperbolic cosine
csinh-c	Complex hyperbolic sine
ctanh-c	Complex hyperbolic tangent

User-callable SLATEC Routines (continued)

ali-s, dli-d	Logarithmic integral
e1-s, de1-d	Exponential integral E1(x)
ei-s, dei-d	Exponential integral Ei(x)
exint-s, dexint-d	Member sequence of exponential integrals
spenc-s, dspenc-d	Form of Spence's integral
algams-s, dlgams-d	Log of Γ function
alngam-s, dlngam-d, clngam-c	Log of Γ function
c0lgmc-c	$(z + 0.5)\log((z + 1)/z) - 1$ with relative accuracy
gamlim-s, dgamlm-d	Min and max bounds for Γ function argument
gamma-s, damma-d, camma-c	Complete Γ function
gamr-s, dgamr-, cgamr-c	Reciprocal of Γ function
albeta-s, dlbeta-d, clbeta-c	Natural logarithm of complete β function
beta-s, dbeta-d, cbeta-c	Complete β function
psi-s, dpsi-d, cpsi-c	Ψ (or digamma) function
psifn-s, dpsifn-d	Derivatives of ψ function
gami-s, dgami-d	Incomplete Γ function
gamic-s, dgamic-d	Complementary incomplete Γ function
gamit-s, dgamit-d	Tricomi's incomplete Γ function
betai-s, dbetai-d	Incomplete β function
erf-s, derf-d	Error function
erfc-s, derfc-d	Complementary error function
daws-s, ddaws-d	Dawson's function
besj0-s, dbesj0-d	Bessel function, 1st kind, $\mathcal{O}$ 0
besj1-s, dbesj1-d	Bessel function, 1st kind, $\mathcal{O}$ 1
besy0-s, dbesy0-d	Bessel function, 2nd kind, $\mathcal{O}$ 0
besy1-s, dbesy1-d	Bessel function, 2nd kind, $\mathcal{O}$ 1
besj-s, dbesj-d	Sequence of Bessel functions
besy-s, dbesy-d	Forward recursion Bessel functions
cbesh-c, zbesh-c	Hankel functions, complex argument
cbesj-c, zbesj-c	Bessel functions, complex argument
cbesy-c, zbesy-c	Bessel functions, complex argument
besi0-s, dbesi0-d	Hyperbolic Bessel function, 1st kind, $\mathcal{O}$ 0
besi0e-s, dbsi0e-d	Scaled modified Bessel function, 1st, $\mathcal{O}$ 0
besi1-s, dbesi1-d	Modified Bessel function, 1st kind, $\mathcal{O}$ 1
besi1e-s, dbsi1e-d	Scaled modified Bessel function, 1st kind, $\mathcal{O}$ 1
besk0-s, dbesk0-d	Modified Bessel function, 3rd kind, $\mathcal{O}$ 0
besk0e-s, dbsk0e-d	Scaled modified Bessel function, 3rd kind, $\mathcal{O}$ 0
besk1-s, dbesk1-d	Modified Bessel function, 3rd kind, of $\mathcal{O}$ 1
besk1e-s, dbsk1e-d	Scaled modified Bessel function, 3rd kind, $\mathcal{O}$ 1
besi-s, dbesi-d	I Bessel functions
besk-s, dbesk-d	Forward recursion. Bessel function
beskes-s, dbskes-d	Scaled modified 3rd kind Bessel functions
besks-s, dbesks-d	Modified 3rd kind Bessel functions, fractional o
cbesi-c, zbesi-c	I Bessel functions, complex argument
cbesk-c, zbesk-c	K Bessel functions, complex argument
aie-s, daie-d	Airy function
bi-s, dbi-d	Bairy function (Airy function of 2nd kind)
bie-s, dbie-d	Bairy function, scaled
cairy-c, zairy-c	Airy function ai, derivative, complex argument
cbiry-c, zbiry-c	Airy function bi, derivative, complex argument
bskin-s, dbskin-d	Repeated integrals of k-zero Bessel function
chu-s, dchu-d	Logarithmic confluent hypergeometric function
rc-s, drc-d	Elliptic integral
rd-s, drd-d	In/complete elliptic integral, 2nd kind
rf-s, drf-d	In/complete elliptic integral, 1st kind
rj-s, drj	In/complete elliptic integral, 3rd kind

Jser-callable SLATEC Routines (continued)

c3jj-s, drc3jj-d	3j symbol
c3jm-s, drc3jm-d	3j symbol
c6j-s, drc6j-d	6j symbol
) (gams)	Various Matrix Operations (see toc)
gedi-s, dgedi-d, cgedi-c	Matrix det & inverse
gefs-s, dgefs-d, cgefs-c	General linear equations
geir-s, cgeir-c	General linear equations
gesl-s, dgesl-d, cgesl-c	Real system AX=B using sgeco or sgefa
qrsl-s, dqrsl-d, cqrsl-c	Coordinate transformations, least-squares solutions
)2A2 (gams)	Banded Matrices
gbsl-s, cgbsl-c	Real AX=B using sgbco or sgbfa
nbfs-s, dnbfs-d, cnbfs-c	Nonsymmetric banded linear equations
nbir-s, cnbir-c	Nonsymmetric banded linear equations
nbsl-s, dnbsl-d, cnbsl-c	Real banded system using snbco or snbfa
gtsl-s, dgtsl-d, cgtsl-c	Tridiagonal linear system
)2A3 (gams)	Triangular
trdi-s, dtrdi-d, ctrdi-c	Det & inverse, triangular matrix
trsl-s, dtrsl-d, ctrsl-c	TX=B, T triangular
)2A4 (gams)	Sparse
bcg-s, scgn-s, dcgn-d	Nonsymmetric AX = B
cgs-s, dcgs-d, sgmres-s, dgmres-d	Nonsymmetric AX = B
ir-s, dir-d	Nonsymmetric AX = B
lpdoc-s, dlpdoc-d	Documentation: sparse linear algebra package
omn-s, domn-d	Orthomin solve AX = B
sdbcg-s, dsdbcg-d	Diagonally scaled AX=B
sdcgn-s, dsdcgn-d, ssdcgs-s, dsdcgs-d; ssdgmr-s,	Diagonally scaled sparse AX=B
dsdgmr-d, ssdomn-s, dsdomn-d	Diagonally scaled sparse AX=B
sgs-s, dsgs-d	Gauss–Seidel AX = B
silur-s, dsilur-d	Incomplete LU sparse AX = B
sjac-s, dsjac-d	Jacobi's sparse AX = B
slubc-s, dslubc-d	Incomplete LU AX=B
slucn-s, dslucn-d, sslucs-s, dslucs-d	Incomplete LU cg sparse AX=B
slugm-s, dslugm-d	Incomplete LU Gmres AX=B
sluom-s, dsluom-d	Incomplete LU Orthomin sparse AX=B
)2B (gams)	Real Symmetric Matrices
sidi-s, dsidi-d, chidi-c, csidi-c	Det, inverse, real symmetric
spdi-s, dspdi-d, chpdi-c, cspdi-c	Det, inverse, real symmetric packed matrix
spsl-s, dspsl-d, chpsl-c, cspsl-c	Real symmetric system
)2B1B (gams)	Positive Definite
chdc-s, dchdc-d, cchdc-c	Cholesky decomposition, + def
podi-s, dpodi-d, cpodi-c	Det, inverse, real symmetric + def
pofs-s, dpofs-d, cpofs-c	+ definite symmetric system
poir-s, cpoir-c	Solve + definite symmetric system
posl-s, dposl-d, cposl-c	Real symmetric + def system
ppdi-s, dppdi-d, cppdi-c	Det,inverse, real symmetric + def
ppfa-s, dppfa-d, cppfa-c	Factor real symmetric packed + def
ppsl-s, dppsl-d, cppsl-c	Real symmetric + definite system
ptsl-s, dptsl-d, cptsl-c	+ definite tridiagonal system
bcg-s, dbcg-d, scg-s, dcg-d	Sparse nonsymmetric AX = B
cgn-s, dcgn-d	Sparse normal AX = B
cgs-s, dcgs-d	Sparse nonsymmetric AX = B
gmres-s, dgmres-d	Gmres sparse AX=B
ir-s, dir-d	Sparse AX = B
lpdoc-s, dlpdoc-d	Documentation, sparse AX = B

User-callable SLATEC Routines (continued)

somn-s, domn-d	Orthomin sparse AX=B
ssdbcg-s, dsdbcg-d, ssdcg-s, dsdcg-d	Diagonally scaled sparse AX=B
ssdcgn-s, dsdcgn-d	Diagonally scaled sparse normal AX=B
ssdcgs-s, dsdcgs-d	Diagonally scaled cgs sparse AX=B
ssdgmr-s, dsdgmr-d	Diagonally scaled nonsymmetric sparse AX=B
ssdomn-s, dsdomn-d	Diagonally scaled orthomin sparse AX=B
ssgs-s, dsgs-d	Gauss–Seidel AX = B
ssiccg-s, dsiccg-d	Incomplete Cholesky sparse AX=B
ssilur-s, dsilur-d	Incomplete LU sparse AX = B
ssjac-s, dsjac-d	Jacobi's sparse AX = B
sslubc-s, dslubc-d	Incomplete LU sparse AX=B
sslucn-s, dslucn-d	Incomplete LU sparse normal AX=B
sslucs-s, dslucs-d	Incomplete LU AX=B
sslugm-s, dslugm-d	Incomplete LU Gmres sparse AX=B
ssluom-s, dsluom-d	Incomplete LU Orthomin sparse AX=B
D2C (gams)	Complex NonHermitian Matrices
cgeco-c, sgeco-s, dgeco-d	Factor, Gaussian elimination
cgedi-c, sgedi-s, dgedi-d	Matrix det
cgefa-c, sgefa-s, dgefa-d	Factor via Gaussian elimination
cgefs-c, sgefs-s, dgefs-d	General complex linear equations
cgeir-c, sgeir-s	General complex linear equations
cgesl-c, sgesl-s, dgesl-d	Complex AX=B
cqrsl-c, sqrsl-s, dqrsl-d	Coordinate transformations, least-squares solutions
csidi-c, ssidi-s, dsidi-d, chidi-c	Det & inverse, complex symmetric matrix
csisl-c, ssisl-s, dsisl-d, chisl-c	Complex symmetric system
cspdi-c, sspdi-s, dspdi-d, chpdi-c	Det & inverse, complex symmetric, packed matrix
cspsl-c, sspsl-s, dspsl-d, chpsl-c	Complex symmetric matrix
cgbsl-c	Complex banded AX=B
cnbfs-c, snbfs-s, dnbfs-d	General nonsymmetric banded AX=B
cnbir-c, snbir-s	General nonsymmetric banded AX=B
cnbsl-c, snbsl-s, dnbsl-d	Complex banded AX=B
cgtsl-c, sgtsl-s, dgtsl-d	Tridiagonal linear system
ctrdi-c, strdi-s, dtrdi-d	Det & inverse, triangular matrix
ctrsl-c, strsl-s, dtrsl-d	Triangular tx=b
D2D (gams)	Complex Hermitian Matrices
chidi-c, ssidi-s, dsisi-d, csidi-c	Det, inverse complex Hermitian
chisl-c, ssisl-s, dsisl-d, csisl-c	Complex Hermitian system
chpdi-c, sspdi-s, dspdi-d, dspdi-c	Det, inverse, packed complex Hermitian matrix
chpsl-c, sspsl-s, dspsl-d, cspsl-c	Complex Hermitian system
D2D1B (gams)	Positive Definite
cchdc-c, schdc-s, dchdc-d	Cholesky decomposition, + definite matrix
cpodi-c, spodi-s, dpodi-d	Det, inverse, complex, Hermitian, + definite matrix
cpofs-c, spofs-s, dpofs-d	+ definite symmetric complex system
cpoir-c, spoir-s	+ definite Hermitian system
cposl-c, sposl-s, dposl-d	Complex Hermitian + definite system
cppdi-c, sppdi-s, dppdi-d	Det, inverse, complex Hermitian + definite matrix
cppsl-c, sppsl-s, dppsl-d	Complex Hermitian + definite system
cpbsl-c, spbsl-s, dpbsl-d	Complex Hermitian + definite banded system
cptsl-c, sptsl-s, dptsl-d	+ definite tridiagonal system
D3 (gams)	Determinants
sgedi-s, dgedi-d, cgedi-c	Det, inverse of matrix
sgbdi-s, dgbdi-d, cgbdi-c	Det, banded matrix
snbdi-s, dnbdi-d, cnbdi-c	Det, banded matrix
strdi-s, dtrdi-d, ctrdi-c	Det, inverse, trigular matrix
ssidi-s, dsidi-d, chidi-c, csidi-c	Det, inverse, real symmetric matrix
sspdi-s, dspdi-d, chpdi-c, cspdi-c	Det, inverse, real symmetric packed matrix

Jser-callable SLATEC Routines (continued)

podi-s, dpodi-d, cpodi-c	Det, inverse, real symmetric + definite matrix
ppdi-s, dppdi-d, cppdi-c	Det, inverse, real symmetric + definite matrix
pbdi-s, dpbdi-d, cpbdi-c	Det, symmetric + def, banded matrix
gedi-c, sgedi-s, dgedi-d	Det, inverse of matrix
sidi-c, ssidi-s, dsidi-d, chidi-c	Det, inverse, complex symmetric matrix
spdi-c, sspdi-s, dspdi-d, chpdi-c	Det, inverse, packed complex symmetric matrix
gbdi-c, sgbdi-s, dgbdi-d	Det, complex banded matrix
nbdi-c, snbdi-s, dnbdi-d	Det, banded matrix
trdi-c, strdi-s, dtrdi-d	Det, inverse, triangular matrix
hidi-c, ssidi-s, dsisi-d, csidi-c	Det, inverse, complex Hermitian matrix
hpdi-c, sspdi-s, dspdi-d, dspdi-c	Det, inverse, packed complex Hermitian matrix
podi-c, spodi-s, dpodi-d	Det, inverse, + def complex Hermitian matrix
ppdi-c, sppdi-s, dppdi-d	Det, inverse, complex Hermitian + definite matrix
pbdi-c, spbdi-s, dpbdi-d	Det, complex Hermitian + definite banded matrix
)4 (gams)	Eigenvalues, Eigenvectors
isdoc-a	Documentation for Eispack, matrix eigenproblems
s-s, ch-c	Eigenvalues, eigenvectors, real symmetric matrix
sp-s	Eigenvalues/vector, real symmetric packed matrix
siev-s, chiev-c	Eigenvalues/vector, real symmetric matrix
spev-s	Eigenvalues/vector, packed real symmetric matrix
g-s, cg-c, sgeev-s, cgeev-c	Eigenvalues/vector, real general matrix
h-c, rs-s	Eigenvalues/vector, complex Hermitian matrix
hiev-c, ssiev-s	Eigenvalues/vector, complex Hermitian matrix
g-c, rg-s	Eigenvalues/vector, complex general matrix
geev-c, sgeev-s	Eigenvalues/vector, complex general matrix
isect-s	Eigenvalues, symmetric tridiagonal matrix, Sturm
mtql1-s	Eigenvalues, symmetric tridiagonal matrix, QL
mtql2-s	Eigenvalues/vector, symmetric tridiagonal matrix, QL
ntqlv-s	Eigenvalues/vector, symmetric tridiagonal matrix, QL
atqr-s	Largest or smallest eigenvalues, symmetric tridiagonal matrix
st-s	Eigenvalues/vector, real symmetric tridiagonal matrix
t-s	Eigenvalues/vector, special real tridiagonal matrix
ql1-s	Eigenvalues, symmetric tridiagonal matrix, QL
ql2-s	Eigenvalues/vector, symmetric tridiagonal matrix
qlrat-s	Eigenvalues, symmetric tridiagonal matrix, rational QL method
ridib-s	Eigenvalues, symmetric tridiagonal matrix, Sturm
sturm-s	Eigenvalues, symmetric tridiagonal matrix, given interval, Sturm
qr-s	Some eigenvalues, real symmetric matrix, qr
sb-s	Eigenvalues/vector, symmetric banded matrix
)4B (gams)	General Eigenvalue Problems, AX = λ BX
sg-s	Eigenvalues/vectors of symmetric general eigenproblem
sgab-s	Eigenvalues/vector, symmetric general eigenproblem
sgba-s	Eigenvalues/vector, symmetric general eigenproblem
alanc-s, cbal-c	Balance real general matrix, isolate eigenvalues
omlr-c	Eigenvalues, complex upper Hessenberg matrix, lr
omlr2-c	Eigenvalues/vector, complex upper Hessenberg matrix
qr-s, comqr-c	Eigenvalues, real upper Hessenberg matrix, qr
qr2-s, comqr2-c	Eigenvalues/vector, real upper Hessenberg matrix, qr
vit-s, cinvit-c	Eigenvectors, real upper Hessenberg matrix, inverse iteration
andv-s	Eigenvectors, real symmetric banded matrix, inverse iteration
nvit-s	Eigenvectors, symmetric tridiagonal matrix, inverse iteration
sia-s, dllsia-d, sglss-s, dglss-d	Linear least-squares problems, qr factorization
ndsol-s, dbndsl-d	Least-squares problems, banded matrix
fti-s, dhfti-d, llsia-s	Linear least-squares problems, qr factorization
llsia-d, sglss-s, dglss-d	Linear least-squares problems, qr factorization
ei-s, dlsei-d	Linearly constrained least-squares problem

User-callable SLATEC Routines (continued)

ulsia-s, dulsia-d	Underdetermined linear equations by LQ factorization
E (gams)	Interpolation
bspdoc-a	Documentation for Bspline, piecewise polynomials
bint4-s, dbint4-d, bintk-s, dbintk-d	B-rep of cubic spline, interpolate data
pchdoc-a	Pchip documentation, cubic Hermite interpolation
bsqad-s, dbsqad-d	Integral of k-th $\mathcal{O}$ b-spline
bvalu-s, dbvalu-d	B-spline for function value, derivatives
chfdv-s, dchfdv-d	Cubic polynomial, derivative, Hermite form
chfev-s, dchfev-d	Cubic polynomial, Hermite form, array of points
pchbs-s, dpchbs-d	Piecewise cubic Hermite to b-spline converter
pchfd-s, dpchfd-d	Piecewise cubic Hermite function, 1st derivative, array
pchfe-s, dpchfe-d	Piecewise cubic Hermite function, array
pchia-s, dpchia-d	Integral, piecewise cubic Hermite function
pchid-s, dpchid-d	Integral, piecewise cubic Hermite function
pfqad-s, dpfqad-d	Integral, product of function and derivative
ppqad-s, dppqad-d	Integral k-th $\mathcal{O}$ b-spline
ppval-s, dppval-d	Derivative of b-spline
f (gams)	Solution of Nonlinear Equations
rpqr79-s, cpqr79-c, rpzero-s, cpzero-c	Zeros of polynomial, real coefficients
cpqr79-c, rpqr79-s, cpzero-c, rpzero-s	Zeros of polynomial, complex coefficients
fzero-s, dfzero-d	Zero of function
snsq-s, dnsq-d	Zero, nonlinear functions, n variables
snsqe-s, dnsqe-d	Easy-to-use zero, nonlinear functions, n variables
sos-s, dsos-d	Least-squares nonlinear equations
splp-s, dsplp-d	Linear programming problems, 1000's constraints
H2 (gams)	Quadrature
qpdoc-a	Quadpack documentation, automatic integration
gaus8-s, dgaus8-d	Integrate, real, adaptive 8-point Legendre–Gauss
qag-s, dqag-d, qage-s, dqage-d	Integrate, specified accuracy
qnc79-s, dqnc79-d	Integrate, 7-point adaptive Newton–Cotes quadrature
qng-s,dqng-d	Integrate, to specified accuracy
H2A1A2 (gams)	Nonautomatic Quadrature
qk15-s, dqk15-d	Integrate f or abs(f) with error estimate
15→21, 31, 41, 51, 61	Integrate f or abs(f) with error estimate
avint-s, davint-d	Integrate tabulated function, parabolas
pchia-s, dpchia-d	Integrate, piecewise cubic Hermite function
pchid-s, dpchid-d	Integrate, piecewise cubic Hermite function, endpoints
bfqad-s, dbfqad-d	Integrate product of function & derivative of b-spline
bsqad-s, dbsqad-d	Integrate k-th $\mathcal{O}$ b-spline
pfqad-s, dpfqad-d	Integrate product, function & derivative of b-spline
ppqad-s, dppqad-d	Integrate k-th $\mathcal{O}$ b-spline
qagp-s, dqagp-d, qagpe-s, dqagpe-d	Automatic integrate, specified accuracy
qawc-s, dqawc-d	Automatic principal value integral, specified accuracy
qawce-s, dqawce-d	Automatic weighted Cauchy principal value integral, specified accuracy
qawo-s, dqawo-d, qawoe-s, dqawoe-di	Automatic integrate trig-weighted
qaws-s, dqaws-d, qawse-s, dqawse-d	Automatic integrate singular-weighted function
qmomo-s, dqmomo-d	Automatic compute k-th modified chebyshev moments
qc25c-s, dqc25c-d	Nonautomtic integrate weighted
qc25f-s, dqc25f-d	Nonautomtic integrate trig-weighted
qc25s-s, dqc25s-d	Integrate singular-weighted, error estimate
qk15w-s, dqk15w-d	Integrate fw or abs(fw), error estimate
qagi-s, dqagi-d, qagie-s, dqagie-d	Integrate, infinite bounds, specified accuracy

User-callable SLATEC Routines (continued)

qawf-s, dqawf-d, qawfe-s, dqawfe-d	Fourier integral, specified accuracy
qk15i-s, dqk15i-d, qagi-s	Integrate, infinite range
I1 (gams)	Ordinary Differential Equations
derkf-s, dderkf-d	Initial value ODE, Runge–Kutta-Fehlberg method
deabm-s, ddeabm-d	Initial value ODE, Adams-Bashforth method
sdriv1-s, ddriv1-d, cdriv1-c	Solve 200 or fewer ODEs
sdriv2-s, ddriv2-d, cdriv2-c	Solve n ODEs, stiff & nonstiff
sdriv3-s, ddriv3-d, cdriv3-c	Solve n ODEs; stiff & nonstiff
steps-s, dsteps-d	Integrate ODEs by one step
debdf-s, ddebdf-d	Initial value ODE, backward differentiation, stiff
sdassl-s, ddassl-d	System of differential/algebraic equations
sdriv1-s, ddriv1-d, cdriv1-c	Solve 200 or fewer ODEs
sdriv2-s, ddriv2-d, cdriv2-c	n ODEs, stiff & nonstiff
sdriv3-s, ddriv3-d, cdriv3-c	n ODEs, stiff & nonstiff
bvsup-s, dbvsup-d	2-pt boundary-value problem, variable-steps
I2 (gams)	Partial Differential Equations
hstcrt-s, hwscrt-s	Helmholtz pde, 5-pt finite difference, cartesian
hstcsp-s, hstssp-s, hwscsp-s, hwsssp-s	Helmholtz eqn, 5-pt finite difference, spherical
hstcyl-s, hwscyl-s	Helmholtz pde, 5-pt finite difference, cylindrical
hstplr-s, hwsplr-s	Helmholtz pde, 5-pt finite difference, polar
hw3crt-s	Helmholtz pde, 7-pt finite difference, cartesian
sepeli-s, sepx4-s	Separable elliptic pde on rectangle
J1 (gams)	Fast Fourier Transforms (see also L)
fftdoc-a	Fftpack documentation, fast Fourier transforms
ezfftb-s	Simplified real, periodic, backward FFT
ezfftf-s	Simplified real, periodic, FFT
rfftf1-s, cfftf1-c	Forward transform, real, periodic sequence
cfftf1-c, rfftf1-s	Forward transform, complex, periodic sequence
cosqb-s	Unnormalized inverse cosine trsform
cosqf-s	Forward cosine trsform, odd wave numbers
cost-s	Cosine transform, real, even sequence
sinqb-s	Unnormalized inverse of sinqf
sinqf-s	Forward sine transform, odd wave numbers
sint-s	Sine trsform, real, odd sequence
qawc-s, dqawc-d, qawce-s, dqawce-d	Cauchy principal value integral, specified accuracy
qc25c-s, dqc25c-d	Integral of f/(x-c) with error estimate
K (gams)	Approximation
bspdoc-a	B spline documentation
efc-s, defc-d, fc-s, dfc-d	Fit piecewise polynomial curve (b-spline) to discrete data
pcoef-s, dpcoef-d	Convert polfit coefficients to Taylor series form
polfit-s, dpolft-d, efc-s	1-D least squares fit to data by polynomials
defc-d, fc-s, dfc-d	1-D least squares fit to data by polynomials
lsei-s, dlsei-d, sbocls-s, dbocls-d	Linearly least-squares fit with constraints
sbols-s, dbols-d, wnnls-s, dwnnls-d	Least-squares fit, with bounds
snls1-s, dnls1-d	Nonlinear minimization, n variables
snls1e-s, dnls1e-d	Easy-to-use nonlinear minimization, n variables
L (gams)	Statistics, Probability
erf-s, derf-d	Error function
erfc-s, derfc-d	Complementary error function
rgauss-s	Normally distributed (Gaussian) random number
rand-s, runif-s	Uniformly distributed random number
efc-s, defc-d, fc-s, dfc-d	Piecewise polynomial (b-splines) fit

SLATEC Documentation

bspdoc-a	Bspline, piecewise polynomial functions
eisdoc-a	Eispack, matrix eigenproblems
fftdoc-a	Fftpack, fast Fourier transforms
fundoc-a	Fnlib, elementary & special functions
pchdoc-a	Pchip, piecewise cubic Hermite interpolation
qpdoc-a	Quadpack, automatic integral evaluation
slpdoc-s, dlpdoc-d	Sparse linear algebra package

16

Bound States in Momentum Space

In Chapter 10, *Quantum Eigenvalues; Zero-Finding and Matching*, we explored how to solve a Schrödinger equation for the energies in coordinate (r) space. This is a differential equation version of the eigenvalue problem. In this chapter we examine the integral equation form of the eigenvalue problem in momentum (k) space.

16.1 PROBLEM: BOUND STATES IN NONLOCAL POTENTIALS

As pictured in Fig. 16.1, a projectile particle interacts with the particles in a medium through which the projectile passes. The multiple-particle nature of the interaction leads to a nonlocal potential in which the potential at $\mathbf{r}$ depends on the wave function at all $\mathbf{r}'$ values. This changes the interaction term in the Schrödinger equation [L 96]:

$$V(r)\psi(r) \to \int dr' V(r, r')\psi(r'). \tag{16.1}$$

The integration in (16.1) leads to a Schrödinger equation that is a combined integral and differential ("integrodifferential") equation:

$$-\frac{1}{2\mu}\frac{d^2\psi(r)}{dr^2} + \int dr' V(r, r')\psi(r') = E\psi(r). \tag{16.2}$$

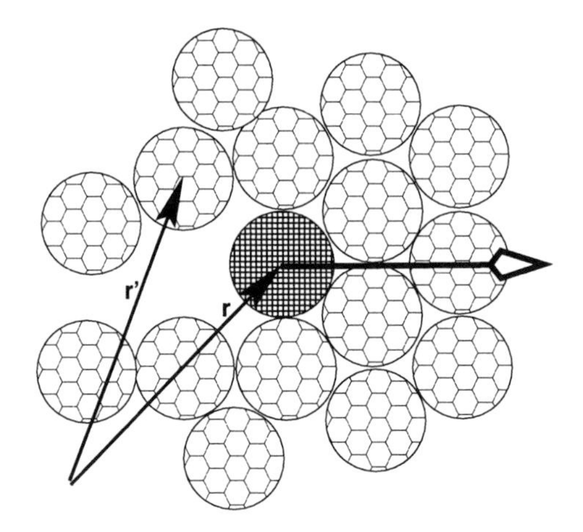

Fig. 16.1 A dark particle moving to the right through a dense multiparticle medium. The nonlocality of the potential felt by the dark particle at $\mathbf{r}$ arises from the particle interactions at all $\mathbf{r}'$.

Your **problem** is to figure out how to find the bound-state energies E_n and wave functions ψ_n for the integral equation in (16.2).[1] In Chapter 17, *Quantum Scattering via Integral Equations*, we indicate how to solve for scattering from this potential.

16.2 THEORY: K-SPACE SCHRÖDINGER EQUATION

One way of dealing with equation (16.2) is by going to momentum space where, in the partial-wave basis, we obtain an integral equation [L 96]:

$$\frac{k^2}{2\mu}\psi_n(k) + \frac{2}{\pi}\int_0^\infty dp p^2 V(k,p)\psi_n(p) = E_n\psi_n(k). \tag{16.3}$$

Here $V(k,p)$ is the momentum-space representation of the potential, $\psi_n(k)$ is the momentum-space wave function (the probability amplitude for finding the particle with momentum k) and the subscript n distinguishes solutions of different energies E_n.

Equation (16.3) is an integral equation for $\psi_n(k)$. It is more than an integral expression for $\psi_n(k)$ because the integral in it can't be evaluated until $\psi_n(p)$ is known for all p; that is, until the equation is solved! Nevertheless, we shall see that we can transform this problem into a matrix eigenvalue

[1] We use "natural" units in which Planck's constant $\hbar \equiv 1$, and so there is no difference between momenta and wave vectors.

problem. The matrix eigenvalue problem is easy to solve on the computer using a mathematical subroutine library.[2]

Those with a mathematical bent may care to observe that the standard formulation of the bound-state problem imposes the boundary condition that the wave function decays exponentially as $r \to \infty$. While the formulation given in this section does not explicitly impose that condition, it assumes that the wave packet is normalizable. The two conditions are equivalent because only an exponentially decaying wave function will be normalizable.

At first, a physicist living in coordinate space might not consider $\psi_n(k)$ as a bona fide wave function. Yet, mathematically, the abstract state vector is $|\psi_n\rangle$, and the usual wave function $\psi_n(r) = \langle r\,|\psi_n\rangle$ is the coordinate-space projection of this state, and $\psi_n(k) = \langle k\,|\psi_n\rangle$ is the momentum-space projection. Both wave functions contain the same amount of information about the state. In any case, $\psi_n(k)$ can always be converted to $\psi_n(r)$ by a Bessel transform,

$$\psi_n(r) = \int_0^\infty dk \psi_n(k) j_l(kr) k^2. \tag{16.4}$$

The momentum-space potential $V(k', k)$ is obtained from the partial-wave potential $V(r', r)$ by a double Bessel transform:

$$\begin{aligned} V(k,p) &= (2\pi)^3 \int_0^\infty dr \int_0^\infty dr' r r' j_l(kr') V(r', r) j_l(pr) && (16.5)\\ &= (2\pi)^3 \int_0^\infty r^2 j_l(kr) V(r) j_l(pr), && (16.6) \end{aligned}$$

where the second form is appropriate for a *local* potential [one for which $V(r', r) = \delta(r' - r)V(r)$]. The $j_l(kr)$'s are spherical Bessel functions; for example

$$j_0(z) = \frac{\sin z}{z}, \qquad j_1(z) = \frac{\sin z}{z^2} - \frac{\cos z}{z}, \qquad \ldots. \tag{16.7}$$

As you may recall, in Chapter 3, *Errors and Uncertainties in Computations*, we developed a technique for computing the spherical Bessel functions.

16.3 METHOD: REDUCING INTEGRAL TO LINEAR EQUATIONS

One technique for solving integral equations is to transform them to linear equations that can be solved as matrix equations. This is also done in our discussion of quantum scattering in §17.4. We convert (16.3) into linear equations by approximating the integral as a weighted sum over N integration

[2]The use of libraries is described in Chapter 15, *Matrix Computing and Subroutine Libraries*.

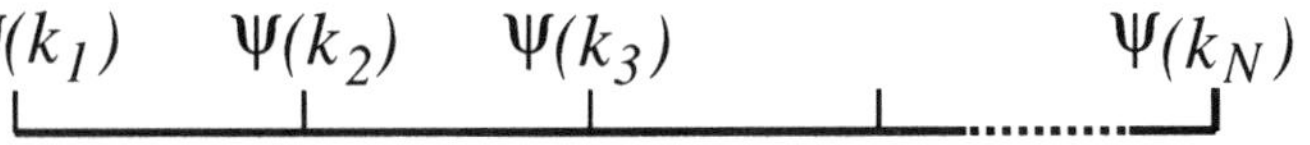

$\Psi(k_1)$ $\Psi(k_2)$ $\Psi(k_3)$ $\Psi(k_N)$

Fig. 16.2 The grid in momentum space on which the integral equation for the wave function is solved.

points (usually Gauss quadrature[3]), $p = k_j,\ j = 1, N$:

$$\int_0^\infty dp p^2 V(k,p)\psi_n(p) \simeq \sum_{j=1}^N k_j^2 V(k,k_j)\psi_n(k_j), \quad (16.8)$$

$$\frac{k^2}{2\mu}\psi_n(k) + \frac{2}{\pi}\sum_{j=1}^N k_j^2 V(k,k_j)\psi_n(k_j) = E_n\psi_n(k). \quad (16.9)$$

For a given value of the label n, (16.9) contains the N unknowns $\psi_n(k_j)$, the single unknown E_n, and the unknown function $\psi_n(k)$. We eliminate the unknown function $\psi_n(k)$ by evaluating the equation on a grid (indicated in Fig. 16.2) composed of the same N k_i values used to approximate the integral. This leads to the set of N coupled linear equations in $N+1$ unknowns:

$$\frac{k_i^2}{2\mu}\psi_n(k_i) + \frac{2}{\pi}\sum_{j=1}^N k_j^2\ V(k_i,k_j)\psi_n(k_j) = E_n\psi_n(k_j), \quad (i = 1, N). \quad (16.10)$$

We write this in matrix form as

$$[H][\psi_n] = E_n[\psi_n], \quad (16.11)$$

$$H_{ij} = \frac{k_i^2}{2\mu}\delta_{ij} + \frac{2}{\pi}V(k_i,k_j)k_j^2 w_j, \quad (i,j = 1, N). \quad (16.12)$$

where δ_{ij} is the Kronecker delta function.

Equation (16.11) is the matrix representation of the Schrödinger equation (16.3). The wave function $\psi_n(k)$ on the grid is the $N \times 1$ vector

$$[\psi_n(k_j)] = \begin{pmatrix} \psi_n(k_1) \\ \psi_n(k_2) \\ \vdots \\ \psi_n(k_N) \end{pmatrix}, \quad (16.13)$$

where the subscript n indicates the energy.

[3]See Chapter 4, *Integration*, for a discussion of numerical integration. See the diskette and the Web for subroutines to supply the points and weights.

The acute reader may be questioning the possibility of solving N equations for $N+1$ unknowns. That reader is wise; only sometimes, and only for certain values of E_n (eigenvalues) will the computer be able to find solutions. To see how this arises, we try to apply the matrix inversion technique (which we will use successfully for scattering in Chapter 17). We rewrite (16.11) as

$$[H - E_n I][\psi_n] = [0]. \tag{16.14}$$

We now multiply both sides of this equation by the inverse of $[H - E_n I]$ to obtain the formal solution

$$[\psi_n] = [H - E_n I]^{-1}[0]. \tag{16.15}$$

This equation tells us that one of two things is happening. If the inverse exists, then we have the *trivial* solution $\psi_n \equiv 0$, which is not very interesting. Alternatively, if nontrivial solutions are to exist, then our assumption that the inverse exists must be incorrect. Yet we know from the theory of linear equations that the inverse fails to exist when the determinant vanishes:

$$\det[H - E_n I] = 0. \quad \text{(bound-state condition)}. \tag{16.16}$$

Equation (16.16) is the additional equation needed to find unique solutions to the eigenvalue problem. There is, as the case may be, no guarantee that solutions of (16.16) can always be found. When they are found, they correspond to the *eigenvalues* of (16.11) and are the bound-state energies of our physical system.

16.4 MODEL: THE DELTA-SHELL POTENTIAL

To keep things simple, and to have an analytic answer to compare with, we consider the local, delta-shell potential:

$$V(r) = \frac{\lambda}{2\mu}\delta(r-b). \tag{16.17}$$

This would be a good model for an interaction that occurs predominantly when two particles are a fixed distance b apart. Because (16.17) is a local potential, we use (16.6) to determine its momentum-space representation:

$$V(k',k) = \int_0^\infty r^2 j_l(k'r')\frac{\lambda}{2\mu}\delta(r-b)j_l(kr)dr \quad = \frac{\lambda b^2}{2\mu}j_l(k'b)j_l(kb), \tag{16.18}$$

where the subscript l indicates the angular momentum state for which we are solving the problem. (*Beware:* This is not a well-behaved function in momentum space. To get stable answers, you need to look at the functional dependence of the integrand, and distribute your integration points on that account.)

If the energy is parameterized in terms of a wave vector κ by

$$E = -\frac{\kappa^2}{2\mu}, \tag{16.19}$$

then for this potential there is one bound state for each value of the angular momentum l, and it satisfies [Gott 66] the transcendental equation

$$1 - \frac{\lambda}{i\kappa}(i\kappa b)^2 j_l(i\kappa b)\left[n_l(i\kappa b) - i j_l(i\kappa b)\right] = 0. \tag{16.20}$$

For $l = 0$, this takes the simple form

$$e^{-2\kappa b} - 1 = \frac{2\kappa}{\lambda}, \quad (l = 0). \tag{16.21}$$

16.5 IMPLEMENTATION: BINDING ENERGIES, BOUND.C (.F)

An actual computation may follow two paths. The first would call subroutines to evaluate the determinant of the $[H - E_n I]$ matrix in (16.16) and then to *search* for those values of energy for which the computed determinant vanishes. This provides E_n, but not wave functions. The other approach would call an eigenproblem solver that may give some or all eigenvalues and eigenfunctions. In both cases, the solution is obtained iteratively, and you may be required to guess starting values for both the eigenvalues and eigenvectors.

1. Write your own, or modify the code on the diskette and Web, so that you can solve the integral equation (16.11) for the delta-shell potential (16.18). You can do this either by evaluating the determinant of $[H - E_n I]$ and then finding the E for which the determinant vanishes, *or* by finding the eigenvalues and eigenvectors for this H.

2. Set the scale by setting $2\mu = 1$ and $b = 10$.

3. Set up the potential and Hamiltonian matrices $V(i,j)$ and $H(i,j)$ for Gaussian quadrature integration with at least $N = 16$ grid points.

4. Adjust the value and sign of λ to find the $l = 0$ bound state. A good approach is to start with a large negative value for λ, and then make it less negative. You should find the eigenvalue moves up in energy.

5. Try increasing the number of grid points in steps of 8, for example; 16, 24, 32, 64, ..., and see how the energy changes.

6. *Note:* Your eigenenergy solver may return several eigenenergies. Only the true bound state will be at negative energy and will be stable as the

number of grid points change.

7. Extract the best value for the bound-state energy and estimate its precision by seeing how it changes with the number of grid points.

8. Verify that, regardless of the potential's strength, there is only a single bound state.

16.6 EXPLORATION: WAVE FUNCTION

1. Determine the momentum-space wave function $\psi_n(k)$. Does it fall off at $k \to \infty$? Does it oscillate? Is it well behaved at the origin?

2. Determine the coordinate-space wave function via the Bessel transform

$$\psi_0(r) = \int_0^\infty dk \psi_0(k) j_0(kr) k^2. \qquad (16.22)$$

   Does $\psi_0(r)$ fall off like you would expect for a bound state? Does it oscillate? Is it well behaved at the origin?

3. Compare the r dependence of your $\psi_0(r)$ to the analytic wave function:

$$\psi_0(r) \propto \begin{cases} e^{-\kappa r} - e^{\kappa r}, & \text{for } r < a, \\ e^{-\kappa r}, & \text{for } r > a. \end{cases} \qquad (16.23)$$

4. Deduce the energy of the $l = 1$ bound state.

17

Quantum Scattering via Integral Equations ⊙

The power and accessibility of high-speed computers has changed the view as to what theories and equations in physics are soluble. In Chapter 9, *Differential Equations and Oscillations,* and Chapter 14, *Differential Chaos in Phase Space*, we see how nonlinear differential equations are solved to give new insight into the physical world. The project in this chapter solves a singular integral equation. After these chapters we hope the reader views both integral and differential equations as easily soluble.

17.1 PROBLEM: QUANTUM SCATTERING IN K SPACE

As shown in Fig. 16.1, a projectile passes through a medium containing particles with which the projectile interacts. The multiple-particle nature of the interaction leads to a nonlocal potential that depends on the wave function over all of space:

$$V(r)\psi(r) \rightarrow \int dr' V(r, r')\psi(r'). \tag{17.1}$$

The integration in (17.1) leads to a Schrödinger equation that is a combined integral and differential ("integrodifferential") equation:

$$-\frac{1}{2\mu}\frac{d^2\psi(r)}{dr^2} + \int dr' V(r, r')\psi(r') = E\psi(r). \tag{17.2}$$

Your **problem** is to solve this equation and deduce the scattering (Fig. 17.1) that occurs when a particle passes through a dense medium.

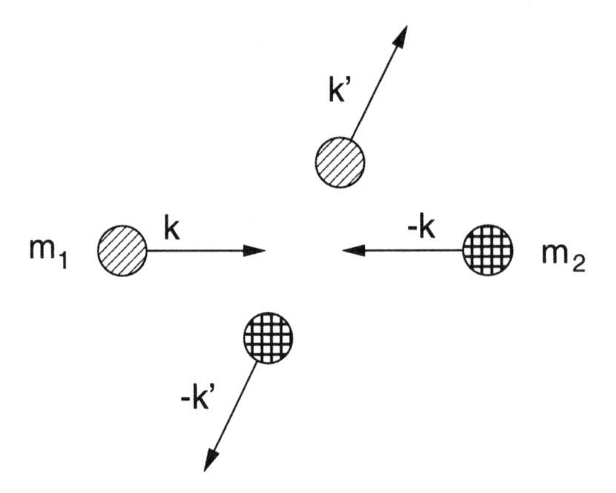

Fig. 17.1 The scattering of mass m_1 and m_2 in their center-of-momentum system.

17.2 THEORY, LIPPMANN–SCHWINGER EQUATION

An escape from the mess of dealing with the integral and the differential aspects of (17.2) is to convert it to an integral equation. This is done by going to momentum space where we deal with the Fourier components of the wave function. In Chapter 16 we did that for the bound-state problem and found that it leads to a homogeneous integral equation. After representing the integral equation as a matrix eigenvalue problem, we can solve it. In this chapter we are concerned with scattering, and the different physics leads to an inhomogeneous integral equation.

Because scattering experiments measure scattering amplitudes, it is convenient to convert the Schrödinger equation into an equation dealing with amplitudes rather than wave functions. An integral form of the Schrödinger equation dealing with the amplitude R (reaction matrix) is the *Lippmann–Schwinger equation:*[1]

$$R(k', k) = V(k', k) + \frac{2}{\pi}\mathcal{P}\int_0^\infty dp \frac{p^2 V(k', p) R(p, k)}{(k_0^2 - p^2)/2\mu}. \tag{17.3}$$

Note that equation (17.3) requires more than just an integral to evaluate. It is an integral equation in which $R(p, k)$ is integrated over, yet since $R(p, k)$ is unknown, the integral cannot be evaluated until after the equation is solved!

The symbol $\mathcal{P}$ in (17.3) indicates that the Cauchy principal-value prescription is used to avoid the singularity arising from the zero of the denominator

[1]To keep the presentation simple, our equations are given in the partial-wave basis but without the l subscripts to indicate it.

(we discuss how to do that in § 17.3). As indicated in Fig. 17.1, this equation describes the scattering of two particles with reduced mass

$$\mu = \frac{m_1 m_2}{m_1 + m_2}, \tag{17.4}$$

center-of-mass energy

$$E = \frac{k_0^2}{2\mu}, \tag{17.5}$$

and initial and final center-of-mass momenta k and k'. The diagonal matrix element $R(k_0, k_0)$ is the experimental scattering amplitude needed to solve your **problem**.

The potential term $V(k', k)$ in (17.3) is the double Bessel transform of the coordinate-space potential $V(r', r)$:

$$\begin{aligned} V(k', k) &= (2\pi)^3 \int_0^\infty dr \int_0^\infty dr' r r' j_l(k'r') V(r', r) j_l(kr) & (17.6) \\ &= (2\pi)^3 \int_0^\infty r^2 j_l(k'r) V(r) j_l(kr). & (17.7) \end{aligned}$$

Here the second form is appropriate for a *local* potential [one for which $V(r', r) = \delta(r' - r)V(r)$]. The $j_l(kr)$'s are spherical Bessel functions; for example

$$j_0(z) = \frac{\sin z}{z}, \quad j_1(z) = \frac{\sin z}{z^2} - \frac{\cos z}{z}, \quad \ldots. \tag{17.8}$$

For a given problem, the momentum-space potential $V(k', k)$ is considered known (it can always be computed from the coordinate-space potential).

17.3 THEORY, MATHEMATICS: SINGULAR INTEGRALS

A *singular* integral

$$\mathcal{G} = \int_a^b g(k) dk, \tag{17.9}$$

is one in which the integrand $g(k)$ is singular at a point k_0 within the interval $[a, b]$, and yet the integral $\mathcal{G}$ is finite. (If the integral itself was infinite, we could not compute it.) Unfortunately, computers are notoriously bad at dealing with infinite numbers, and if an integration point gets too near to the singularity, severe subtractive cancellation or overflow occurs. Consequently, we apply some results from complex analysis before evaluating singular integrals numerically.[2]

[2] [S&T 93] describe a different approach using *Maple* and *Mathematica*.

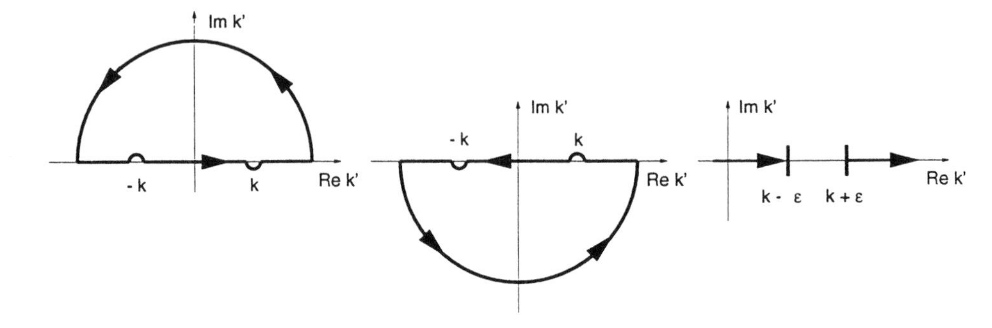

Fig. 17.2 Three different paths in the complex k' plane used to evaluate line integrals when there are singularities. Here the singularities are at k and $-k$, and the integration variable is k'.

In Fig. 17.2 we show three ways in which the singularity of an integrand can be avoided. The paths in A and B move the singularity slightly off the real k axis by giving the singularity a small imaginary part $\pm i\epsilon$. The Cauchy principal-value prescription $\mathcal{P}$ is indicated in the right-most portion of Fig. 17.2 and is seen to "pinch" both sides of the singularity at k_0, but not to pass through it:

$$\mathcal{P}\int_{-\infty}^{+\infty} f(k)dk = \lim_{\epsilon\to 0}\left[\int_{-\infty}^{k_0-\epsilon} f(k)dk + \int_{k_0+\epsilon}^{+\infty} f(k)dk\right]. \tag{17.10}$$

The preceding three prescription are related by

$$\int_{-\infty}^{+\infty} \frac{f(k)dk}{k-k_0\pm i\epsilon} = \mathcal{P}\int_{-\infty}^{+\infty} \frac{f(k)dk'}{k-k_0} \mp i\pi f(k_0), \tag{17.11}$$

which follows from Cauchy's residue theorem and some contour distortions.

17.3.1 Numerical Principal Values

A numerical principal value limit (17.10) is awkward because computers have limited precision. A better prescription for computers follows from the calculus relation

$$\mathcal{P}\int_{-\infty}^{+\infty} \frac{dk}{k-k_0} = 0. \tag{17.12}$$

This equation means that the curve of $1/(k-k_0)$ as a function of k has equal and opposite areas on both sides of the singular point k_0. If we break the integral up into one over positive k and one over negative k, a change of variable $k \to -k$ permits us to express (17.12) as

$$\mathcal{P}\int_0^{+\infty} \frac{dk}{k^2 - k_0^2} = 0. \tag{17.13}$$

We observe that the principal-value exclusion of the singular point's contribution is equivalent to a simple subtraction of the zero integral (17.13):

$$\mathcal{P}\int_0^{+\infty} \frac{f(k)dk}{k^2 - k_0^2} = \int_0^{+\infty} \frac{[f(k) - f(k_0)]dk}{k^2 - k_0^2}. \tag{17.14}$$

We notice that there is no $\mathcal{P}$ on the RHS of (17.14) because the integrand is no longer singular at $k = k_0$ (it's proportional to the df/dk) and can therefore be evaluated numerically as can any other integral! The integral (17.14) is called the *Hilbert transform* of f and also arises in inverse problems.

17.4 METHOD: CONVERTING INTEGRAL TO MATRIX EQUATIONS

Now that we know how to handle singular integrals, we return to our problem of a singular integral equation. One technique for solving integral equations is to reduce them to linear equations that are then solved with matrix operations. We need to solve the integral equation (17.3) with the potential (17.7). The momentum[3] k_0 is related to the energy E and the reduced mass μ by (17.5). The experimental observable that results from a solution of (17.3) is the amplitude $R(k_0, k_0)$, or equivalently, the scattering phase shift δ_l:

$$R(k_0, k_0) = -\frac{\tan \delta_l}{\rho}, \qquad \rho = 2\mu k_0. \tag{17.15}$$

The procedure for the computer solution of (17.3) [H&T 70] uses (17.14) to rewrite the principal-value prescription as a definite integral:

$$R(k', k) = V(k', k) + \frac{2}{\pi}\int_0^{\infty} dp \frac{p^2 V(k', p)R(p, k) - k_0^2 V(k', k_0)R(k_0, k)}{(k_0^2 - p^2)/2\mu}. \tag{17.16}$$

We convert the integral equation (17.16) into linear equations by approximating the integral as a sum over N integration points (usually Gauss quadrature) $\{k_j;\ j = 1, N\}$ with weights w_j:

[3] We are formulating this problem with "natural" units in which Planck's constant $\hbar \equiv 1$. This means that there is no difference between momentum and wave vectors.

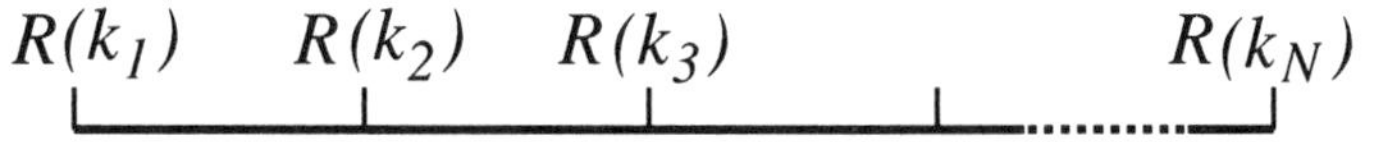

R(k1) R(k2) R(k3) R(kN)

Fig. 17.3 The grid in momentum space on which the integral equation for the R is solved.

$$R(k,k_0) \simeq V(k,k_0) + \frac{2}{\pi}\sum_{j=1}^{N} \frac{k_j^2 V(k,k_j) R(k_j,k_0) w_j}{(k_0^2 - k_j^2)/2\mu} - \frac{2}{\pi} k_0^2 V(k,k_0) R(k_0,k_0) \sum_{m=1}^{N} \frac{w_m}{(k_0^2 - k_m^2)/2\mu}. \tag{17.17}$$

We note that the last term in (17.17) implements the principal-value prescription and cancels the singular behavior of the first term.

Equation (17.17) contains the $N+1$ unknowns $R(k_j, k_0)$ for $j = 1, N$, and $R(k_0, k_0)$. As indicated in Fig. 17.3, we turn it into $N+1$ simultaneous equations by evaluating it for $N+1$ k values on a grid consisting of the observable and integration points:

$$k = k_i = \begin{cases} k_j, & j = 1, N \quad \text{(quadrature points)}, \\ k_0, & i = 0 \quad \text{(observable point)}. \end{cases} \tag{17.18}$$

There are now $N+1$ unknowns $R(k_i, k_0) \equiv R_i$, and $N+1$ linear equations for them:

$$R_i = V_i + \frac{2}{\pi}\sum_{j=1}^{N} \frac{k_j^2 V_{ij} R_j w_j}{(k_0^2 - k_j^2)/2\mu} - \frac{2}{\pi} k_0^2 V_{ii} R_0 \sum_{m=1}^{N} \frac{w_m}{(k_0^2 - k_m^2)/2\mu} \quad (i = 1, N+1). \tag{17.19}$$

We express these equations as matrix equations by combining the denominators and weights into a single denominator vector D:

$$D_i = \begin{cases} +\frac{2}{\pi}\frac{w_i k_i^2}{(k_0^2 - k_i^2)/2\mu}, & \text{for } i = 1, N, \\ -\frac{2}{\pi}\sum_{j=1}^{N} \frac{w_j k_0^2}{(k_0^2 - k_j^2)/2\mu}, & \text{for } i = N+1. \end{cases} \tag{17.20}$$

The linear equations (17.19) now assume that the matrix form

$$R - DVR = [1 - DV]\, R = V, \tag{17.21}$$

where R and V are *column vectors* of length $N_1 \equiv N + 1$:

$$[R] \stackrel{\text{def}}{==} [R_{i,N_1}] = \begin{pmatrix} R_{1,N_1} \\ R_{2,N_1} \\ \vdots \\ R_{N_1,N_1} \end{pmatrix}, \tag{17.22}$$

$$[V] \stackrel{\text{def}}{==} [V_{i,N_1}] = \begin{pmatrix} V_{1,N_1} \\ V_{2,N_1} \\ \vdots \\ V_{N_1,N_1} \end{pmatrix}. \tag{17.23}$$

The matrix in brackets in (17.21) if called the wave matrix Ω or F,

$$F_{ij} = \delta_{ij} - D_j V_{ij}, \tag{17.24}$$

and the integral equation is then the matrix equation:

$$[F][R] = [V]. \tag{17.25}$$

With R the unknown vector, (17.25) is in the standard form $AX = B$, which can be solved by the mathematical subroutine libraries discussed in Chapter 15, *Matrix Computing and Subroutine Libraries.*

17.4.1 Solution via Inversion or Elimination

An elegant (but alas not efficient) solution to (17.25) is by matrix inversion:

$$[R] = [F]^{-1}[V]. \tag{17.26}$$

Because the inversion of even complex matrices is a standard routine in mathematical libraries, (17.26) is a *direct solution* for the R matrix. A more efficient approach is to find an $[R]$ that solves $[F][R] = [V]$ without computing the inverse. This is accomplished by Gaussian *elimination.*

17.4.2 Solving $i\epsilon$ Integral Equations ⊙

The integral equation most commonly encountered in quantum mechanics corresponds to outgoing wave boundary conditions. This means that the singularity is handled by giving the energy $k_0^2/2\mu$ a small positive imaginary part $i\epsilon$. This procedure leads to the Lippmann–Schwinger equation for the T matrix:

$$T(k', k) = V(k', k) + \frac{2}{\pi} \int_0^\infty dp \frac{p^2 V(k', p) T(p, k)}{(k_0^2 - p^2 + i\epsilon)/2\mu}. \tag{17.27}$$

Solving this equation is essentially the same as solving (17.3) for the $R(k', k)$ matrix. We use the identity (17.11) and decompose the $i\epsilon$ integral into a principal-value part and an on-shell term:

$$T(k', k) = V(k', k) + \frac{2}{\pi}\mathcal{P}\int_0^\infty dp\frac{p^2 V(k', p)T(p, k)}{(k_0^2 - p^2)/2\mu} - 2i\mu k_0 V(k', k_0)T(k_0, k). \tag{17.28}$$

Now the last term is incorporated into the numerical analysis by adding an imaginary term to the D matrix (17.20):

$$D_{N+1} = -\frac{2}{\pi}\sum_{j=1}^{N}\frac{w_i k_0^2}{(k_0^2 - k_j^2)/2\mu} - 2\mu i k_0. \tag{17.29}$$

The solution proceeds as before, only now with complex matrices arising from the new definition of D. The resulting on-shell T matrix element is related to the same experimental phase shift as before, only now through the complex expression

$$T(k_0, k_o) = -\frac{e^{i\delta_l}\sin\delta_l}{\rho}, \qquad \rho = 2\mu k_0. \tag{17.30}$$

17.5 IMPLEMENTATION: DELTA-SHELL POTENTIAL, SCATT.F

Although the integral equation techniques we have been discussing come into their own when dealing with nonlocal potentials, they also work for local potentials. As a specific implementation of these techniques, we consider a potential for which the analytic solution is known, the *delta-shell potential*:

$$V(r) = \frac{\lambda}{2\mu}\delta(r - b). \tag{17.31}$$

Physically, the delta-shell potential is a good model for an interaction that occurs almost entirely when two particles are a fixed distance b apart. We use (17.7) to determine the momentum-space representation of this V:

$$V(k', k) = \int_0^\infty r^2 j_l(k'r')\frac{\lambda}{2\mu}\delta(r - b)j_l(kr)dr \tag{17.32}$$

$$= \frac{\lambda b^2}{2\mu}j_l(k'b)j_l(kb), \tag{17.33}$$

where the subscript l indicates the angular momentum state for which we are solving the problem.

The analytic solution of the Lippmann–Schwinger integral equation (17.3)

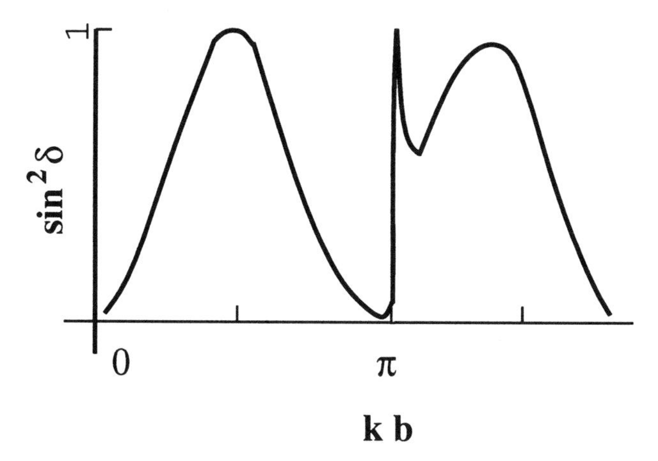

Fig. 17.4 The energy dependence of that part of the scattering cross section arising from the $l = 0$ phase shift.

for the delta-shell potential is [Gott 66]

$$R(k_0, k_0) = -\frac{\tan \delta_l}{\rho} = -\frac{\lambda (kb)^2 j_l(kb) \rho}{k - \lambda (kb)^2 j_l(kb) n_l(kb)}, \tag{17.34}$$

where n_l is the spherical Neumann function; for example

$$n_0(z) = -\frac{\cos z}{z}, \qquad n_1(z) = -\left[\frac{\cos z}{z^2} + \frac{\sin z}{z}\right]. \tag{17.35}$$

With these equations we can easily calculate the $l = 0$ and $l = 1$ phase shifts and compare to those obtained from your numerical solution of the integral Schrödinger equation. In Fig. 17.4 we give a plot of $\sin^2 \delta_0$ versus kb. This is proportional to the scattering cross section arising from the $l = 0$ phase shift. It is seen to reach its maximum value at energies corresponding to resonances. Your numerical results should be similar to this, although it may be difficult to reproduce the very sharp energy dependence.

1. Set up the matrices $V(i, j)$, $V(i)$, $D(j)$, and $F(i, j)$ according to (17.24), (17.22), (17.23), and (17.33). Use Gaussian quadrature points with at least $N = 16$ for your grid.

2. Employ a matrix inversion routine you have obtained from a library to calculate F^{-1}.

3. Calculate the vector R by matrix multiplication:

$$R = F^{-1} V. \tag{17.36}$$

4. Deduce the S-wave phase shift δ by using the last number in your R vector:

$$R(k_0, k_0) = R_{N1,N1} = -\frac{\tan\delta_l}{\rho}, \quad \rho = 2\mu k_0. \tag{17.37}$$

5. Estimate the precision of your solution by increasing the number of grid point in steps of 4. If your phase shift changes in the second or third decimal place, you probably have that much precision.

6. Plot $\sin^2\delta_0$ versus energy $E = k_0^2/2\mu$ starting at zero energy and ending at energies where the phase shift is again small. Your results should be similar to those in Fig. 17.4 (calculated from the analytic result). Note that a *resonance* occurs when δ_l increases rapidly through $\pi/2$; that is, when $\sin^2\delta_0 = 1$.

7. Check your answer against the analytic results (17.34).

17.6 EXPLORATION: SCATTERING WAVE FUNCTION

1. The F^{-1} matrix that occurred in our solution to the integral equation,

$$R = F^{-1}V = (1 - VG)^{-1}V \tag{17.38}$$

is actually quite useful. In scattering theory it is known as the *wave matrix* because it is used in the expansion of the wave function:

$$u_l(r) = N_0 \sum_{i=1}^{N} j_l(k_i r) F(k_i, k_0)^{-1}. \tag{17.39}$$

Here N_0 is a normalization constant and standing-wave boundary conditions are built into u_l if the R matrix is used to calculate F. Plot this wave function and compare it to a free wave.

2. Solve for the part of the scattering cross section arising from the $l = 1$ phase shift for $0 \le kb \le 2\pi$.

18

Computing Hardware Basics: Memory and CPU

18.1 PROBLEM: SPEEDING UP YOUR PROGRAM

Your **problem** is to make a numerically intensive program run faster, but not by porting it to a faster computer. We assume that you are already running your programs on a scientific computer. By running the short implementations given in this chapter, you may discover how to speed them up. In the process you will experiment with your computer's memory and experience some of the concerns, techniques, and rewards of high-performance computing (HPC).[1]

In HPC, you generally modify or "tune" your program to take advantage of a computer's architecture. Often the real problem is to determine which parts of your program get used the most and to decide whether they would run significantly faster if you modified them to take advantage of a computer's architecture. In this chapter we mainly discuss the theory of an high-performance computer's memory and central processor design. In Chapter 19, *High-Performance Computing: Profiling and Tuning*, we concentrate on how you determine the most numerically intensive parts of your program, and how specific hardware features affect them.

Be warned, there is a negative side to high-performance computing. Not only does it take mental concentration and time to tune a program, but as you optimize a program for a specific piece of hardware and its special software features, you make your program less portable and probably less readable. One school of thought says it's the compiler's, and not the scientist's, job

[1]This chapter is derived from [L&F 93, Chapters 12 and 13].

to worry about computer architecture, and it is old-fashioned for you to tune your programs. Yet many computational scientists who run large and complex programs on a variety of machines frequently obtain a 300–500% speedup when they tune their programs for the CPU and memory architecture of a particular machine. You, of course, must decide whether it is worth the effort for the problem at hand; for a program run only once, it is probably not, for an essential tool used regularly, it probably is.

18.2 THEORY: COMPONENTS OF A HIGH-PERFORMANCE COMPUTER

By definition, supercomputers are the fastest and most powerful computers available, and are the superstars of the high-performance class of computers. At this instant, "supercomputers" almost always refer to parallel machines. Workstations, which are small enough in size and cost to be used by a small group or an individual, yet powerful enough for large-scale scientific and engineering applications, can also be high-performance computers. We define high-performance computers as machines with good balance among the following major elements:

- Multistaged (pipelined) functional units
- Multiple central processing units (for parallel machines)
- Fast, central registers
- Very large and fast memories
- Very fast communication speed among functional units
- Vector or array processors
- Software that uses all the above effectively

True HPCs get major computing jobs done quickly by having all their parts working together and by doing several things at one time.

18.2.1 Memory Hierarchy

An idealized model of computer architecture is a CPU sequentially executing a stream of instructions and reading from a continuous block of memory. For example, in Fig.18.1 we see a vector and an array loaded in memory and about to be processed. The real world is more complicated than this. First, matrices are not stored in blocks, but rather in linear order. For example, in

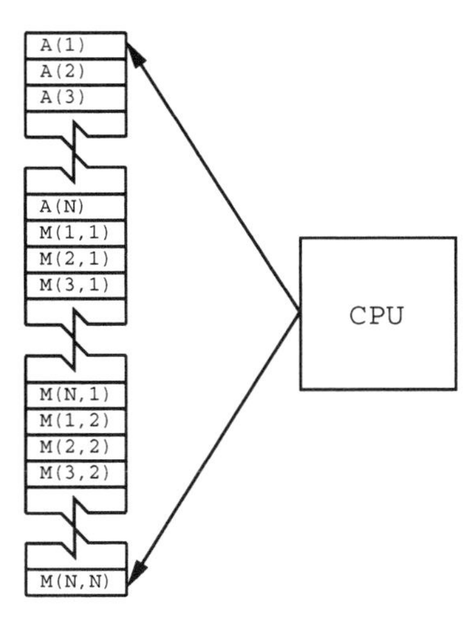

Fig. 18.1 The logical arrangement of CPU and memory showing a Fortran array, $A(N)$, and matrix $M(N,N)$ loaded into memory.

Fortran it would be

$$\mathtt{m}(1,1)\mathtt{m}(2,1)\mathtt{m}(3,1)\mathtt{m}(1,2)\mathtt{m}(2,2)\mathtt{m}(3,2)\mathtt{m}(1,3)\mathtt{m}(2,3)\mathtt{m}(3,3), \tag{18.1}$$

and in C it would be

$$\mathtt{m}(0,0)\mathtt{m}(0,1)\mathtt{m}(0,2)\mathtt{m}(1,0)\mathtt{m}(1,1)\mathtt{m}(1,2)\mathtt{m}(2,0)\mathtt{m}(2,1)\mathtt{m}(2,2). \tag{18.2}$$

Second, as indicated in Figs.18.2 and 18.3, the values for the matrix elements may not even be in the same physical memory. Some may be in RAM, some on the disk, some in cache, and some in the CPU. To understand the meaning of this last statement, we show in Fig.18.3 a simple model of the complex memory architecture of a high-performance computer. This hierarchical arrangement arises from an effort to balance speed and cost, with fast, expensive memory, supplemented by slow, less expensive memory. It may include the

CPU: Central processing unit, the fastest part of the computer. The CPU consists of a number of very high-speed memory units called *registers*, containing the *instructions* sent to the hardware to do things like fetch, store, and operate on data. There are usually separate registers for instructions, addresses, and *operands* (current data).

FPU: Floating point (or Arithmetic) unit, a piece of fast hardware designed for floating-point arithmetic. It may be a separate unit called a *math coprocessor* and is usually on the same board or chip as the CPU.

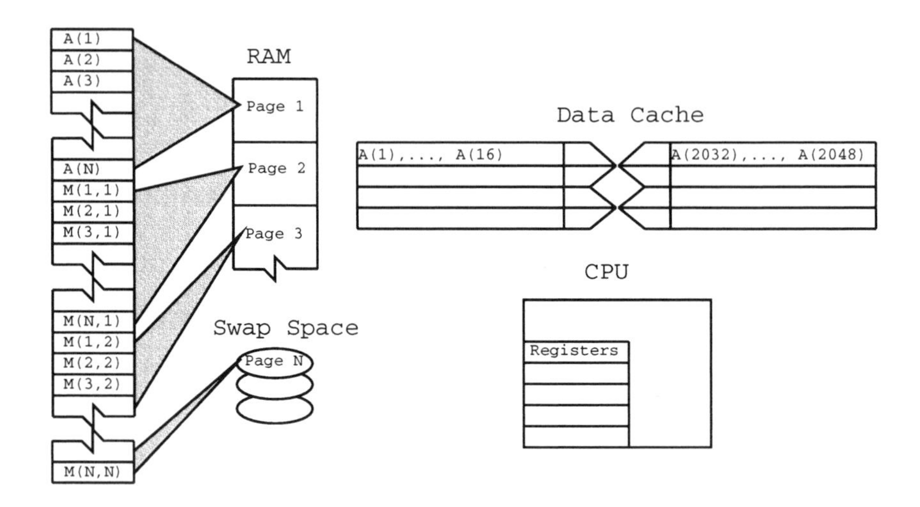

Fig. 18.2 The elements of a computer's memory architecture.

Cache: A small, very fast bit of memory also called the *high-speed buffer* holds instructions, addresses, and data in their passage between the very fast CPU registers and the slower main RAM memory. This is seen in the next level down the pyramid in Fig.18.3. The main memory is also called *dynamic RAM* (DRAM), while the cache is *static RAM* (SRAM). If the cache is used properly, it eliminates the need for the CPU to wait for data to be fetched from memory.

Cache and data lines: The data transferred to and from the cache or CPU are grouped into *cache lines* or *data lines.* The time it takes to bring data from memory into cache is called *latency.*

RAM: Random access memory or central memory is in the middle memory hierarchy in Fig.18.3. RAM can be accessed directly; that is, in random order, and it can be accessed quickly; that is, without mechanical devices. It is where your program resides while it is being processed.

Pages: Central memory is organized into *pages*, which are blocks of memory of fixed length. The operating system labels and organizes its memory pages much like we do the pages of books; they are numbered and kept track of with a *table of contents.*

Hard disk: Finally, at the bottom of the memory pyramid is the permanent storage on magnetic disks or optical devices. Although disks are very slow compared to RAM, they can store vast amounts of data and sometimes compensate for their slower speeds by using a cache of their own, the *paging storage controller.*

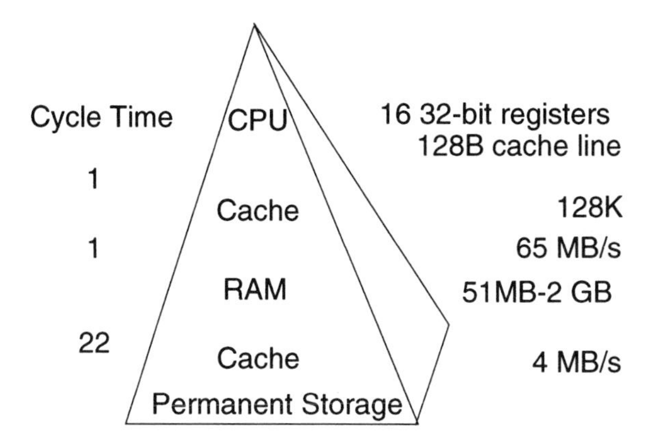

Fig. 18.3 Typical memory hierarchy for a single-processor high-performance computer (where B = bytes, K = kilobytes, MB = megabytes, and GB = gigabytes).

Virtual memory: True to its name, this is a part of memory you will not find in our figures because it is *virtual*. Virtual memory permits your program to use more pages of memory than will physically fit into RAM at one time. Pages not currently in use are stored in slower memory and brought into fast memory only when needed. The separate memory location for this switching is known as *swap space*, and it is shown in Fig.18.2. Thanks to virtual memory, it is possible to run programs on small computers that otherwise would require larger machines (or extensive reprogramming). The price you pay for virtual memory is an order-of-magnitude slowdown of your program's speed when virtual memory is actually invoked. But this may be cheap compared to the time you would have to spend to rewrite your program so it fits into RAM.

When we speak of "fast" and "slow" memory we are using a time scale set by the clock in the CPU. For example, a modern workstation may have a clock speed or cycle time of 5 ns. This means that it could execute 200 million instructions per second (200 Mips), if it could get its hands on the needed data quickly enough. While it usually takes 1 cycle to transfer data from the cache into the CPU, the other memory is much slower, and so you can speed your program up by minimizing transfers among different levels of memory. Compilers try to do this for you, but their success is affected by your programming style.

18.2.2 The Central Processing Unit

How does the CPU get to be so fast? Often, it is *pipelined*; that is, it has the ability to begin the steps necessary to execute the next instruction before the current one has finished. It is like an assembly line or a bucket brigade in which the person filling the buckets at one end of the line does not wait for each bucket to arrive at the other end before filling another bucket. In this same way a processor fetches, reads, and decodes an instruction while another instruction is executing. Consequently, even though it may take more than one cycle to perform some operations, it is possible for data to be entering and leaving the CPU on each cycle. For example, consider how the operation $c = (a + b)/(d * f)$ is handled:

Computation of $c = (a + b)/(d * f)$

Arithmetic unit	Step 1	Step 2	Step 3	Step 4
A1:	Fetch a	Fetch b	Add	—
A2:	Fetch d	Fetch f	Multiply	—
A3:	—	—	—	Divide

Here the pipelined arithmetic units A1 and A2 are simultaneously doing their jobs of fetching and operating on operands, yet arithmetic unit A3 must wait for the first two units to complete their tasks before it gets something to do (during which time the other two sit idle).

18.2.3 CPU Design: RISC

RISC is an acronym for reduced instruction set computer. It is an approach to CPU architecture that increases arithmetic speed by decreasing the number of instructions within the CPU. In contrast, a *CISC* or complex instruction set computer, has some 1000 or more basic machine language instructions on the CPU chip as well as complex codes written in them. Having the instructions on the chip is useful, but because a typical instruction takes more than 10 cycles to execute, this slows down the CPU. Because the instructions are not on the chip, RISC relies on a high-level compiler to translate a high-level language such as Fortran or C into efficient machine instructions. The RISC philosophy is to have the compiler utilize fully the few instructions available on the chip. The theory is that a simpler scheme is cheaper to design and produce, that the processor runs at increased speeds, and that the space saved in the silicon by eliminating the large basic code is used to attain greater arithmetic power by including more registers and some instruction-level parallelism.

18.2.4 CPU Design: Vector Processing

Often the most demanding part of a scientific computation involves matrix operations. On a classic (von Neumann) scalar computer, the addition of two vectors of physical length 99 to form a third ultimately requires 99 sequential additions:

Computation of $[A] + [B] = [C]$

Step 1	Step 2	...	Step 99
$a(1) + b(1) = c(1)$	$a(2) + b(2) = c(2)$	...	$a(99) + b(99) = c(99)$

There is actually much behind-the-scene work here. For each element i there is the *fetch* of $a(i)$ from its location in memory, the *fetch* of $b(i)$ from its location in memory, the addition of the numeric values of these two elements in a CPU register, and the *storage* in memory of the sum into $c(i)$. This fetching uses up time and is wasteful in the sense that the computer is being told again and again to do the same thing.

When we speak of a computer doing "vector" processing, we mean that there are hardware components that perform mathematical operations on entire rows or columns of "matrices" as opposed to individual elements. (This hardware also can handle single-subscripted matrices, that is, mathematical vectors.) In *vector* processing of $[A] + [B] = [C]$, the successive fetching and additions of the elements of A and B get grouped together and overlaid, and $Z \simeq 64 - 256$ elements (the *section size*) are processed with one command:

Vector processing of $[A] + [B] = [C]$

Step 1	Step 2	Step 3	...	Step Z
$a(1) + b(1) = c(1)$	al			
	$a(2) + b(2) = c(2)$			
		$a(3) + b(3) = c(3)$		
			...	
				$a(Z) + b(Z) = c(Z)$

Typically, this vector processing speeds up the processing of vectors by a factor of about 10. If all Z elements were processed in the same step, then the speedup would be $\sim$ 64–256.

18.2.5 Virtual Memory

Virtual memory extends the effective size of a computer's RAM by using the relatively plentiful (and cheap) hard-disk space. In this way an application program effectively sees a continuous span of memory addresses that is greater than the physical size of central memory. A combination of operating system and hardware *maps* this virtual memory into pages with typical lengths of 4 KB. As indicated in Fig. 18.2, when the application accesses the memory location for `m(i,j)`, the number of the page of memory holding this address is found, and the location of `m(i,j)` within this page is noted. A *page fault* occurs if the needed page resides on disk rather than in RAM. In this case the entire page must be read into memory while the least-recently used page in RAM is swapped onto the disk.

Virtual memory not only permits the computer to run *one* program that otherwise would be too big for RAM but also allows *multitasking*, the simultaneous loading into memory of more programs than physically fit into RAM. Although the ensuing switching among applications uses computing cycles, by avoiding long waits while an application is loaded into memory, multitasking increases the total throughout and permits an improved computing environment for users.

For example, it is multitasking that permits windows system to provide us with multiple windows. Even though each window application uses a fair amount of memory, only the single application currently receiving input must actually reside in memory; the rest are *paged out* to disk. This explains why you may notice a slight delay when switching to an idle window; the pages for the now-active program are being placed into RAM and simultaneously the least-used application still in memory is paged out.

18.3 METHOD: PROGRAMMING FOR VIRTUAL MEMORY

While paging makes little appear big, you pay a price because your program's run time increases with each page fault. If your program does not fit into RAM all at once, it will run significantly slower. If virtual memory is shared among multiple programs that run simultaneously, they all can't have the entire RAM at once, and so there will be memory access *conflicts*, in which case the performance of all programs suffer.

The basic rules for programming for virtual memory are

1. Worry about reducing the amount of memory used (the *working set size*) only if your program is large. In that case apply *global optimization*.

2. Avoid page faults by organizing your programs to successively perform its calculations on subsets of data, each fitting completely into RAM.

3. Avoid simultaneous calculations in the same program to avoid competition for memory and consequent page faults. Complete each major calculation before starting another.

4. Group data elements close together in memory blocks if they are going to be used together in calculations.

18.4 IMPLEMENTATION: GOOD AND BAD VIRTUAL MEMORY USE

To see the effect of virtual memory use, try out these simple examples on your machine. Use a command such as `time` to measure the time being used for each example. These examples call functions `force12` and `force21`. You should write these functions and make them have significant memory requirements for both local and global variables.

BAD Program, Too Simultaneous

```
Do j = 1, n
  Do i = 1, n
       f12(i,j) = force12(pion(i), pion(j))                      Fill f12.
       f21(i,j) = force21(pion(i), pion(j))                      Fill f21.
       ftot = f12(i,j) + f21(i,j)                               Fill ftot.
   EndDo
EndDo
```

You see that each iteration of the Do loop requires the data and code for all the functions and access to all elements of the matrices and arrays. The working set size of this calculation is the sum of the sizes of the arrays `f12(N,N)`, `f21(N,N)` and `pion(N)` plus the sums of the sizes of the functions `force12` and `force21`.

A better way to do this is to break the calculation into separate components:

GOOD Program, Two Loops

```
Do j = 1, n
  Do i = 1, n
    f12(i,j) = force12(pion(i), pion(j))                   Fill just f12.
  EndDo
EndDo
Do j = 1, n                                                  Second nest.
  Do i = 1, n
    f21(i,j) = force21(pion(i), pion(j))                   Fill just f21.
  EndDo
EndDo
```

```
Do j = 1, n                                             Third nest.
  Do i = 1, n
    ftot = f12(i,j) + f21(i,j)                          Compute ftot.
  EndDo
EndDo
```

Here the separate calculations are independent and the working set size is reduced. Because the working set size of the first Do loop is the sum of the sizes of the arrays `f12(N,N)` and `pion(N)`, and of the function `force12`, we have approximately half the previous size. The size of the last Do loop is the sum of the sizes for the two arrays. The working set size of the entire program is the larger of the working set sizes for the different Do loops.

As an example of the need to group data elements close together in memory or Common blocks if they are going to be used together in calculations, consider the following code:

BAD Program, Discontinuous Memory

```
Common zed, ylt(9), part(9), zpart1(49123), zpart2(49123), med2(9)
Do j = 1, n
  ylt(j) = zed*part(j)/med2(9)                 Discontinuous variables.
```

Here the variables `zed`, `ylt`, and `part` are used in the same calculations and are adjacent in memory because the programmer grouped them together in Common. Later, when the programmer realized that the array `med2` was needed, it got tacked onto the end of Common. All the data comprising the variables `zed`, `ylt`, and `part` fit into one page, but the `med2` variable is on a different page because the large array `zpart2(49123)` separates it from the other variables. In fact, the system may be forced to make the entire 4-KB page available in order to fetch the 72 bytes of data in `med2`. While it is difficult for the Fortran or C programmer to assure the placement of variables within page boundaries, you will improve your chances by grouping data elements together:

GOOD Program, Continuous Memory

```
Common zed, ylt(9), part(9), med2(9), zpart1(49123), zpart2(49123)
Do j = 1, n
  ylt(j) = zed*part(j)/med2(9)                    Continuous variables.
```

18.5 METHOD: PROGRAMMING FOR DATA CACHE

Data caches are small, very fast memory used as temporary storage between the ultrafast CPU registers and the fast main memory. They have grown in importance as high-performance computers have become more prevalent. On systems that use a data cache, this may well be the single most important

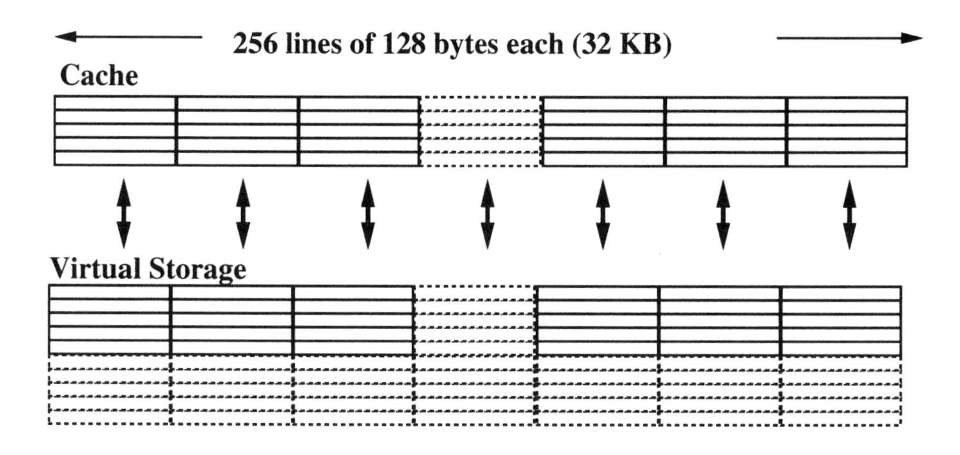

Fig. 18.4 The cache manager's view of RAM. Each 128-byte cache line is read into one of four lines in cache.

programming consideration; continually referencing data that are not in the cache (*cache misses*) may lead to an order-of-magnitude increase in CPU time.

As indicated in Fig.s 18.2 and 18.4, the data cache holds a copy of some of the data in memory. The basics are the same for all caches but the sizes are manufacturer-dependent. When the CPU tries to address a memory location, the *cache manager* checks to see if the data are in cache. If they are not, the manager reads the data from memory into cache and then the CPU deals with the data directly in cache. The cache manager's view of RAM is shown in Fig. 18.4.

When considering how some matrix operation uses memory, it is important to consider the *stride* of that operation; that is, the number of array elements that get stepped through as an operation repeats. For example, summing the diagonal elements of a matrix to form the trace

$$\mathrm{Tr}A = \sum_{i=1}^{N} a(i,i) \tag{18.3}$$

involves a large stride because the diagonal elements are stored far apart for large `N`. However, the sum

$$c(i) = x(i) + x(i+1) \tag{18.4}$$

has stride 1 because adjacent elements of `x` are involved. The basic rule in programming for cache is

- Keep the stride low, preferably at 1.

In practice, this means

- Vary the leftmost index first on Fortran matrices.
- Vary the rightmost index first on C matrices.

18.6 IMPLEMENTATION 1: CACHE MISSES

For high-performance computing, you should write programs that keep as much of the data being processed as possible in cache. To do this you should recall that Fortran matrices are stored in successive memory locations with the row index varying most rapidly (column-major order), and C matrices are stored in successive memory locations with the column index varying most rapidly (row-major order). While it is difficult to isolate the effects of cache from other elements of the computer's architecture, you should now estimate its importance by comparing the time it takes to step through matrix elements row by row, to the time it takes to step through matrix elements column by column.

By actually running on machines available to you, check that these two simple codes with the same number of arithmetic operations will take significantly different times to run because one of them must make large jumps through memory with the memory locations addressed not yet read into cache:

Sequential Column Reference

```
Do j = 1, 9999
    x(j) = m(1,j)                              Sequential column reference.
```

Sequential Row Reference

```
Do j = 1, 9999
    x(j) = m(j,1)                              Sequential column reference.
```

18.7 IMPLEMENTATION 2: CACHE MISSES

Test the importance of cache flow on your machine by comparing the time it takes to run these two simple programs. Run for increasing column size `idim` and compare the times for loop A versus those for loop B. A machine with a very small cache may be most sensitive to stride.

GOOD f77/BAD C Program, Minimum/Maximum Stride

```
Dimension Vec(idim,jdim)                                          Loop A.
```

```
Do j = 1, jdim
  Do i=1, idim
    Ans = Ans + Vec(i,j)*Vec(i,j)                    Stride 1 fetch (f77).
EndDo
```

BAD f77/GOOD C Program, Maximum/Minimum Stride

```
Dimension Vec(idim, jdim)                                       Loop B.
Do i = 1, idim
  Do j=1, jdim
    Ans = Ans + Vec(i,j)*Vec(i,j)                 Stride jdim fetch (f77).
EndDo
```

Loop A steps through the matrix `Vec` in column order. Loop B goes in row order. By changing the size of the columns (the leftmost index for Fortran), we change the size of the step (*stride*) we take through memory. Both loops take us through all elements of the matrix, but the stride is different. By increasing the stride, we use fewer elements already present in cache, require additional swapping and loading of cache, and thereby slow down the whole process.

18.8 IMPLEMENTATION 3: LARGE MATRIX MULTIPLICATION

The penultimate example of memory usage is large matrix multiplication:

$$[C] = [A] \times [B]. \tag{18.5}$$

This involves all the concerns with the different kinds of memory. The natural way to code (18.5) follows from the definition of matrix multiplication:

$$c_{ij} = \sum_{k=1}^{N} a_{ik} \times b_{kj}. \tag{18.6}$$

The sum is over a column of A times a row of B.

Try out these two codes on your computer. In Fortran, the first code has B with stride 1, but C with stride N. This is cured in the second code by performing the initialization in another loop. In the C language the problems are reversed. On one of our machines, we found a factor of 100 difference even though the number of operations is the same!

BAD f77/GOOD C Program, Maximum/Minimum Stride

```
Do i = 1, N                                                        Row.
  Do j = 1, N                                                   Column.
    c(i,j) = 0.0                                            Initialize.
```

```
      Do k = 1, N
        c(i,j) = c(i,j) + a(i,k)*b(k,j)              Accumulate sum.
EndDo
```

GOOD f77/BAD C Program, Minimum/Maximum Stride

```
Do j = 1, N                                          Initialization.
  Do  i = 1, N
    c(i,j) = 0.0
    EndDo
  Do k = 1, N
    Do 20 i = 1, N
      c(i,j) = c(i,j) + a(i,k)*b(k,j)
EndDo
```

19

High-Performance Computing: Profiling and Tuning

19.1 PROBLEM: EFFECT OF HARDWARE ON PERFORMANCE

Your **problem** is the same as in the last chapter, namely, to make a numerically intensive program run faster. In this chapter you run an experiment in which you run a full program on as many computers as you can get your hands on. In this way you explore how a computer's architecture affects the program's performance. Specifically, first you determine which parts of the program are most demanding of computing time ("profile for hot spots") so that you will know which parts may be "tuned." Then you see the effects of running tuned programs on different architectures.

19.2 METHOD: TABULATING SPEEDUPS

We start with a simple Fortran program called *tune* and show how to profile and tune it for vector and superscalar (RISC) architectures. The program is shorter than any normally worth the effort of tuning, but the steps are the ones you would follow for an industrial-strength program. You should run all versions of *tune* on the machine available to you, ensure that the program returns the same answers on each, and then compare the relative speedups. The results from some of our runs are in Table 19.1. You should:

1. To set the scale properly, normalize the times relative to the CPU time consumed by the baseline program (*tune*) on each machine.

Table 19.1 Effect of tuning on platforms with different architectures

Platform	base (no opt)	tune (-O)	tune1 (sqrt)	tune2 (vector)	tune3 (index)	tune4 (unroll)
Sun SparcStation	1	1.7	1.8	2.1	2.5	2.8
IBM RS/6000	1	8.0	8.1	0.9	8.1	10.5
IBM ES Vector	1	2.3	2.3	18.0	5.1	5.4

2. The baseline program should have no compiler-option optimizations (no *-O*, *-O1*, or *-O2*).

3. Next run the unmodified *tune* again, this time using the optimize options of the compiler. As you see in the *tune* column of Table 19.1, this can have a tremendous effect on running times.

19.3 IMPLEMENTATION 1: BASELINE PROGRAM, TUNE.F

The program *tune* solves for the eigenvalues and eigenvectors of the matrix $[H]$ by using an iterative perturbation expansion, effectively, a modification of the power or Davidson method:

$$H\mathbf{c} = E\mathbf{c}, \tag{19.1}$$

$$c(k) \simeq c(k) + \frac{H-E}{E-H(k,k)}c(k), \tag{19.2}$$

$$E \simeq \frac{\langle \mathbf{c}|H|\mathbf{c}\rangle}{\langle \mathbf{c}|\mathbf{c}\rangle}. \tag{19.3}$$

Here is a listing of the program:

```
c        tune: a Fortran example program.
c
      Program  tune
      Parameter (ldim = 2050)
      Implicit Double Precision (a-h,o-z)
      Dimension ham(ldim,ldim), coef(ldim), sigma(ldim)
c                                     set up Hamiltonian and starting vector
      Do 10 i = 1,ldim
         Do 11 j = 1,ldim
            If (Abs(j-i) .gt. 10) Then
               ham(j,i) = 0.0
            Else
               ham(j,i) = 0.3**Abs(j-i)
            EndIf
 11      Continue
         ham(i,i) = i
```

```
          coef(i) = 0.0
  10  Continue
      coef(1) = 1.0
c                                      start iterating towards the solution
      err = 1.0
         iter = 0
  20     If (iter .lt.15 .and. err. gt. 1.0e-6) Then
            iter = iter + 1
c                                 compute current energy \& norm, \& normalize
      ener = 0.0
      ovlp = 0.0
      Do 21   i = 1,ldim
         ovlp = ovlp+coef(i)*coef(i)
         sigma(i) = 0.0
         Do 30   j = 1,ldim
            sigma(i) = sigma(i) + coef(j)*ham(j,i)
  30     Continue
         ener = ener + coef(i)*sigma(i)
  21  Continue
      ener = ener/ovlp
      Do 22   I = 1,ldim
         coef(i) = coef(i)/Sqrt(ovlp)
         sigma(i) = sigma(i)/Sqrt(ovlp)
  22  Continue
c                                                 compute update and error norm
      err = 0.0
      Do 23 i = 1,ldim
         If (i.eq.1) GoTo 23
         step = (sigma(i) - ener*coef(i))/(ener-ham(i,i))
         coef(i) = coef(i) + step
         err = err + step**2
  23  Continue
      err = Sqrt(err)
      Write(*,'(1x,i2,7f10.5)') iter, ener, err, coef(1)
      GoTo 20
      EndIf
      Stop
      End
```

This program computes `ener`, the lowest eigenvalue, and `coef`, the corresponding eigenvector of a predominantly diagonal Hamiltonian matrix `ham`. The eigenvector c has elements `coef(i)` and norm `ovlp`, the latter used in an intermediate normalization of the eigenvector.

19.4 METHOD: PROFILING

To determine the time spent in different parts of the program and the number of times all subprograms are called, compile and then profile `tune`. We used the Unix profiling command `gprof`:

```
> f77 -O -pg -o  tune  tune.f     Compile, profile & optimize options.
```

```
> tune                                        Run and produce gmon.out
> gprof  tune                                              Do profile.
```

In spite of `tune` being rather simple, the resulting output is interesting. For example, part of it tells us about the `sqrt` function:

```
                                    called/total       parents
 index %time self descendants  called+self    name          index
                                    called/total       children
 -------------------------------------------------------------
             0.02          0.00   23111/23111        _MAIN_ [3]
[11]     0.1 0.02          0.00   23111              _sqrt [11]
 -------------------------------------------------------------
```

This says that the `sqrt` function is called some 23,111 times by `tune`, which is somewhat of a concern because `sqrt` is a rather slow function. Yet we also see that only 10% of the code's time is spend evaluating that function, so it's by no means a hot spot for this code. Nevertheless, in the interests of pedagogy we tune for the `sqrt` function as if it were worthy of the effort.

19.5 IMPLEMENTATION 2: BASIC OPTIMIZATION, TUNE1.F

We apply some basic optimizations and produce `tune1`. Explicitly, we notice that `sqrt(ovlp)` is called twice in the main Do loop, which clearly is a waste of time because `ovlp` and the corresponding `sqrt(ovlp)` are constant throughout the loop. Consequently, we speed up the code by calculating `sqrt(ovlp)` only once in the loop. Next, we convert the division by `sqrt(ovlp)` into a multiplication by `1/Sqrt(ovlp)` because most computers multiply faster than they divide.

We also notice that `fact*coef(i)` is unnecessarily repeated in the loop. We substitute a local scalar variable `t` for it and thereby calculate it only once. This should speed up the code somewhat because the compiler now can store the scalar variable in a register for fast access, whereas before it had to go into ROM to retrieve the array element.[1]

```
c       tune1: a Fortran example program, basic optimization.
c
      Program  tune1
      PARAMETER (ldim = 2050)
      Implicit Double Precision (a-h,o-z)
      Dimension ham(ldim,ldim),coef(ldim),sigma(ldim)
c                                     set up Hamiltonian and starting vector
      Do 10 i = 1,ldim
```

[1]Really, we shouldn't have to do this at all because a good optimizing compiler should recognize repeated expressions within a loop, calculate them only once, and hold the value in a register. We are being conservative to help get the point across.

```
         Do 11 j = 1,ldim
             If (Abs(j-i) .gt. 10) Then
                ham(j,i) = 0.0
             Else
                ham(j,i) = 0.3**Abs(j-i)
             EndIf
 11      Continue
         ham(i,i) = i
         coef(i) = 0.0
 10   Continue
      coef(1) = 1.0
c                                   start iterating towards the solution
      err = 1.0
      iter = 0
 20      if(iter.lt.15 .and. err.gt.1.0e-6) Then
          iter = iter+1
c                   compute energy, norm of current approximation,\&  normalize
      ener = 0.0
      ovlp = 0.0
      Do 21   i = 1,ldim
         ovlp = ovlp+coef(i)*coef(i)
         sigma(i) = 0.0
         Do 30    j = 1,ldim
            sigma(i) = sigma(i)+coef(j)*ham(j,i)
 30      Continue
         ener = ener+coef(i)*sigma(i)
 21     Continue
      ener = ener/ovlp
      fact  = 1.0/Sqrt(ovlp)
      coef(1)  =  fact*coef(1)
      err  =  0.0
      Do 22   i = 2,ldim
         t         =   fact*coef(i)
         u         =   fact*sigma(i) - ener*t
         step      =   u/(ener - ham(i,i))
         coef(i)   =   t + step
         err       =   err + step*step
 22   Continue
      err = Sqrt(err)
      Write(*,'(1x,i2,7f10.5)') iter,ener,err,coef(1)
      GoTo 20
      EndIf
      Stop
      End
```

Problems may arise if a repeated expression is *not* written exactly the same each time. You improve the compiler's ability to recognize repeated expressions, and make the code clearer, by watching the order of variables:

Exactly Repeated Expressions; Will Recognize

```
H(i) = sin(phi) + alpha*scale + offset
E(i) = cos(phi) + alpha*scale + offset       Repeat last 3 variables.
```

Not Exactly Repeated Expressions; Won't Recognize

```
H(i) = sin(phi) + offset + scale*alpha
E(i) = cos(phi) + alpha*scale + offset              Repeated, yet mixed.
```

You may be wondering why the compiler is smart enough to recognize repetitions of `fact*coef(i)` but not of `sqrt(ovlp)`. The reason is that `sqrt` is a function and not an expression to evaluate; calling a subprogram may have side effects beyond returning a value, and the compiler knows that it does not know those possible side effects. Consequently, compilers seldom will move a function outside of a loop even if it is an intrinsic function. There are preprocessors that rewrite source code for you and will move intrinsic functions; unfortunately they may also leave you with a hard-to-read code.

Now that we have done some tuning, we run `gprof` on our revised code `tune1`. The part analyzing `sqrt()` is

```
                                   called/total      parents
index %time self descendants called+self   name         index
                                   called/total      children
-------------------------------------------------------------
                0.00        0.00      22/22          _MAIN_ [2]
[153]           0.0 0.00        0.00      22          _Sqrt [153]
-------------------------------------------------------------
```

Notice that there are now only 22 calls to `sqrt` where previously there were 23,111, and that the program spends 0.0% doing it (yet it only spent 10% of its time there in the first place). The overall savings would be greater in a realistic case with large matrices or where the tuned code is the inner loop of a large program. The lesson is twofold: writing clean code saves CPU time, but dramatic savings requires you to spend your efforts where the program spends most of its time.

For the next step in the tuning to produce `tune1`, we remove the `If (i.eq.1) GoTo 23` condition from the loop. Whereas some compilers do permit `If` statements inside vector loops, `If` statements slow down the program ("are not free"). As general practice, we remove all `If` statements, I/O calls, and user-written subroutine calls from loops to be vectorized.

The final change is to replace `err = err + step**2` by `err = err + step*step` because multiplication is faster than exponentiation. Once again, while this optimization is handled automatically by some compilers, we also want to ensure that the expression has the form of `add + multiply` because this form is often handled as one floating point calculation on vector and advanced RISC machines.[2] As we see from the `tune1` column in Table 19.1, these modifications lead to a relatively small speedup (compared to what the compiler did when

[2]The careful reader will notice that the definition coef(i)= coef(i)+step can be sped up if written as a scalar add + multiply form: coef(i)=t+step.

asked to optimize).

19.6 IMPLEMENTATION 2: VECTOR TUNING, TUNE2.F

In `tune2` we optimize the code for vector processing and for the intermediate memory cache found on the IBM ES/9000. (You may not have a vector processor to try out these modifications, but you can explore what effects they have on your nonvector computer.) Specifically, we are assessing the effect of not having all the variables needed in a calculational step fitting into cache. We note that the definition of `step` uses only the diagonal values of the matrix `ham`, that is, `ham(i,i)`. That being so, there is a very large *stride* for accessing the matrix, and to speed up the processing, we created an array `diag` to hold the diagonal elements. The data are then stride 1.

We also tried to speed up the code by *strip mining* or *blocking* the major loop for the vectors. Rather than splitting the loop into separate pieces, we simply interchanged the index on the `ham` matrix so that the vector processor can grab a full stride at one time.

```
c       tune2: a Fortran example program, vector tuned.
c
      Program  tune2
      PARAMETER (ldim = 2050)
      Implicit Double Precision (a-h,o-z)
      Dimension  ham(ldim,ldim),coef(ldim),sigma(ldim),diag(ldim)
c                                set up Hamiltonian and starting vector
      Do 10 i = 1,ldim
         Do 11 j = 1,ldim
             If( Abs(j-i) .gt. 10) Then
                ham(j,i) = 0.0
             Else
                ham(j,i) = 0.3**Abs(j-i)
             EndIf
 11      Continue
      ham(i,i) = i
      coef(i) = 0.0
 10   Continue
      coef(1) = 1.0
c                                      start iterating towards the solution
      Do 15 i = 1,ldim
         diag(i) = ham(i,i)
 15    Continue
      err = 1.0
      iter = 0
 20   If (iter.lt.15 .and. err.gt.1.0e-6) Then
          iter = iter+1
c                   compute energy, norm of current approximation,\&  normalize
      ener = 0.0
      ovlp = 0.0
      Do 21   i = 1,ldim
         ovlp = ovlp+coef(i)*coef(i)
```

```
          t = 0.0
          Do 30   j = 1,ldim
             t = t + coef(j)*ham(i,j)
 30       Continue
          sigma(i) = t
          ener = ener + coef(i)*t
 21    Continue
       ener = ener/ovlp
       fact  = 1.0/Sqrt(ovlp)
       coef(1)  =  fact*coef(1)
       err  =  0.0
       Do 22   i = 2,ldim
           t          =    fact*coef(i)
           u          =    fact*sigma(i) - ener*t
           step       =    u/(ener - diag(i))
           coef(i)    =    t + step
           err        =    err + step*step
 22    Continue
       err = Sqrt(err)
       Write(*,'(1x,i2,7f10.5)') iter,ener,err,coef(1)
       GoTo 20
       EndIf
       Stop
       End
```

As we see from Table 19.1 under tune2, this speeds up the calculation by a factor of 9 on the ES/9000, but slows it down on almost all other machines!

19.7 IMPLEMENTATION 3: VECTOR CODE ON RISC, TUNE3.F

As we see from Table 19.1, the modifications tune2 for vector hardware work very well on the IBM 3090 but slows the code down on RISC workstations. To pinpoint the reason for this, we create tune3, in which we change the matrix indices back. As we see from the table, this removes most of the speedup attained on the 3090, but also removes most of the slowdown for the other platforms.

```
c         tune3:
c
       Program  tune3: Running Vector Code on RISC
       PARAMETER (ldim = 2050)
       Implicit Double Precision (a-h,o-z)
       Dimension ham(ldim,ldim),coef(ldim),sigma(ldim),diag(ldim)
c                                           set up Hamiltonian and starting vector
       Do 10 i = 1,ldim
          Do 10 j = 1,ldim
              If (Abs(j-i) .gt. 10) Then
                 ham(j,i) = 0.0
              Else
                 ham(j,i) = 0.3**Abs(j-i)
              EndIf
```

```
 10  Continue
c                                 start iterating towards the solution
     Do 15 i = 1,ldim
        ham(i,i) = i
        coef(i) = 0.0
        diag(i) = ham(i,i)
 15  Continue
     coef(1) = 1.0
     err = 1.0
     iter = 0
 20  If(iter.lt.15 .and. err.gt.1.0e-6) Then
         iter = iter+1
c                compute energy, norm of current approximation,\&  normalize
     ener = 0.0
     ovlp = 0.0
     Do 21   i = 1,ldim
        ovlp = ovlp+coef(i)*coef(i)
        t = 0.0
        Do 30   j = 1,ldim
           t = t + coef(j)*ham(j,i)
 30     Continue
        sigma(i) = t
        ener = ener + coef(i)*t
 21  Continue
     ener = ener/ovlp
     fact  = 1.0/Sqrt(ovlp)
     coef(1)  =  fact*coef(1)
     err  =  0.0
     Do 22   i = 2,ldim
         t          =   fact*coef(i)
         u          =   fact*sigma(i) - ener*t
         step       =   u/(ener - diag(i))
         coef(i)    =   t + step
         err        =   err + step*step
 22  Continue
     err = Sqrt(err)
     Write(*,'(1x,i2,7f10.5)') iter,ener,err,coef(1)
     GoTo 20
     EndIf
     Stop
     End
```

19.8 IMPLEMENTATION 4: SUPERSCALAR TUNING, TUNE4.F

This time we tune for a RISC or superscalar architecture. We unroll the Do loops in order to ensure that the RISC floating-point arithmetic unit is not idle for too long. We are conservative in our changes and go only two levels down in unrolling the loops. Greater unrolling and blocking would produce better results, but also less clear code.

```
     Program  tune4
     PARAMETER (ldim = 2050)
```

```
      Implicit Double Precision (a-h,o-z)
      Dimension ham(ldim,ldim),coef(ldim),sigma(ldim),diag(ldim)
c       set up Hamiltonian and starting vector
      Do 10 i = 1,ldim
         Do 10 j = 1,ldim
             If(Abs(j-i) .gt. 10) Then
                ham(j,i) = 0.0
             Else
                ham(j,i) = 0.3**Abs(j-i)
             EndIf
 10   Continue
c       start iterating towards the solution
      Do 15 i = 1,ldim
         ham(i,i) = i
         coef(i)  = 0.0
         diag(i)  = ham(i,i)
 15   Continue
      coef(1) = 1.0
      err = 1.0
      iter = 0
 20   If(iter.lt.15 .and. err.gt.1.0e-6) Then
        iter = iter+1
      ener = 0.0
      ovlp1 = 0.0
      ovlp2 = 0.0
      Do 21   i = 1,ldim-1,2
         ovlp1 = ovlp1+coef(i)*coef(i)
         ovlp2 = ovlp2+coef(i+1)*coef(i+1)
         t1    = 0.0
         t2    = 0.0
         Do 30   j = 1,ldim
            t1 = t1 + coef(j)*ham(j,i)
            t2 = t2 + coef(j)*ham(j,i+1)
 30      Continue
      sigma(i)   = t1
      sigma(i+1) = t2
      ener       = ener + coef(i)*t1 + coef(i)*t2
 21   Continue
      ovlp = ovlp1 + ovlp2
      ener = ener/ovlp
      fact  = 1.0/Sqrt(ovlp)
      coef(1)  =  fact*coef(1)
      err  =  0.0
      Do 22   i = 2,ldim
         t          =   fact*coef(i)
         u          =   fact*sigma(i) - ener*t
         step       =   u/(ener - diag(i))
         coef(i)    =   t + step
         err        =   err + step*step
 22 Continue
    err = Sqrt(err)
    Write(*,'(1x,i2,7f10.5)') iter,ener,err,coef(1)
    GoTo 20
    EndIf
    Stop
    End
```

In Table 19.1 under tune4 we see a 10–25% speedup.

19.9 ASSESSMENT

The results presented in Table 19.1 show four versions of `tune` compared to the baseline with no optimizations and no programming assistance. After the baseline run, all versions of `tune` were compiled with full compiler optimization. You may have a similar table resulting from your explorations. Check whether your results agree with our conclusions that

1. Compiler optimization helps significantly in all cases (`base` versus `tune`), with a range of values 2–8 found.
2. Simple improvements to the `sqrt` made in `tune1` produced very little speedup. This is expected because `gprof` shows that the program spends little time there.
3. The program `tune2` optimized for vector processing produced a factor-of-8 speedup (and that's a factor of 18 over nonoptimized code) on the vector ES/9000. Nonetheless, the program runs slower on most workstations. When we produced `tune3` by removing one of the vector changes, namely, switching the matrix indexes, this version eliminates much of the speedup on the vector mainframe, but restored the performance on the workstations. This shows that the improvement in stride provided by using an array to hold diagonal elements of the large matrix was important for the vector machine.
4. Unrolling the loops in `tune4` led to a marked improvement on all machines.

20

Parallel Computing and PVM

20.1 PROBLEM: SPEEDING UP YOUR PROGRAM

Your **problem** is to take the program you wrote to generate the bifurcation plot for bug populations in Chapter 13, *Unusual Dynamics of Nonlinear Systems*, and run different ranges of μ values simultaneously on several CPUs. While this may lead to a faster turnaround time, the real purpose is for you to get the needed experience in parallel computing so that you can apply it to programs that otherwise would run for hours or days on one CPU.

In this chapter we discuss the types and language of parallel computing, and then end with an application using the Parallel Virtual Machine (PVM) package.[1] Although the PVM package is free, it must be installed right on a number of machines before you can use it, and this is not a job for the casual user. That being so, we recommend that the reader take their implementation from the associated Web materials. In this way the latest details are there for those who have a system to use them on, while realistic working examples are there for readers who may not have access to PVM.

[1]Some of this chapter is based on [L&F 93] and some of it on materials developed by Hans Kowallik for the Web tutorials [NACSE].

20.2 THEORY: PARALLEL SEMANTICS

We have seen in Chapter 18, *Computing Hardware Basics: Memory and CPU*, that many of the tasks on a high-performance computer occur in parallel as a consequence of internal structures such as pipelined and segmented CPUs, hierarchical memory, and independent I/O processors. While this processing is "in parallel," in modern terminology "parallel computing" denotes a computing environment in which some number of CPUs are running asynchronously and communicating with each other to avoid conflicts or to exchange intermediate results.

20.2.1 Parallel Instruction and Data Streams

The processors in a parallel computer are placed at the *nodes* of a communication network (usually, but not necessarily, residing within the computer). One way of categorizing parallel systems is by how they handle instructions and data. From this viewpoint there are four types of machines:

Single instruction, single data (SISD): These are the traditional (and least expensive) serial computers in which a single instruction is executed and acts on a single data stream before the next instruction and next data stream are encountered.

Single instruction, multiple data (SIMD): Here instructions are processed from a single stream, but they act concurrently on multiple pieces of data. Generally the nodes are very simple and relatively slow processors, but are very large in number.

Multiple instructions, multiple data (MIMD): In this category each processor runs independently of the others with independent instructions and data. These are the type of machines that employ *message passing*. They can be a collection of workstations linked via a network or more integrated machines with thousands of processors.

It is possible for the multiple data streams to be extracted from a common memory and for the different central processors to be running completely independent programs. In fact, the running of independent programs is not very different from the multitasking feature familiar on mainframes, workstations, and PCs. In multitasking (for example, Unix) several independent programs reside in the computer's memory simultaneously and share the processing time in a round-robin or priority order. In multiprocessing, these jobs may all be running at the same time. Clearly, this gets more complicated if separate processors are operating on different parts of the *same* program because then synchronization and load balance (keeping all processors equally busy) are a concern.

20.2.2 Granularity

In addition to instruction and data streams, another way to categorize both the hardware and software in parallel computing is by *granularity*. A *grain* is defined as a measure of the computational work to be done, more specifically, the ratio of computation work to communication work.

Coarse-grain parallel: Here there are separate programs running on separate computer systems with the systems coupled via a conventional communication network. For example, three Sun SPARC Stations or six IBM RS/6000's sharing the same files across a network but with a different central memory system for each workstation. Each computer could be operating on a different and independent part of one problem at the same time.[2]

Medium-grain parallel: This approach may have several processors executing (possibly different) programs simultaneously while accessing a common memory. The processors are usually placed on a common *bus* (communication channel) and communicate with each other through the memory system. Medium-grain programs have different, independent, *parallel subroutines* running on different processors. Because the compilers seldom often are smart enough to figure out which parts of the program to run like that, the user must hand-program in multitasking routines.

Fine-grain parallel: As the granularity decreases and the number of nodes increases, there is an increased requirement for fast communication among the nodes. For this reason fine-grain systems tend to be custom-designed machines. The communication may be via a central bus or shared memory for a small number of nodes (less than eight) or through some form of high-speed network for massively parallel machines. In this latter case, the compiler divides the work among the processing nodes. For example, different *Do* loops of a program may be running on different nodes.

20.2.3 Parallel Performance

In general, the slowest step in a complex process determines the overall process rate because the other steps must wait around. In parallel processing, the issuing of instructions, occurring approximately once every 6 cycles, is usually the slowest step, and this one slow step may eliminate the advantage of multitasking. In general, the speedup of a program will not be great unless ~90% of the program is run in parallel. An infinite number of processors give

[2]Some experts define our medium as coarse, yet this fine point changes with time.

a speedup of only 2 when half the code is not parallelized. Yet even that theoretical speedup is not attainable in practice because as you divide up a problem into smaller and smaller pieces in order to run on more and more processors, the finite communication time limits the speedup, and adding more processors only increases the time they have to wait around.

20.3 METHOD: MULTITASKING PROGRAMMING

It only makes sense to run the most numerically intensive codes on parallel machines, and frequently these are very large programs assembled over a number of years or decades. It may come as no surprise, then, that the programming languages for the parallel machines are primarily Fortran with explicit parallel extensions (Fortran 90) and to a lesser extent C. While the language may be familiar, parallel programming becomes more demanding as the number of processors increases, and it may be best to rewrite a code written for one CPU rather than to try convert it to parallel.

20.3.1 Method: Multitask Organization

A typical organization of a program for multitasking is

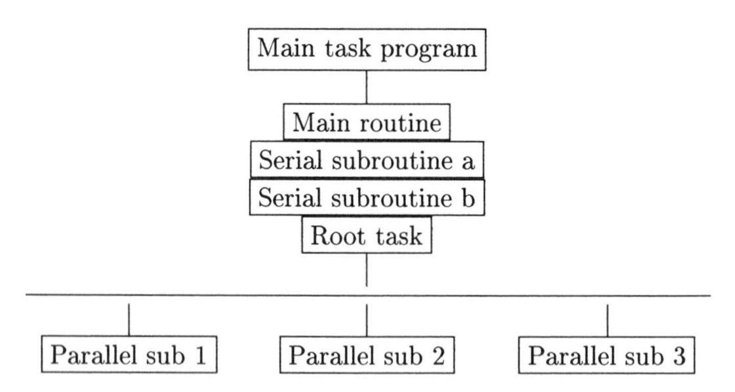

The operating system or the user organizes the work into units called *tasks*, and the tasks assign work to each processor. There is a main task to control the overall execution as well as subtasks to run independent parts of the program (called *parallel subroutines*, *slaves*, or *subtasks*). These parallel subroutines can be distinctive subprograms or multiple copies of the same subprogram. The main-task program does its own computations as well as actually calling and scheduling the parallel subroutines.

The programmer helps speed up the running by keeping as many processors as possible as simultaneously busy as possible and by avoiding storage conflicts from different parallel subprograms. You do this *load balancing* by dividing your program into subtasks of approximately equal numerical intensity that will run simultaneously on different processors. The rule of thumb is to make the task with the largest granularity (workload) dominant by forcing it to execute first, and to keep all the processors busy by having the number of tasks be an integer multiple of the number of processors.

While vector processors will usually work on the innermost loop of a program, the parallel subroutines should be of broad scope—which usually means the outermost loop. For example, vectorizing an outer matrix of dimension 3 would not produce much, if any, speedup. But if many floating-point operations are needed to calculate the elements of this matrix, then this may be a good choice for three-way parallel.

To avoid storage conflicts, design your program so that parallel subtasks use data that are independent of the data in the main task and in other parallel tasks. This means that these data should not be modified *or even examined* by different tasks simultaneously. In organizing these multiple tasks, some concern about *overhead costs* is appropriate. These costs tend to be high for fine-grain programming and to vary for different scheduling commands.

20.4 METHOD: DISTRIBUTED MEMORY PROGRAMMING

An approach to concurrent processing that has gained wide acceptance for coarse- and medium-grain systems is *distributed memory.* In it, each processor has its own memory and the processors exchange data among themselves over a (preferably high-speed) network with a fast switch. The data exchanged or *passed* among processors have encoded forward and return addresses and are called *messages.*

For a messages-passing program to be successful, the data must be divided among nodes so that, at least for a while, each node has all the data it needs to run an independent subtask. When a program begins execution, data are sent to all nodes. When all nodes have completed their subtasks, they exchange data again in order for each node to have a complete new set of data to perform the next subtask. This repeated cycle of data exchange followed by processing continues until the full task is completed. Message-passing MIMD programs are also *single-program multiple-data* programs, which means that the programmer writes a single program that is executed on all of the nodes. Often a separate host program, which starts the programs on the nodes, reads the input files and organizes the output.

Although MIMD systems are becoming very popular, a number of standards, such as a standard programming language, have yet to be adopted. Some popular systems are *Express*, a commercial product from ParaSoft Corporation, *PVM*, and *MPI.* All hide the messy details of passing messages and synchronization, and are available for use on single MIMD machines or for clusters of workstations. The existence and utility of these packages mean that parallel systems do not have to consist of only a set of dedicated processors; there can also be a number of workstations from various manufactures connected by some network. And while the top priority of these workstations may be the work of their owners, when there are no local demands on them they will automatically switch over to helping someone else's big problem get done concurrently. The integrated sum is tremendous computing power that might otherwise go wasted.

20.5 IMPLEMENTATION: PVM BUG POPULATIONS, WWW

PVM is a software package that allows a number of computers connected over a network to be combined into a single parallel virtual machine. This virtual machine can consist of computers with different architectures, running different operating systems.

To learn more about PVM and to use it to solve your problem, we recommend that you work through the Web tutorial at

http://nacphy.physics.orst.edu/PVM/

That tutorial computes different regions of Fig. 20.1 on different processors. The Web tutorial supplies details for those who can apply them, as well as realistic working examples for those readers who may not have access to PVM.

20.5.1 The Plan

The most important part in using PVM is probably the program that sets up the communications, starts the programs on remote hosts, and sends messages to other programs. The most straightforward model for writing parallel programs with PVM is with a *master* and a *slave*. The master process is the only one started by the user on one of the machines. It then starts and controls processes on the other machines that perform the actual work.

The **master**

- Determines which physical machines are part of the virtual machine.
- Starts a slave process on every physical machine.
- Collects the results that are sent back by the slaves.
- Prints the results to standard output.

The **slave**

- Determines the names of the machines on which the process is running.
- Determines the time on these machines.

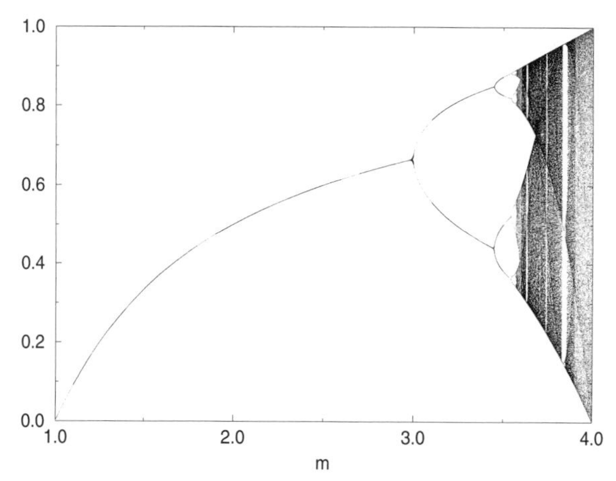

Fig. 20.1 The population of bugs versus growth rate for the logistics map. The shades of the different regions indicate that different PVM machines were used to calculate them.

- Sends a message with this information back to the master.

The bug population problem is essentially perfect for parallel processing, because the initial population value is independent for each value of the growth parameter. This means that the population calculation for each μ value can be done independently on separate machines with no message passing necessary. Sample programs are given in the Web tutorial, and some sample results are in Fig. 20.1.

21

Object-Oriented Programming: Kinematics ⊙

In this chapter we provide examples of object-oriented programming (OOP) using the C++ language. Even though this subject fits in well with our earlier discussion of programming principles, we have put it off until now and marked it as optional (the ⊙) since we do not use explicit OOP in the projects and because it is difficult for the new programmer. However, we suggest that everyone read through §21.2.1.

21.1 PROBLEM: SUPERPOSITION OF MOTIONS

The isotropy of space implies that motion in one direction is independent of motion in other directions. So, for example, when a soccer ball is kicked, we have acceleration in the vertical direction and simultaneous, yet independent, uniform motion in the horizontal direction. In addition, Galilean invariance (velocity independence of Newton's laws of motion) tells us that when an acceleration is added in to uniform motion, the distance covered due to the acceleration adds to the distance covered by uniform motion.

Your **problem** is to use the computer to describe motion in such a way that velocities and accelerations in each direction are treated as separate entities or objects, independent of motion in other directions. In this way the problem is viewed consistently by both the programming philosophy and the basic physics.

21.2 THEORY: OBJECT-ORIENTED PROGRAMMING

We will analyze this problem from an object-oriented programming viewpoint. While the *objects* in OOP are often graphical ones on a computer screen, the objects in our problem are the motions in each dimension. By reading through and running the programs in this project, you should become familiar with the concepts of OOP.

21.2.1 OOP Fundamentals

Object-oriented programming (OOP) has a precise definition. The concept is general and the *object* can be a component of a program with the properties of *encapsulation*, *abstraction*, *inheritance*, and *polymorphism* (to be defined shortly). Of interest to us is OOP's programming paradigm, which aims to simplify writing large programs by providing a framework for reusing components developed and tested in previous problems. A true object-oriented language has four characteristics [Smit 91, P&W 91]:

Encapsulation: The data and the *methods* used to produce or access the data are encapsulated into an entity called an *object.* For our 1-D problem, we take the data as the initial position and velocity of the soccer ball, and the object as the solution $x(t)$ of the equations of motion that gives the position of the ball as a function of time. As part of the OOP philosophy, data are manipulated only via distinct *methods.*

Abstraction: Operations applied to objects must give expected results according to the nature of the objects. For example, summing two matrices always gives another matrix. The programmer can in this way concentrate on solving the problem rather than on details of implementation.

Inheritance: Objects inherit characteristics (including code) from their ancestors, yet may be different from their ancestors. In our problem, motion in two dimensions inherits the properties of 1-D motion in each of two dimensions, and accelerated motion inherits the properties of uniform motion.

Polymorphism: Different objects may have *methods* with the same name, yet the method may differ for different objects. Child objects may have *member* functions with the same name but properties differing from those of their ancestors. In our problem, a member function *archive*, which contains the data to be plotted, will be redefined depending on whether the motion is uniform or accelerated.

21.3 THEORY: NEWTON'S LAWS, EQUATION OF MOTION

Newton's second law of motion relates the force vector $\mathbf{F}$ acting on a mass m to the acceleration vector $\mathbf{a}$ of the mass:

$$\mathbf{F} = m\mathbf{a}. \tag{21.1}$$

When the vectors are resolved into Cartesian components, each component yields a second-order differential equation:

$$F_i = m\frac{d^2 x_i}{dt^2}, \quad (i = 1, 2, 3). \tag{21.2}$$

If the force in the x direction vanishes, $F_x = 0$, the equation of motion (21.2) has a solution corresponding to uniform motion in the x direction with a constant velocity v_{0x}:

$$x = x_0 + v_{0x}t. \tag{21.3}$$

Equation (21.3) is the *base* or *parent* object in our project. If the force in the y direction also vanishes, then there also will be uniform motion in the y direction:

$$y = y_0 + v_{0y}t. \tag{21.4}$$

In our project we consider uniform x motion as a parent and view the y motion as a child.

Equation (21.2) tells us that a constant force in the x direction causes a constant acceleration a_x in that direction. The solution of the x equation of motion with uniform acceleration is

$$x = x_0 + v_{0x}t + \tfrac{1}{2}a_x t^2. \tag{21.5}$$

For projectile motion without air resistance, we usually have no x acceleration and a negative y acceleration due to gravity, $a_y = -g = -9.8\text{m/s}^2$. The y equation of motion is then

$$y = y_0 + v_{0y}t - \tfrac{1}{2}gt^2. \tag{21.6}$$

We define this accelerated y motion to be a child to the parent uniform x motion.

21.4 OOP METHOD: CLASS STRUCTURE

The *class structure* we use to solve our problem contains the objects:

Parent class Um1D:	1-D uniform motion for given initial conditions,
Child class Um2D:	2-D uniform motion for given initial conditions,
Child class Am2d:	2-D accelerated motion for given initial conditions.

The *member functions* include

x:	position after time t,
archive:	creator of a file of position versus time.

For our projectile motion, *encapsulation* is the combination of the initial conditions (x_0, v_{x0}) with the member functions used to compute $x(t)$. Our member functions are the creator of the class of uniform 1-D motion `Um1D`, its destructor `~Um1D`, and the creator `x(t)` of a file of x as a function of time t. *Inheritance* is the child class `Um2D` for uniform motion in both x and y directions, it being created from the parent class `Um1D` of 1-D uniform motion. *Abstraction* is present (although not used powerfully) by the simple addition of motion in the x and y directions. Polymorphism is present by having the member function that creates the output file be different for 1-D and 2-D motions. In our implementation of OOP, the class `Am2D` for accelerated motion in two dimensions inherits uniform motion in two dimensions (which, in turn, inherits uniform 1-D motion), and adds to it the attribute of acceleration.

21.5 IMPLEMENTATION: UNIFORM 1-D MOTION, UNIM1D.CPP

For 1-D motion we need a program that outputs positions along a line as a function of time, (x, t). For 2-D motion we need a program that outputs positions in a plane as a function of time (x, y, t). Time varies in discrete steps of $\Delta t =$`delt` $= T/N$, where the total time T and the number of steps N are input parameters. We give here program fragments that can be pasted together into a complete C++ program (the complete program is on the diskette).

Our parent class `Um1D` of uniform motion in one dimension contains

x00:	the initial position,
delt:	the time step,
vx0:	the initial velocity,
time:	the total time of the motion,
steps:	the number of time steps.

To create it, we start with the C++ headers:

```
#include <stdio.h>                                    /* Input-output libe */
#include <stdlib.h>                                           /* Math libe */
```

The encapsulation of the data and the member functions is achieved via `Class Um1D`:

21.5.1 Uniform Motion in 1-D, Class Um1D

```
class Um1D                                          /* Create base class */
  {
  public:
  double x00,delt,vx,time;        /* Initial position and velocity, dt */
  int steps;                                              /* Time steps */
  Um1D(double x0,double dt,double vx0,double ttot);  /* Class Constructor
*/
  ~ Um1D(void);                                     /* Class Destructor */
  double x(double tt);                                          /* x(t) */
  void archive();                                 /* send  x vs t to file */
  };
```

Next, the variables `x0`, `delt`, `vx`, and `time` are initialized by the constructor of the class `Um1D`:

```
Um1D::Um1D(double x0,double dt,double vx0,double ttot)
{                                                 /* Constructor Um1D */
x00 = x0;
delt = dt;
vx = vx0;
time = ttot;
steps = ttot/delt;
}
```

After that, we make the destructor of the class, which also prints the time when the class is destroyed:

```
Um1D::~Um1D(void)                          /* Destructor of class Um1D */
{
printf("Class Um1D destroyed \ n");
}
```

Given the initial position x_0, the member function returns the position after time dt:

```
double Um1D::x(double tt)                              /* x=x0+dt*v */
{ return x00+tt*vx;  }
```

The algorithm is implemented in a member routine, and the positions and times are written to the file `Motion1D.dat`:

```
void Um1D::archive()                          /* Produce x vs t file */
{
FILE  *pf;
int i;
```

```
double xx,tt;
if((pf = fopen("Motion1D.dat","w+"))==NULL)
{
printf("Could not open file\ n");
exit(1);
}
tt = 0.0;
for(i = 1;i<=steps;i++)
{
xx = x(tt);                                   /* Computes x=x0+t*v */
fprintf(pf,"%f  %f\ n",tt,xx);
tt = tt+delt;
}
fclose(pf);
}
```

The main program defines an object (class) `unimotx` of type `Um1D` and gives initial numeric values to the data:

```
main()
{
double inix,inivx,dtim,ttotal;
inix = 5.0;
dtim = 0.1;
inivx = 10.0;
ttotal = 4.0;
Um1D unimotx(inix,dtim,inivx,ttotal);          /* Class constructor */
unimotx.archive();                            /* Produce y vs x file */
}
```

21.5.2 Implementation: Uniform Motion in 2-D, Child Um2D, unimot2d.cpp

The first part of the program is the same as before. We now make the child class `Um2D` from the class `Um1D`.

```
#include <stdio.h>
#include <stdlib.h>
class Um1D                                     /* Base class created */
{
public:
double x00,delt,vx,time;          /* Initial conditions, parameters */
int steps;                                            /* Time step */
Um1D(double x0,double dt,double vx0, double ttot);    /* constructor */
~Um1D(void);                                   /* Class destructor */
double x(double tt);                                  /* x=xo+v*t */
void archive();                               /* Send x vs t to file */
};
```

```
   Um1D::Um1D(double x0,double dt,double vx0,double ttot)          /* Um1D
Constructor */
   {
   x00 = x0;
   delt = dt;
   vx = vx0;
   time = ttot;
   steps = ttot/delt;
   }
   Um1D::~Um1D(void)                                 /* Class Um1D destructor */
   {
   printf("Class Um1D destroyed\ n");
   }
   double Um1D::x(double tt)                                   /* x=x0+dt*v */
   {
   return x00+tt*vx;
   }
   void Um1D::archive()                               /* Produce x vs t file */
   {
   FILE  *pf;
   int i;
   double xx,tt;
   if((pf = fopen("Motion1D.dat","w+"))==NULL)
   {
   printf("Could not open file\ n");
   exit(1);
   }
   tt = 0.0;
   for(i = 1;i<=steps;i++)
   {
   xx = x(tt);                                        /* computes x=x0+t*v */
   fprintf(pf,"%f  %f\ n",tt,xx);
   tt = tt+delt;
   }
   fclose(pf);
   }
```

21.5.3 Class Um2D: Uniform Motion in 2-D

To include another degree of freedom, we define a new class that inherits the x component of uniform motion from the parent `Um1D`, as well as the y component of uniform motion from the parent `Um1D`. The new data for the class are the initial y position `y00` and the velocity in the y direction `vy0`. A new member `y` is included to describe the y motion. Note, in making the constructor of the `Um2D` class, that our interest in the data y versus x leads to the member `archive` being redefined. This is polymorphism in action.

```
class Um2D : public Um1D                     /* Child class, parent Um1D */
{
public:                                      /* Data accessible to other code */
double y00,vy;                       /* member functions  accessible to all */
Um2D(double x0,double dt,double vx0,double ttot,double y0,double vy0);
~Um2D(void);                                 /* destructor of Um2D class */
double y(double tt);                            /* Added motion for 2-D */
void archive();                                 /* redefinition for 2-D */
};
```

Observe the way the constructor for the class `Um2D` initializes the data in `Um1D`, `y00`, and `vy`:

```
Um2D::Um2D(double x0,double dt,double vx0,double ttot,double y0,double vy0)
/* Constructor of class Um2D */
{
y00 = y0;
vy = vy0;
}
```

The destructor of the new class is

```
Um2D::~Um2D(void)
{ printf("Class Um2D is destroyed\ n");  }
```

The new member of class `Um2D` accounts for the y motion:

```
double Um2D::y(double tt)
{ return y00+tt*vy;  }
```

The new member function for two dimensions contains the data of the y and x positions versus time, and incorporates the polymorphism property of the objects:

```
void Um2D::archive()                            /* Uniform motion in 2D */
{
FILE  *pf;
int i;
double xx,yy,tt;
if((pf = fopen("Motion2D.dat","w+"))==NULL)
{
printf("Could not open file\ n");
exit(1);
}
tt = 0.0;
for(i = 1;i<=steps;i++)                        /* uses member function x */
{
xx = x(tt);
yy = y(tt);                                  /* add the second dimension */
fprintf(pf,"%f  %f\ n",yy,xx);
tt = tt+delt;
```

```
}
fclose(pf);
}
```

The differences with the previous `main` program is the inclusion of the y component of the motion and the constructor `unimotxy` of class type `Um2D`:

```
main()
{
double inix,iniy,inivx,inivy,dtim,ttotal;
inix = 5.0;
dtim = 0.1;
inivx = 10.0;
ttotal = 4.0;
iniy = 3.0;
inivy = 8.0;
Um2D unimotxy(inix,dtim,inivx,ttotal,iniy,inivy);  /* class constructor
*/
unimotxy.archive();                         /* To obtain file of y vs x */
}
```

21.5.4 Implementation: Projectile Motion, Child Accm2D, accm2d.cpp

Consider now the problem of the motion of a soccer ball kicked at $(x, y) = (0, 0)$ with initial velocity (v_{0x}, v_{0y}) =(14,14) m/s. This is, of course, motion of a projectile in a uniform gravitational field which we know is described by a parabolic trajectory. We define a new child class `Accm2D` derived from the parent class `Um2D` that adds acceleration to the motion. The first part of the program is the same as the one for uniform motion in two dimensions, but now there is an extension with the class `Accm2D`:

```
#include <stdio.h>
#include <stdlib.h>
class Um1D                                   /* Base class is created */
{
public:                            /* Initial conditions.  parameters */
double x00,delt,vx,time;
int steps;                             /* Time steps to write in file */
Um1D(double x0,double dt,double vx0,double ttot);     /* Constructor */
~Um1D(void);                                          /* Destructor */
double x(double tt);                                  /* x=xo+v* dt */
void archive();                                 /* send x vs t to file */
};
Um1D::Um1D(double x0,double dt,double vx0,double ttot)   /* Constructor
*/
{
x00 = x0;
delt = dt;
```

```
vx = vx0;
time =  ttot;
steps = ttot/delt;
}
Um1D::~Um1D(void)                       /* Destructor of the class Um1D */
{
printf("Class Um1D destroyed\ n");
}
double Um1D::x(double tt)                                 /* x=x0+dt*v */
{
return x00+tt*vx;
}
void Um1D::archive()                        /* Produce  file of  x vs t */
{
FILE  *pf;
int i;
double xx,tt;
if((pf = fopen("Motion1D.dat","w+"))==NULL)
{
printf("Could not open file\ n");
exit(1);
}
tt=0.0;
for(i = 1;i< = steps;i++)         /* computes x=x0+dt*v, changing x0 */
{
xx = x(tt);
fprintf(pf,"%f  %f\ n",tt,xx);
tt = tt+delt;
}
fclose(pf);
}
class Um2D : public Um1D              /* Child class Um2D, parent Um1D */
{
public:    /* Data accessible:  other code; functions:  all members */
double y00,vy;
Um2D(double x0,double dt,double vx0,double ttot, double
y0,double vy0);                          /* constructor of Um2D class */
~Um2D(void);                              /* destructor of Um2D class */
double y(double tt);
void archive();                              /* redefine member for 2D */
};
Um2D::Um2D(double x0,double dt,double vx0,double ttot,
double y0,double vy0):Um1D(x0,dt,vx0,ttot)   /* Construct class Um2D */
{
y00 = y0;
vy = vy0;
}
```

```
Um2D::~Um2D(void)
{
printf("Class Um2D is destroyed\ n");
} double Um2D::y(double tt)
{
return y00+tt*vy;
}
void Um2D::archive()
{
FILE  *pf;
int i;
double xx,yy,tt;                          /*  now in 2D, still uniform */
if((pf = fopen("Motion2D.dat","w+"))==NULL)
{
printf("Could not open file\ n");
exit(1);
}
tt = 0.0;
for(i = 1;i<=steps;i++)                        /* uses member function x */
{
xx = x(tt);                                     /* adds second dimension */
yy = y(tt);
fprintf(pf,"%f  %f\ n",yy,xx);
tt = tt+delt;
}
fclose(pf);
}
```

21.5.5 Accelerated Motion in Two Directions

A child class `Accm2D` is created from the parent class `Um2D`. It will inherit uniform motion in two dimensions and add acceleration in both the x and y directions. It has member functions:

1. Constructor of class
2. Destructor of class
3. A new member `xy` that gives the x and y components of acceleration
4. An *archive* to override the member function of same name in the parent class for 2-D uniform motion

```
class Accm2D : public Um2D
{
public:
double ax,ay;
```

```
Accm2D(double x0,double dt,double vx0,double ttot,
double y0,double vy0,double accx,double accy);
~Accm2D(void);
void xy(double *xxac, double *yyac,double tt);
void archive();
};
```

Observe the method used to initialize the classes `Am2d` and `Um2D`:

```
Accm2D::Accm2D(double x0,double dt,double vx0,double ttot, double y0,double
vy0,double accx,doubleaccy):
Um2D(x0,dt, vx0,ttot,y0, vy0)
{
ax = accx;
ay = accy;
}
Accm2D::~Accm2D(void)
{
printf(" Class Accm2D destroyed\ n");
}
```

Next we introduce a member function to show the inheritance of the parent class functions `xpox` and `y` and to include the two components of acceleration:

```
void Accm2D::xy(double *xxac, double *yyac,double tt)
{
double dt2;
dt2 = 0.5*tt*tt;
*xxac = x(tt)+ax*dt2;
*yyac = y(tt)+ay*dt2;
}
```

To override *archive* (which creates the data file), we redefine *archive* to take into account the acceleration:

```
void Accm2D::archive()
{
FILE  *pf;
int i;
double tt,xxac,yyac;
if((pf = fopen("Motion.dat","w+"))==NULL)
{
printf("Could not open file\ n");
exit(1);
}
tt = 0.0;
for(i = 1;i<=steps;i++)
{
xy(&xxac,&yyac,tt);
fprintf(pf,"%f  %f\ n",xxac,yyac);
```

```
tt = tt+delt;
}
fclose(pf);
}
```

Next a file is produced with the y and x positions of the ball as functions of time:

```
main()
{
double inix,iniy,inivx,inivy,aclx,acly,dtim,ttotal;
inix = 0.0;
dtim = 0.1;
inivx = 14.0;
ttotal = 4.0;
iniy = 0.0;
inivy = 14.0;
aclx = 0.0;
acly = -9.8;
Accm2D acmo2d(inix,dtim,inivx,ttotal,iniy,inivy,aclx,acly);
printf("  \ n");
printf("    \ n");
acmo2d.archive();
}
```

21.6 ASSESSMENT: EXPLORATION, SHMS.CPP

The superposition of independent simple harmonic motion in each of two dimensions can be studied with OOP. Define a class `ShmX` for harmonic motion in the x direction:

$$x = A_x \sin(\omega_x t + \phi_x), \tag{21.7}$$

where A_x is the amplitude, ω_x the angular frequency, and ϕ_x the phase. Define another class `Shmy` for independent harmonic motion in the y direction:

$$y = A_y \sin(\omega_y t + \phi_y). \tag{21.8}$$

Now employ the concept of *multiple inheritance* to define a child class `ShmXY` of both `ShmX` and `ShmY`. It should have a member function to write a file with the x and y positions at several times (which can then be used to plot Lissajous figures). To obtain multiple inheritance use `class ShmXY : public ShmX, public ShmY`. To obtain the constructor, use something like this:

```
ShmXY::ShmXY(double Axi,double delx,double wx,double tx,
double dtx,double Ayi,double dely,double wy,double ty,double dty)
:  ShmX(Axi,delx,wx,tx,dtx),ShmY(Ay,dely,wy,ty,dty)
```

22

Thermodynamic Simulations: The Ising Model

22.1 PROBLEM: HOT MAGNETS

Ferromagnetic materials such as bar magnets contain *domains* that are magnetized even in the absence of an external magnetic field. When an external magnetic field is applied, the different domains align and the internal fields become very high. Yet, as the temperature of the ferromagnet is raised, the magnetism decreases and in some cases a *phase transition* occurs in which the magnetism decreases precipitously. Your **problem** is to explain the thermal behavior and the phase transitions of ferromagnets.

We will solve this problem with the quantum mechanical Ising model and the simulated annealing (Metropolis) algorithm. The model is simple but contains much physics, and the simulation gives a visualization of thermal equilibrium that is absent from formal studies. The same algorithm and theory is used for lattice quantum mechanics, and so the present project should be completed before attempting Chapter 23, *Functional Integration on Quantum Paths.*

22.2 THEORY: STATISTICAL MECHANICS

When we say that an object is at a temperature T, we mean that the atoms composing this object are in a state of thermodynamic equilibrium with an environment at temperature T. While this may be an equilibrium state, it is also a dynamic state (it is thermo*dynamics* after all). The system's energy is continually changing as it exchanges energy with the environment. If the system is at temperature T, then its atoms have an average kinetic energy proportional to T, with larger and larger fluctuations from this average occurring as the temperature increases.

An example of how the equilibrium state changes with temperature is the annealing process (it's one we will simulate on the computer). Let's say that we are making a blade for a sword and are hammering away at it while it's red hot to get its shape just right. At this high a temperature there is a lot of internal motion and not the long-range order needed for a stiff blade. So, as part of the process, we *anneal* the blade; that is, we heat and slow cool it in order to reduce brittleness and increase strength. Too rapid a cooling would not permit long-range equilibration (and ordering) of the blade, and this would lead to brittleness.

In the present problem we deal with the thermal properties of magnetized materials. The magnetism arises from the alignment of the quantum mechanical spins of the atoms. The spin of each atom, in turn, arises from its electrons. When the number of electrons is large, the problem is too difficult to really solve, and so statistical methods are used to obtain average quantities (in most cases that is all we can measure, anyway). If the system is described microscopically by classical or quantum mechanics, then this method is called *statistical mechanics.*

Statistical mechanics starts with the elementary interactions among a system's particles and constructs the macroscopic thermodynamic properties such as temperature T and internal energy U. The essential assumption is that all microscopic configurations of the system consistent with the constraints are equally probable. This leads to a distribution for the states of the system in which state α_j with energy $E(\alpha_j)$ occurs with a probability given by the Boltzmann factor:

$$P(\alpha_j) = \frac{e^{-E(\alpha_j)/kT}}{Z(T)}, \tag{22.1}$$

$$Z(T) = \sum_{\alpha_j} e^{-E_j/kT}. \tag{22.2}$$

Here k is Boltzmann's constant, T is the temperature, and the partition function $Z(T)$ is a weighted sum over states.

Notice that the Boltzmann distribution (22.1) does not require a thermal system to be in the state of lowest energy. Rather, it states that it is less likely for the system to have a high energy. Of course, as $T \to 0$ only the $E = 0$ state has a nonvanishing probability, yet for finite temperatures we expect the system's energy to have fluctuation on the order of kT.

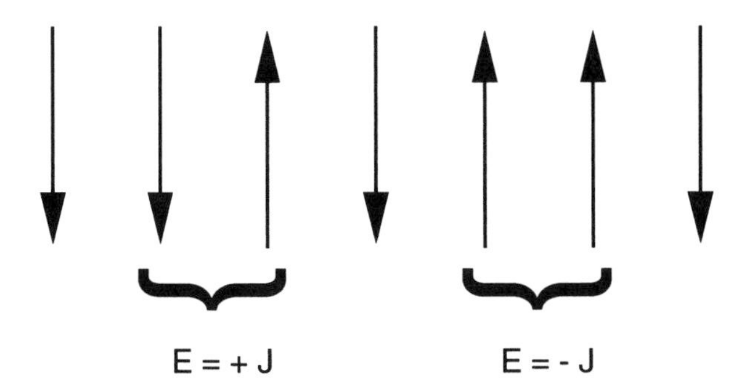

Fig. 22.1 A 1-D lattice of N spins. The interaction energy $V = \pm J$ between nearest-neighbor pairs is shown for aligned and opposing spins.

22.3 MODEL: AN ISING CHAIN

As our model, we consider N magnetic dipoles fixed on the links of a linear chain as shown in Fig. 22.1. (It is a straightforward generalization to handle two- and three-dimensional lattices, and indeed we do it as an exploration.) Because the particle are fixed, their positions and momenta are not dynamical variables, and we need to only worry about their spins.

We assume that the particle at site i has spin s_i, which can be either up or down:

$$s_i \equiv s_{z,i} = \pm\tfrac{1}{2}. \tag{22.3}$$

Each possible configuration or state of the N particles is described by the quantum state vector

$$|\alpha_j\rangle = |s_1, s_2, \ldots, s_N\rangle \tag{22.4}$$

$$= \{\pm\ \tfrac{1}{2}\ , \pm\ \tfrac{1}{2}\ , \ldots\}, \quad j = 1, 2^N. \tag{22.5}$$

Since the spin of each particle can assume that any one of *two* values, there are 2^N different possible states of the N particles in the system. We do not worry about the effect on the state vector of interchanging identical particles because they are all fixed in place and so can't be confused with each other.

The energy of the system arises from the interaction of the spins with each other and with the external magnetic field B. We know from quantum mechanics that an electron's spin and magnetic moment are proportional to each other, so a "dipole–dipole" interaction is equivalent to a "spin–spin" interaction. We assume that each dipole interacts with the external magnetic field and with its nearest neighbor through the potential:

$$V_i = -J\mathbf{s}_i \cdot \mathbf{s}_{i+1} - g\mu\mathbf{s}_i \cdot \mathbf{B}. \tag{22.6}$$

Here the constant J is called the *exchange energy* and is a measure of the strength of the spin–spin interaction. The constant g is the gyromagnetic ratio; that is, the proportionality constant between the angular momentum and magnetic moment. The constant μ is the Bohr magneton, the unit for magnetic moments.

In our model with N particles, there are 2^N possible configurations. Since $2^{20} > 10^6$, we see that even a small number of particles can lead to a prohibitively large number of configurations to consider. For a small number of particles, the computer can examine all possible spin configurations, yet as N gets larger, or as we go to two and three dimensions, a less exact statistical approach is used. We will consequently apply equations (22.1)–(22.2). In addition, we know from our previous studies that Monte Carlo techniques work well on the computer, and so we expect that the simulation may be realistic. Just how large N must be for this to occur depends somewhat on how good a description is needed; for our purposes $N > 200$ should appear statistical, with $N \simeq 2000$ being reliable.

The energy of the system to be used in the Boltzmann distribution (22.1) is the expectation value of the sum of V over the spins of the particles:

$$E(\alpha) = \langle\alpha|\sum_i V_i\,|\alpha\rangle \tag{22.7}$$

$$= -J\sum_{i=1}^{N-1} s_i s_{i+1} - B\mu\sum_{i=1}^{N} s_i. \tag{22.8}$$

For simplicity of notation, and to be able to compare with an analytic result, we now turn off the magnetic field, that is, assume that$B = 0$. Nonetheless, it is easy to add it back into the final expression for the energy, and we recommend that you do it because giving a preferred direction to space stabilizes the configuration.

The equilibrium alignment of the spins depends critically on the sign of the exchange energy J. If $J > 0$, the lowest energy state will tend to have neighboring spins aligned. If the temperature is low enough, the ground state will be a *ferromagnet* with essentially all spins aligned. If $J < 0$, the lowest energy state will tend to have neighbors with opposite spins. If the temperature is low enough, the ground state will be a *antiferromagnet* with alternating spins.

Unfortunately, a simple model such as this has its limits. Its approach to thermal equilibrium is qualitatively but not quantitatively correct. Further, when $B = 0$ there is no preferred direction in space, which means that the average magnetization may vanish, and this may lead to instabilities in which the spins spontaneously reverse. These instabilities are a type of Bloch-wall transitions in which regions of different spin orientations change size. In addition, the phase transition at the Curie temperature, a characteristic of magnetic materials, does not occur in the 1-D version of the model.

As you will verify in your simulation, a system described by the Boltzmann distribution (22.1) does not have a single configuration. Rather, there is a continual and random interchange of thermal energy with the environment that leads to fluctuations in the total energy. Even at equilibrium, the system fluctuates with the fluctuations getting larger as the temperature rises.

22.4 SOLUTION, ANALYTIC

For very large numbers of particles, the thermodynamic properties of the 1-D Ising model can be calculated analytically [P&B 89]. This tells us that the average energy (in J units) is

$$\frac{U}{J} = -N \tanh \frac{J}{kT} = -N \frac{e^{J/kT} - e^{-J/kT}}{e^{J/kT} + e^{-J/kT}}, \tag{22.9}$$

$$\Rightarrow \begin{cases} N, & \text{for } kT \to 0, \\ 0, & \text{for } kT \to \infty. \end{cases} \tag{22.10}$$

The analytic result for the specific heat per particle is

$$C(kT) = \frac{1}{N}\frac{dU}{dT} = \frac{(J/kT)^2}{\cosh^2(J/kT)}. \tag{22.11}$$

While the specific heat for the 1-D model does have a maximum as a function of temperature, it does not exhibit the characteristic discontinuity of a phase transition (the 2-D and 3-D models do). For the 1-D Ising model, the magnetization is

$$M(kT) = \frac{N e^{J/kT} \sinh(B/kT)}{\sqrt{e^{2J/kT} \sinh^2(B/kT) + e^{-2J/kT}}}. \tag{22.12}$$

22.5 SOLUTION, NUMERICAL: THE METROPOLIS ALGORITHM

We need an algorithm to evaluate the sums that appear in the partition function (22.2). This is analogous to a 2^N-dimensional numerical integration, and we know from §7.13 that a Monte Carlo approach is best for such high dimensions. Yet we do not want to generate random configurations uniformly for a system in thermal equilibrium because the Boltzmann factor essentially vanishes for those configurations whose energies are not close to the minimum energy. In other words, the vast majority of the terms we sum over to determine the thermodynamic properties of the system make hardly any contribution, and it would be quicker to weight the random numbers we generate such that most of the sum is over large terms.

In their simulation of neutron transmission through matter, Metropolis,

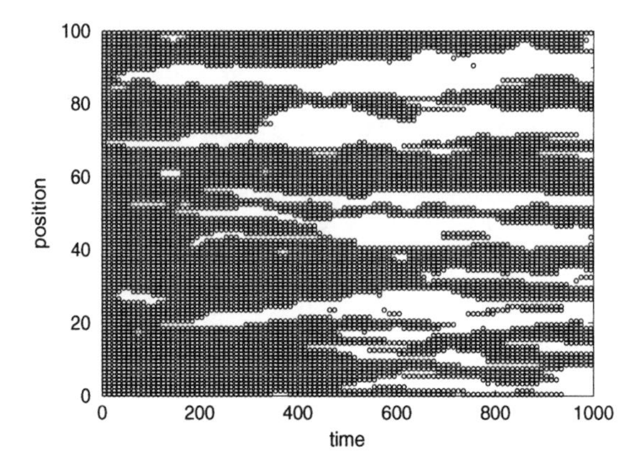

Fig. 22.2 A 1-D lattice of 100 spins. As indicated by the dark circles, initially all the spins are pointing up (a "cold" start). Magnetic domains are seen to form as the system heats up.

Rosenbluth, Teller, and Teller [Metp 53] found an algorithm to improve the Monte Carlo calculation of averages. This *Metropolis algorithm* has now become a cornerstone of computational physics. The sequence of configurations it produces is an example of a *Markov chain*, and in our case simulates the fluctuations in spin configurations that occur during thermal equilibrium. Although not simple or particularly illuminating to prove, this algorithm changes randomly the individual spins in such a way that on the average, the probability of any one configuration occurring follows a Boltzmann distribution.

The Metropolis algorithm involves two steps. First we start off with an arbitrary initial condition and repeatedly apply the algorithm until thermal equilibrium for a specific temperature is reached. For example, in Fig. 22.2 we have an initially "cold" system, whereas the spins would be random for have an initially "hot" system. Once the system reaches thermal equilibrium, the algorithm generates the statistical fluctuations about equilibrium that determine the thermodynamic quantities.

Because the 2^N configurations of N spins can be a lot, the amount of computer time needed can be very long (yet the program is simple). The hope is that the configurations obtained after a small number ($\simeq 10N$) of iterations will be close to those that produce minimum energy. While the answers do get better if we sample more and more configurations, there is a limit to improvement because roundoff error increases as more calculations are made.

Explicitly, the Metropolis algorithm is used to generate a nonuniform, random distribution of spin configurations α_j values, (22.5), with each α_j having probability

$$\mathcal{P}(\alpha_j) = \frac{1}{Z} e^{-E(\alpha_j)/kT}. \tag{22.13}$$

The technique is a variation of von Neumann rejection (stone throwing of §7.8) in which we start with an initial configuration and vary it randomly to obtain a *trial* configuration. The Boltzmann factor tells us that the relative probability of this trial configuration is proportional to $\Delta\mathcal{P} = \exp(-\Delta E/kT)$, where ΔE is the difference in energy between the previous and the trial configuration. If the trial configuration has a lower energy, ΔE will be negative, the relative probability will be greater than one, and we accept the trial configuration as the new configuration with no further ado. If, on the other hand, the trial configuration has a higher energy, we do not reject it out of hand, but accept it with the relative probability $\Delta\mathcal{P} = \exp(-\Delta E/kT) < 1$.

To accept a configuration with a probability, we pick a uniform random number between 0 and 1, and if the relative probability is greater than this number, we accept the trial configuration; if the Boltzmann factor is smaller than the chosen random number, we reject it. When the trial configuration is not accepted, the next configuration is identical to the preceding one.

The key aspects of the Metropolis algorithm is that the weight given to a trial configuration depends on how far it is from the minimum-energy configuration. Those configurations that stray far from the minimum-energy configuration are deemphasized but not completely discarded. By permitting the system to deviate away from the minimum-energy configuration (go "uphill" for a while), this technique is successful at finding a global extremum for situations in which other techniques are successful at finding only local ones. Its success relies on it not being too quick in "cooling" to the minimum-energy configuration; for this reason the algorithm is sometimes called *simulated annealing.* The algorithm is expected to be good at simulating the fluctuations about minimum energy, and gives explicit results for the thermodynamic quantities like those in Fig. 22.3.

The explicit **rules** for the Metropolis algorithm are

1. Start with an arbitrary spin configuration $\alpha_k = \{s_1, s_2, \ldots, s_N\}$.
2. To generate a new configuration α_{k+1}:
   (a) Pick particle i randomly.
   (b) Reverse i's spin direction to create trial configuration α_{tr}.
   (c) Calculate the energy $E(\alpha_{tr})$ of the trial configuration.
   (d) If $E(\alpha_{tr}) \leq E(\alpha_k)$, accept the trial; that is, set $\alpha_{k+1} = \alpha_{tr}$.
   (e) If $E(\alpha_{tr}) > E(\alpha_k)$, accept with probability $\mathcal{P} = \exp(-\Delta E/kT)$:
      i. Choose a uniform random number: $0 \leq r \leq 1$,
      ii. Let $\alpha_{k+1} = \begin{cases} \alpha_{tr}, & \text{if } \mathcal{P} \geq r \quad \text{(accept)}, \\ \alpha_k, & \text{if } \mathcal{P} < r \quad \text{(reject)}. \end{cases}$

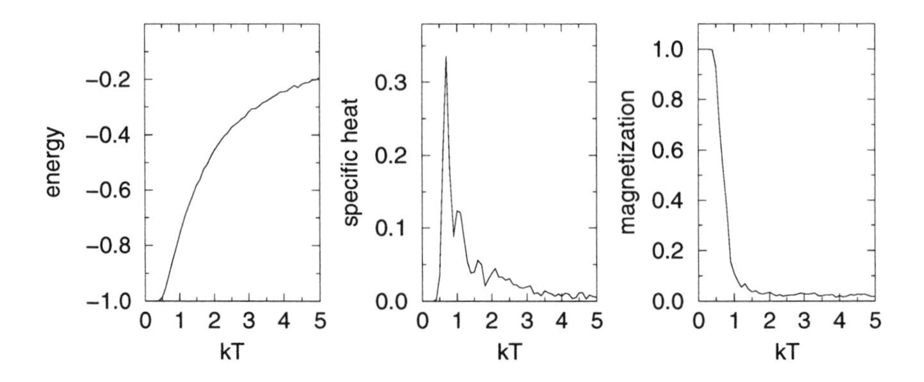

Fig. 22.3 Simulated results for the energy, specific heat, and magnetization of a 1-D lattice of 100 spins as a function of temperature.

How do you start? One possibility, clearly a good choice for high temperatures, is to start with random values of the spins. Another possibility, clearly a good choice for low temperatures, is to start with a ferromagnetic or antiferromagnetic configuration (for positive and negative J, respectively). In general, one tries to remove the importance of the starting configuration by letting the calculation "run a while" ($\simeq N$ rearrangements) before calculating averages. Then you repeat the calculation for different starting configurations, and take the average of the results.

22.6 IMPLEMENTATION, METROPOLIS ALGORITHM, ISING.F (.C)

1. Write a program that implements the Metropolis algorithm, that is, that produces a new configuration α_{k+1} from the present configuration α_k. (Alternatively, use the program supplied on the diskette or the Web.)

2. Make the key data structure in your program an array `s(1:N)` containing the values of s_i. For debugging, print out + and - to give the spins at each lattice point and the trial number.

3. The value for the exchange energy J fixes the scale for energy. Keep it fixed at $J = 1$ (or -1 for an antiferromagnet).

4. The thermal energy kT in units of J is the independent variable that your program should treat as an input parameter. Use $kT = 1$ for debugging.

5. Use periodic boundary conditions on your chain to minimize end effects. (This means that make the first and last spins the same.)

6. Try $N \simeq 20$ for debugging and larger values for production runs.
7. Use the printout to check that the system equilibrates for
   (a) A totally ordered initial configuration (cold start).
   (b) A random initial configuration (hot start).

Your cold start simulation should resemble Fig.22.2.

22.7 ASSESSMENT: APPROACH TO THERMAL EQUILIBRIUM

1. Watch a chain of N atoms attain thermal equilibrium when in contact with a heat bath. At high temperatures, or for small numbers of atoms, you should see large fluctuations, while at lower temperatures you should see smaller fluctuations.
2. The largest kT may be unstable as the system can absorb enough energy to flip all its spin. This is related to the fact that we have eliminated the magnetic field and in this way have no preferred direction to space. Introducing an external magnetic field B will stabilize the system but will also change the total energy and the analytic results.
3. Note how at thermal "equilibrium" the system is still quite dynamic with spins flipping all the time. It is the energy exchange that determines the thermodynamic properties.
4. You may well find that simulations at small kT (say that $kT \sim 0.1$ for $N = 200$) are slow to equilibrate. Higher kT values equilibrate faster yet have larger fluctuations.

22.8 ASSESSMENT: THERMODYNAMIC PROPERTIES

For a given spin configuration α_j, the energy and magnetization are given by

$$E_j = -J \sum_{i=1}^{N-1} s_i s_{i+1}, \tag{22.14}$$

$$\mathcal{M}_j = \sum_{i=1}^{N} s_i. \tag{22.15}$$

At high temperatures we expect a random assortment of spins and so a vanishing magnetization. At low temperature we expect $\mathcal{M}$ to approach $N/2$ as all the spins get aligned.

While the specific heat can be computed from the elementary definition

$$C = \frac{1}{n}\frac{dU}{dT}, \tag{22.16}$$

doing a numerical differentiation of a fluctuating variable is not expected to be accurate. A better way is to first calculate the fluctuations in energy occurring during a number of simulations

$$U_2 = \frac{1}{M}\sum_{t=1}^{M}(E_t)^2, \tag{22.17}$$

and then determine the specific heat from the energy fluctuations:

$$C = \frac{U_2 - (U)^2}{kT^2}. \tag{22.18}$$

1. Extend your program to calculate the total internal energy U (22.14) and the magnetization $\mathcal{M}$ (22.15) for the chain. Notice that you do not have to recalculate entire sums for each new configuration because only one spin changes.

2. Make sure you wait for your system to attain thermal equilibrium before you calculate thermodynamic quantities. (You can check that U is fluctuating about its average.) Your results should resemble those shown in Fig.22.3.

3. The large statistical fluctuations are reduced by running the simulation a number of times with different seeds and taking the average of the results.

4. The simulations you run for small N may be realistic but may not agree with statistical mechanics, which assumes $N \simeq \infty$ (you may assume that $N \simeq 2000$ is close to infinity). Check that agreement with the analytic results for the thermodynamic limit is better for large N than small N.

5. Check that the simulated thermodynamic quantities are independent of initial conditions (within statistical uncertainties). In practice, your cold and hot start results should agree.

6. Make a plot of the internal energy U as a function of kT and compare to the analytic result (22.9).

7. Make a plot of the magnetization $\mathcal{M}$ as a function of kT and compare to the analytic result. Does this agree with how you expect a heated magnet to behave?

8. Compute the fluctuations of the energy U_2 (22.17), and the specific heat C (22.18). Make a graph of your simulated specific heat compared to the analytic result (22.11).

22.9 EXPLORATION: BEYOND NEAREST NEIGHBORS

Extend the model so that the spin–spin interaction (22.6) extends to next-nearest neighbors as well as nearest neighbors. For the ferromagnetic case, this should lead to less fluctuation because we have increased the couplings among spins and thus an increased thermal inertia.

22.10 EXPLORATION: 2-D AND 3-D ISING MODELS

Extend the model so that the ferromagnetic spin–spin interaction (22.6) extends to nearest neighbors in two and then three dimensions. Continue using periodic boundary conditions and keep the number of particles small, at least to start [G&T 96].

1. Form a square lattice and place $\sqrt{N}$ spins on each side.
2. Examine the mean energy and magnetization of the system as it equilibrates.
3. Is the temperature dependence of the average energy qualitatively different from the 1-D model?
4. Print out the spin configurations for small N ($\simeq$ 16–25) and identify the domains.
5. Once your system appears to be behaving properly, calculate the heat capacity of the 2-D Ising model with the same technique used for the 1-D model. Use a total number of particles $100 \leq N \leq 1000$.
6. Look for a phase transition from an ordered to unordered configuration by examining the heat capacity as a function of temperature. It should diverge at the phase transition (you may get only a peak).

23

Functional Integration on Quantum Paths⊙

This optional chapter deals with Feynman's path integral formulation of quantum mechanics [F&H 65]. It is hardest material in this book. In recent times this path integral formulation has been applied to field theory calculations (quantum chromodynamics) and, in the process, has become a major consumer of the world's high-performance computer time. The calculations we present are based on those of other authors [Mann 83, MacK 85, M&N 87]. Different approaches and further references can be found in an article in *Computers in Physics* [Potv 93].

23.1 PROBLEM: RELATION OF QUANTUM TO CLASSICAL TRAJECTORIES

Consider a particle in a harmonic oscillator potential. We know from our nearly endless studies of oscillators that a classical particle moves back and forth periodically through the potential. Your **problem** is to show that the quantum wave function of this particle is a fluctuation about the classical trajectory.

23.2 THEORY: FEYNMAN'S SPACETIME PROPAGATION

An examination of the motion of quantum wave functions leads [L 96] to a form of Huygen's wavelet principle for propagating a free particle from

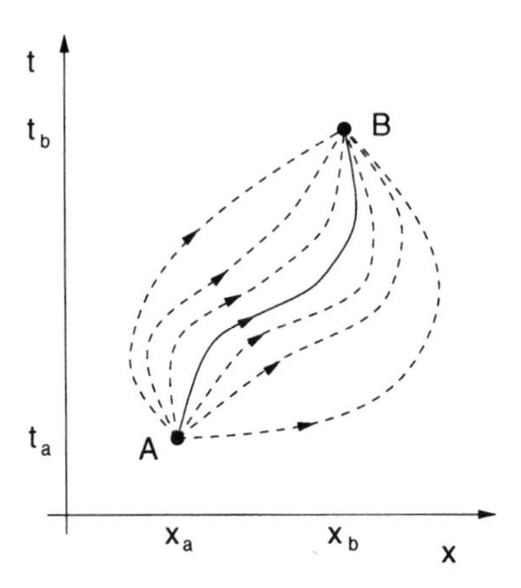

Fig. 23.1 A collection of paths connecting the initial and final spacetime points. The solid line represents the classical trajectory. A classical particle somehow "knows" ahead of time to travel along just that trajectory for which the action S is an extremum.

spacetime point (x_a,t_a) to point (x_b,t_b):

$$\psi(x_b,t_b) = \int dx_a G(x_b,t_b;x_a,t_a)\psi(x_a,t_a), \tag{23.1}$$

where the spacetime Green's function (propagator) is

$$G(x_b,t_b;x_a,t_a) = \sqrt{\frac{\mu}{2\pi i(t_b-t_a)}}\exp\left\{\frac{i\mu(x_b-x_a)^2}{2(t_b-t_a)}\right\}. \tag{23.2}$$

The idea is that each point on the wavefront emits a spherical wavelet that propagates forward in space and time. By interference with all the other wavelets, a new wavefront is created.

One way of interpreting (23.1) is illustrated in Fig. 23.1. We visualize the probability amplitude (wave function ψ) for the particle to be at b as the sum over all *paths* through spacetime originating at the time t_a. In this view, the statistical nature of quantum mechanics is implicit because different paths are possible.

Path integrals arose originally in Feynman's quest for a least-action principle of quantum mechanics in which classical mechanics occurs as the special case of quantum mechanics for vanishingly small values of $\hbar$. The basis is the *Hamilton's principle of least action:*

> *The most general motion of a physical particle moving along the classical trajectory $\bar{x}(t)$ from time t_a to t_b is along a path such that the action $S[\bar{x}(t)]$ is an extremum:*

$$\delta S[\bar{x}(t)] \;=\; S[\bar{x}(t)+\delta x(t)] - S[\bar{x}(t)] = 0, \tag{23.3}$$

where the paths are constrained to pass through the endpoints:

$$\delta(x_a) = \delta(x_b) = 0. \tag{23.4}$$

Here the action S is a line integral of the Lagrangian along the path:

$$S[\bar{x}(t)] = \int_{t_a}^{t_b} dt L\left[x(t), \frac{dx}{dt}(t)\right], \tag{23.5}$$

$$\text{where} \quad L\left[x(t), \frac{dx}{dt}(t)\right] = T\left[x(t), \frac{dx}{dt}(t)\right] - V[x(t)]. \tag{23.6}$$

Here T is the kinetic energy, V is the potential energy, and the square brackets in (23.5) indicate that S is a *functional*[1] of the function $x(t)$.

Feynman observed that because the classical action for a free ($V = 0$) particle has the value

$$S[x(t)] = \frac{\mu}{2}\left(\frac{dx}{dt}\right)^2 (t_b - t_a) = \frac{\mu}{2}\frac{(x_b - x_a)^2}{t_b - t_a}, \tag{23.7}$$

S is related to the free propagator (23.2) by

$$G(x_b, t_b; x_a, t_a) = \sqrt{\frac{\mu}{2\pi i(t_b - t_a)}} \exp\left\{iS[x(t)]/\hbar\right\}. \tag{23.8}$$

This led Feynman to a reformulation of quantum mechanics in which the propagator $G(b, a)$ for a particle making a transition from a to b is postulated to be the sum over all paths connecting a to b, with the phase of each path equal to its classical action:

$$G(b, a) = \sum_{\text{paths}} e^{iS[b,a]/\hbar}. \tag{23.9}$$

This sum over paths is called a *path integral,* in part because the classical action $S[b, a]$ is a line integral, and in part because sums and integrals are much the same.

The key connection between classical- and quantum mechanics is the realization that in units of $\hbar$, a classical action is a very large number, $S/\hbar \simeq \infty$. This means that even though all paths enter into the sum in (23.9) with a weight of equal magnitude, because S is a constant to first order in the variation of paths, (23.3), those paths adjacent to the classical trajectory $\bar{x}$, have phases that vary smoothly and relatively slowly. In contrast, those paths

[1] A *functional* is a number whose value depends on the complete behavior of some function and not just its behavior at one point. For example, the derivative $f'(x)$ depends on the value of f at x, yet the integral $\int_a^b dx f(x)$ depends on the entire function and is therefore a functional of f.

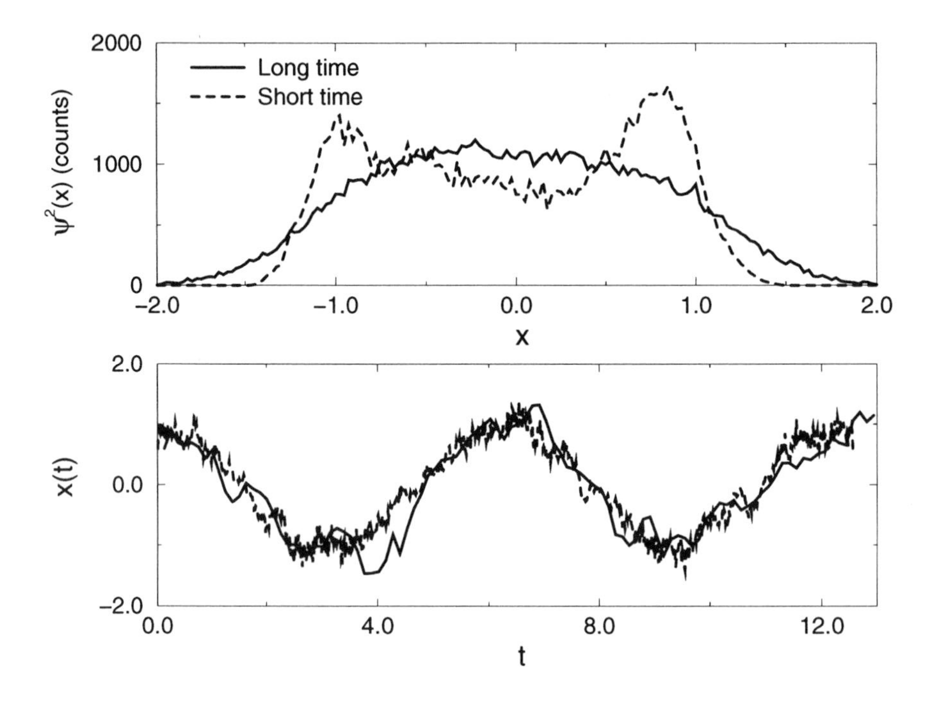

Fig. 23.2 The ground-state wave function of the harmonic oscillator as determined with a path-integral calculation. (*Upper*) The dashed curve is the wave function for a short time $t_b - t_a$ (twice the classical period) and the solid curve for a long time (20 times the classical period). The long time yields a wave function closer to the Gaussian form expected for the ground state. (*Lower*) The long- and short-time trajectories in spacetime used last in the solutions for the wave function. The oscillator has initial and final amplitudes of $x = 1$, $m = k = 1$, and, consequently, a period of $T = 2\pi$.

far from the classical trajectory enter with rapidly varying phases, and when many are included, they tend to cancel each other out. In the classical limit, $\hbar \to 0$, only the classical trajectory contributes and (23.9) becomes Hamilton's principle of least action! In Fig. 23.2 we make these abstract concepts more concrete by showing trajectories used in actual path-integral calculations.

23.3 METHOD, ANALYTIC: THE BOUND-STATE WAVE FUNCTION

The general eigenfunction expansion of the wave function leads to the eigenfunction expansion of the propagator:

$$G(x, t; x_0, t_0 = 0) = \sum_n \psi_n^*(x_0)\psi_n(x)e^{-iE_n t}. \tag{23.10}$$

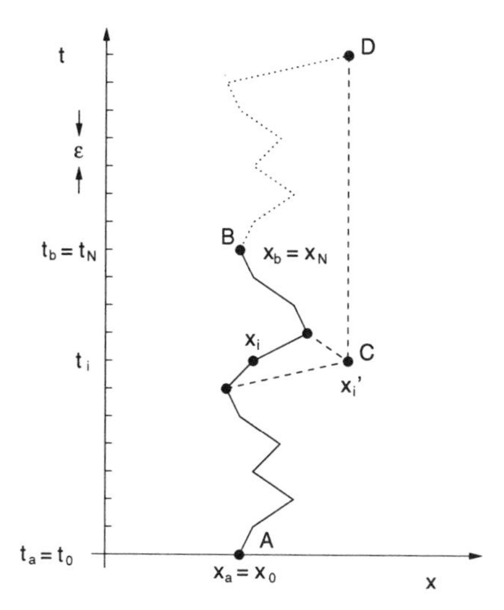

Fig. 23.3 A path through the spacetime lattice that starts and ends at $x = x_a = x_b$. The action is an integral over this path, while the *path integral* is a sum of integrals over all paths. The dotted path BD is a transposed replica of the path AC.

Here n labels the eigenvalues and we leave off the $\hbar$'s we included previously only for pedagogical purposes. We relate (23.10) to the bound-state wave function [recall that our **problem** is to calculate that] by being so bold as to evaluate (23.10) at a large negative imaginary time. If we wait a long imaginary time, the parts of ψ with higher energies decay away more quickly and leave only the ground state ψ_0. This gives us our final expression for the wave function:

$$\boxed{|\psi_0(x)|^2 = \lim_{\tau\to\infty} e^{E_0\tau} G(x, t = -i\tau; x_0 = x, 0).} \tag{23.11}$$

The $x = x_0 \equiv x_a$ appearing in (23.11) is the initial and final positions of the system at the time $t = t_0 \equiv 0$. While the general paths shown in Fig. 23.1 do not have to start and end at the same space position, this expression for $|\psi_0(x)|^2$ requires that the paths return to the same x value. This is as shown in Fig. 23.3.

Equation (23.11) provides a closed-form solution for the ground-state wave function directly in terms of the propagator G. If we evaluate the propagator using Feynman's postulates, we obtain a closed-form solution for the ground-state wave function.

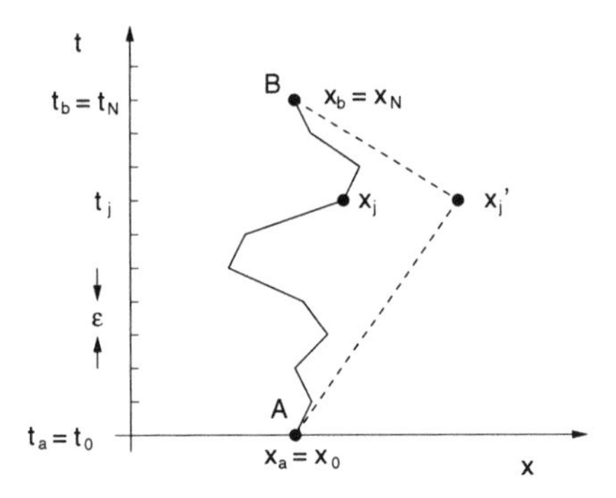

Fig. 23.4 Spacetime with time as a discrete variable. The dashed path joins the initial and final times in two equal time steps, the solid curve uses N steps each of size ε. The position of the curve at time t_j defines the position x_j.

23.4 METHOD, NUMERICAL: PATH INTEGRATION ON A LATTICE

Path integration on a lattice is based on the *composition law* for propagators:

$$G(x_b, t_b; x_a, t_a) = \int dx_j G(x_b, t_b; x_j, t_j) G(x_j, t_j; x_a, t_a), \quad (t_a < t_j, t_j < t_b). \tag{23.12}$$

For a free particle, substitution into (23.12) yields

$$\begin{aligned} G(x_b, t_b; x_a, t_a) &= \sqrt{\frac{\mu}{2\pi i(t_b - t_j)}} \sqrt{\frac{\mu}{2\pi i(t_j - t_a)}} \int dx_j e^{i(S[b,j]+S[j,a])} \\ &= \sqrt{\frac{\mu}{2\pi i(t_b - t_a)}} \int dx_j e^{iS[b,a]}, \end{aligned} \tag{23.13}$$

where we add the actions together because line integrals combine as

$$S[b,j] + S[j,a] = S[b,a]. \tag{23.14}$$

Equation (23.13) tells us that we combine propagators by multiplying and integrating.

In Fig. 23.4 we illustrate how we have "discretized" the time between a and b into N equal steps of size ε, and labeled them with the index j:

$$\varepsilon \stackrel{\text{def}}{=} \frac{t_b - t_a}{N}, \tag{23.15}$$

$$t_j = t_a + j\varepsilon, \quad (j = 0, N). \tag{23.16}$$

In (23.16) we give the fixed endpoints a discrete index even though they will not vary as we change paths. Next we draw the solid curve in Fig. 23.4 as a representation of a single path through spacetime. The position of the curve at time t_j defines the position x_j:

$$x_j \overset{\text{def}}{=} x(t_j). \tag{23.17}$$

Although only a crude numerical approximation, in drawing paths we assume that successive x_j values are connected by straight lines. By considering all curves that connect points a and b, the associated x_j values and t_j values map out a *lattice* in spacetime on which time and space are discrete. Presumably, the solution obtained on the lattice approaches the true solution in the limit of infinitesimal lattice spacing ε.

The lattice construct is convenient. We can replace time differentials dt by the discrete value $\Delta t \equiv \varepsilon = t_{j+1} - t_j$, and space differentials dx by $\Delta x = x_{j+1} - x_j$. In this way, if we use Euler's rule for the first derivative,[2] the velocity and action for one link on the lattice take the following simple forms:

$$\frac{dx_j}{dt} \simeq \frac{x_j - x_{j-1}}{t_j - t_{j-1}} = \frac{x_j - x_{j-1}}{\varepsilon}, \tag{23.18}$$

$$S_j \simeq L_j \Delta t \simeq \frac{1}{2}\mu \frac{(x_j - x_{j-1})^2}{\varepsilon} - V(x_j)\varepsilon, \tag{23.19}$$

where we assume that the Lagrangian is constant on each link of the lattice.

Next we generalize (23.12) and (23.13) for the N-linked path of Fig. 23.4:

$$G(x_b, t_b; x_a, t_a) = \int dx_1 \cdots dx_{N-1} e^{iS[b,a]}, \tag{23.20}$$

$$S[b,a] = \sum_{j=1}^{N} S_j. \tag{23.21}$$

We use here the expression (23.19) for the value of the action for each segment. Consequently, we see that even without a path variation, the integral over the *single* path in Fig. 23.4 is an N-term sum that becomes an infinite sum as the time step ε approaches zero.

Feynman's postulate means that we should sum over all paths connecting a to b to obtain the transition amplitude $G(b,a)$. This means that we must sum not only over the links in one path but *also* over all paths in order to produce the variation in paths that is required by Hamilton's principle. The sum is constrained such that paths must pass through a and b and cannot double back on themselves. This is the essence of *path integration.* Because

[2] Even though, of all the rules we have studied for the derivative, Euler's has the largest approximation error, it is usually used in lattice calculations because it makes the calculation so much simpler. Nonetheless, if the Lagrangian involves a second derivative, you may need the more precise central-difference method to avoid infinities.

we are integrating over functions as well as along paths, the technique is also known as *functional integration.*

23.5 METHOD: LATTICE COMPUTATION OF PROPAGATORS

The propagator $G(b, a)$ is the sum (23.9) over all paths connecting a to b, with each path weighted by the exponential of the action evaluated along that path:

$$G(x, t; x_0, t_0) = \sum\!\!\!\!\!\!\int dx_1 dx_2 \cdots dx_{N-1} e^{iS[x,x_0]}. \tag{23.22}$$

We take the total action to be the sum of the action for N links

$$S[x, x_0] = \sum_{j=1}^{N-1} S[x_{j+1}, x_j] \simeq \sum_{j=1}^{N-1} L\left(x_j, \frac{dx_j}{dt}\right) \varepsilon, \tag{23.23}$$

where $L(x_j, dx_j/dt)$ is the average value of the Lagrangian on link j (which corresponds to time $t = j\varepsilon$). To keep the computation simple, we assume that a local and velocity-independent potential $V(x)$.

Next we observe that in the expression (23.11) for the ground-state wave function, G is evaluated with a negative imaginary time. This means that t in the Lagrangian should change to $-i\tau$:

$$L\left(x, \frac{dx}{dt}\right) = T - V(x) \quad = \quad +\tfrac{1}{2} m \left(\frac{dx}{dt}\right)^2 - V(x), \tag{23.24}$$

$$L\left(x, \frac{dx}{-id\tau}\right) \quad = \quad -\tfrac{1}{2} m \left(\frac{dx}{d\tau}\right)^2 - V(x). \tag{23.25}$$

Yet this reversal in the sign of the kinetic energy term in $L(t = -\tau)$ means that now L equals the negative of the Hamiltonian evaluated at a real positive time $t = \tau$:

$$H\left(x, \frac{dx}{d\tau}\right) \quad = \quad \tfrac{1}{2} m \left(\frac{dx}{d\tau}\right)^2 + V(x) = E, \tag{23.26}$$

$$\Rightarrow \quad L\left(x, \frac{dx}{-id\tau}\right) \quad = \quad -H\left(x, \frac{dx}{d\tau}\right). \tag{23.27}$$

We in this way rewrite the t-path integral of L as a τ-path integral of H along a trajectory and thereby express the Green's function completely in terms of the Hamiltonian:

$$S[x_{j+1}, x_j] \quad = \quad \int_{t_j}^{t_{j+1}} L(x, t) dt = -i \int_{\tau_j}^{\tau_{j+1}} H(x, \tau) d\tau, \tag{23.28}$$

$$\Rightarrow G(x, -i\tau; x_0, 0) = \int dx_1 \cdots dx_{N-1} e^{-\int_0^\tau H(\tau) d\tau}, \tag{23.29}$$

where this line integral of H is over an entire trajectory.

To make the expressions simpler, we express the path integral in terms of an average energy of the particle on each link, $E_j = T_j + V_j$, and then sum over links to obtain the summed energy $\mathcal{E}$:

$$\int H(\tau) d\tau \simeq \sum_j \varepsilon E_j = \varepsilon \mathcal{E}(\{x_j\}), \tag{23.30}$$

$$\mathcal{E}(\{x_j\}) \stackrel{\text{def}}{=} \sum_{j=1}^{N} \left[\frac{m}{2} \left(\frac{x_j - x_{j-1}}{\varepsilon} \right)^2 + V \left(\frac{x_j + x_{j-1}}{2} \right) \right]. \tag{23.31}$$

In (23.31) we have approximated each path link as a *straight line*, we have used Euler's derivative rule $dx/dt \simeq \Delta x / \Delta t$ for the velocity on that link, and we have evaluated the potential at the midpoint of each link.[3]

We now substitute this expression for G into our solution (23.11) for the ground-state wave function, which, you may recall, requires that the initial and final points in space be the same, $x_0 = x$:

$$|\psi_0(x)|^2 = \lim_{\tau \to \infty} \frac{G(x, t = -i\tau, x_0 = x, t = 0)}{\int_{-\infty}^{\infty} dx G(x, t = -i\tau, x_0 = x, t = 0)} \tag{23.32}$$

$$= \frac{\int dx_1 \cdots dx_{N-1} \exp\left[-\int_0^\tau H d\tau'\right]}{\int dx dx_1 \cdots dx_{N-1} \exp\left[-\int_0^\tau H d\tau'\right]} \tag{23.33}$$

$$= \frac{1}{Z} \lim_{\tau \to \infty} \int dx_1 \cdots dx_{N-1} e^{-\varepsilon \mathcal{E}}, \tag{23.34}$$

$$Z = \lim_{\tau \to \infty} \int dx dx_1 \cdots dx_{N-1} e^{-\varepsilon \mathcal{E}}. \tag{23.35}$$

The similarity of these expressions to thermodynamics, even with a partition function Z, is no accident; by making the time parameter of quantum mechanics imaginary, we have converted the time-dependent Schrödinger equation to the heat-diffusion equation. This similarity leads to the integrand in (23.34) being called a probability function f or $\mathcal{P}$, as in statistical mechanics. The $\mathcal{P}$ is identified as a Boltzmann distribution function with a temperature proportional to the inverse of the time step:

$$\mathcal{P} = e^{-\varepsilon \mathcal{E}} = e^{-\mathcal{E}/k_B T}, \tag{23.36}$$

[3]We could also have used the linear average, $V \simeq [V(x_j) + V(x_{j-1})]/2$. For more precision we could use Simpson's rule for the potential, $V \simeq [V(x_j) + 4V((x_j + x_{j-1})/2) + V(x_{j-1})]/6$, and the central difference rule for the derivative, $dx_j/dt \simeq (x_{j+1} - x_{j-1})/(2\varepsilon)$.

$$\Rightarrow \quad k_B T \quad = \quad \frac{1}{\varepsilon} \equiv \frac{\hbar}{\varepsilon}. \tag{23.37}$$

Consequently, the $\varepsilon \to 0$ limit, which makes time continuous, is a "high-temperature" limit. The $\tau \to \infty$ limit, which is required to project out the ground-state wave function, means that we must integrate over a path that is long in imaginary time (i.e., long compared to a typical time $\hbar/\Delta E$).[4] Once the system equilibrates, the remaining thermal fluctuations will simulate Feynman's quantum fluctuations about the classical trajectory in spacetime. At last, the solution to our **problem**.

The evaluation of path integrals like (23.34) requires the integration of the Hamiltonian along each trajectory and the summation over all trajectories. We evaluate this path integral as the sum of 1 over all space and time points in our lattice, with the points chosen with a Monte Carlo technique so that they are random but weighted by the Boltzmann factor. We generate this weighted distribution with the Metropolis (simulated annealing) algorithm described in §22.5.

In general, Monte Carlo Green's function techniques work best if we start off with a good guess at the correct answer and have the algorithm calculate corrections to our guess. For the present problem this means that if we start off with a path close to the classical trajectory in spacetime, the algorithm may be expected to do a good job at simulating the quantum fluctuations about our initial guess. It may, therefore, not be very good at finding the classical trajectory.

As we have formulated our computation, we would pick a value of x and perform a rather lengthy computation of line integrals over all space and time to obtain the modulus of the wave function at this x. To obtain the wave function at another value of x, the entire simulation would have to be repeated from scratch. Rather than go through all that trouble again and again, there is a trick that permits us to compute the entire x dependence of the wave function in a single run. The trick is to insert in the probability integral (23.34) a delta function that fixes the initial position x_0, and then to integrate over all initial positions:

$$|\psi_0(x)|^2 \quad = \quad \frac{1}{Z}\int dx_1 \cdots dx_{N-1} e^{-\varepsilon\mathcal{E}(x,x_1,\ldots,x_N)} \tag{23.38}$$

$$= \quad \frac{1}{Z}\int dx_0 \cdots dx_{N-1}\delta(x-x_0) e^{-\varepsilon\mathcal{E}(x_0,x_1,\ldots,x_N)}, \tag{23.39}$$

where the limit $\tau \to \infty$ is understood.

Equation (23.39) expresses the wave function as an average of a delta func-

[4]Just like our simulation of the Ising model, we must wait around a really long real time while the algorithm repeats in order for our system to equilibrate (find the classical trajectory).

tion over all paths, a procedure that might appear totally inappropriate for numerical computation because there is tremendous error in representing a singular function on a finite-word-length computer. Yet when we simulate the sum over all paths with (23.39), there will always be some x value for which the integral is nonzero, and we need to only accumulate the solution for various (discrete) x values to determine $|\psi_0(x)|^2$ for all x.

To understand how this works, consider path AB in Fig. 23.3 for which we have calculated the summed energy. We next let one point on the chain jump to point C (which changes two links) to form a new path. If we replicate section AC and use it as the extension AD to form the top path, we see that the path CBD has the same summed energy (action) as path ACB and in this way can be used to determine $|\psi(x'_j)|^2$. Accordingly, once the system is equilibrated, we determine new values of the wave function at new locations x'_j by flipping links to new values and calculating new actions. The more frequently some x_j is accepted, the greater is the wave function at that point.

23.6 IMPLEMENTATION: LATTICE PROGRAMMING, QMC.F (.C)

The program on the diskette and the Web evaluates the integral (23.9) by finding the average of the integrand $\delta(x_0 - x)$ with paths distributed according to the weighting function $\exp[-\varepsilon\mathcal{E}(x_0, x_1, \ldots, x_N)]$. The physics enters via (23.41), the calculation of the summed energy $\mathcal{E}(x_0, x_1, \ldots, x_N)$. We evaluate the action integral for the harmonic oscillator potential

$$V(x) = \frac{1}{2}x^2, \tag{23.40}$$

and for a particle of mass $\mu = 1$. A convenient set of natural units is to measure lengths in $\sqrt{1/\mu\omega} \equiv \sqrt{\hbar/\mu\omega} = 1$, and times in $1/\omega = 1$. Correspondingly, the oscillator has a period $T = 2\pi$.

Fig. 23.2 shows results from an application of the Metropolis algorithm. In this computation we start off with an initial path close to the classical trajectory and then examine one-half million variations about this path. All paths are constrained to begin and end at $x = 1$ (which turns out to be somewhat less than the amplitude of the classical oscillation).

When the time difference $t_b - t_a$ equals a short time like $2T$, the system does not have enough time to equilibrate to its ground state and, as we see in the top of Fig. 23.2, the wave function looks like the probability distribution of an excited state (nearly classical, in fact, with the probability highest for the particle to be near its turning points where its velocity vanishes). However, when the time difference $t_b - t_a$ equals the long time $20T$, the system has enough time to decay to its ground state and the wave function looks like the Gaussian probability distribution of the ground state. In either case we see in the bottom part of Fig. 23.2 that the trajectory through spacetime fluctuates about the classical trajectory. This fluctuation is a consequence of the Metropolis algorithm occasionally going uphill in its search; if you modify the program so that searches go only downhill, the spacetime trajectory would be a very smooth trigonometric function (the classical trajectory), but the wave function, which is a measure of the fluctuations about the classical trajectory, would vanish!

The steps of the calculation are

1. Construct a time grid of N time steps of length ε as in Fig. 23.3. Start at $t = 0$ and extend to time $\tau = N\varepsilon$ [this means N time intervals and $(N+1)$ lattice points in time]. Note that time always increases monotonically along a path.

2. Construct a space grid of M points separated by steps of size δ. Use a range of x values several time larger than the characteristic size or range of the potential being used, and start with $M \simeq N$.

3. When calculating the wave function, any x or t value falling between lattice points should be assigned to the closest lattice point.

4. Associate a position x_j with each time τ_j, subject to the boundary conditions that the initial and final positions always remain the same, $x_N = x_0 = x$.

5. Choose an *arbitrary path* of straight-line links connecting the lattice points. For the most realistic simulation, it may be best to start with something close to the classical trajectory as otherwise the simple numerical procedures may not converge. Note that the x values for the links of the path may have values that increase, decrease, or remain unchanged (in contrast to time, which always increases).

6. Evaluate the energy $\mathcal{E}(x_0, x_1, \ldots, x_N)$ by summing the kinetic and potential energies for each link of the path starting at $j = 0$:

$$\mathcal{E}(x_0, x_1, \ldots, x_N) \simeq \sum_{j=1}^{N} \left[\frac{m}{2} \left(\frac{x_j - x_{j-1}}{\varepsilon} \right)^2 + V\left(\frac{x_j + x_{j-1}}{2} \right) \right]. \tag{23.41}$$

7. Begin the first of a sequence of repetitive steps in which a random position x_j associated with time t_j is changed to the position x'_j. As shown by point C in Fig. 23.3, this changes *two* links in the path.

8. For the coordinate that gets changed, weigh the change with the Boltzmann distribution (23.36) by using the Metropolis algorithm.

9. For each lattice point establish a running sum to represent the value of the wave function squared at that point.

10. After each single-link change (or decision not to change), increase the running sum for the new x value by 1. After a sufficiently long running time, the sum divided by the number of steps is the simulated value for $|\psi(x_j)|^2$ at each lattice point x_j.

11. Repeat the entire link-changing simulation using a different seed for the Metropolis algorithm. The average wave function from a number of intermediate-length runs should be better than that from one very long run.

23.7 ASSESSMENT AND EXPLORATION

1. For a more continuous picture of the wave function, make the x lattice spacing smaller; for a more precise value of the wave function at any particular lattice site, sample more points (run longer) and use a smaller time step ε.

2. Because there are no nodes in a ground-state wave function, you can ignore the phase and assume

$$\psi(x) = \sqrt{\psi^2(x)}. \tag{23.42}$$

You then can estimate the energy via

$$E = \frac{\omega}{2} \int_{-\infty}^{+\infty} \psi^*(x) \left(-\frac{d^2}{dx^2} + x^2 \right) \psi(x) dx, \tag{23.43}$$

where the space derivative is evaluated numerically.

3. Modify your program so that the search always goes downhill, that is, to a state of lower summed energy. Verify that this leads to the classical trajectory but zero wave function.

4. Test the wave function computation for the gravitational potential

$$\begin{aligned} V(x) &= mgx, \qquad (23.44) \\ x(t) &= x_0 + v_0 t + \tfrac{1}{2} g t^2. \qquad (23.45) \end{aligned}$$

You may want to set the initial positions to be close to the classical trajectory to ensure convergence.

24

Fractals

24.1 PROBLEM : FRACTALS

It is fairly common in nature to notice objects that do not have well-defined geometric shapes, but appear to be constructed according to some simple mathematical rule. Examples are to be found in plants, sea shells, polymers, thin films, colloids, and aerosols. Often these structures have unusual, but pleasing, shapes.

If we describe a line as having dimension 1, a square as having dimension 2, and a cube as having dimension 3, then we will see that these unusual figures have a dimension that is a fraction. Benoit Mandelbrot, who first studied these fractional-dimension figures with the supercomputers at IBM Research, gave them the name *fractal* [Mand 82]. Some geometric objects, such as the Koch curves, are exact fractals with the same dimension for all their parts. Other objects, such as bifurcation curves, are statistical fractals in which the dimension can be defined only locally or on the average.

We will not study the scientific theories that lead to fractal geometry, but rather will look at how some simple models and rules produce fractals. To the extent that these models generate structures like those occurring in nature, it is reasonable to assume that the natural processes must be following similar rules. The rules, presumably, arise from the basic physics of the process. For example, if we look at the bifurcation plot of the logistics map, Fig. 13.2 in Chapter 13, we see a self-similarity of structure that is characteristic of fractals; in this case we know the structure arises from the equation we used for the map. Detailed applications of fractals in physics can be found in many literature sources [Arm 91, E&P 88, Sand 94, PhT 88].

24.2 THEORY: FRACTIONAL DIMENSION

Consider an abstract "object" such as the density of charge within an atom. There are an infinite number of ways to measure the "size" of this object, for example; each moment of the distribution provides a measure of the size, and there are infinite numbers of moments. Likewise, when we deal with complicated objects that have fractional dimensions, there are different definitions of dimension, and each may give a somewhat different answer. In addition, the fractal dimension is often defined by using a measuring box whose size approaches zero. In realistic applications there may be numerical difficulties in approaching such a limit, and for this reason, too, the exact value of the dimension for complicated empirical objects may not be absolutely well defined.

Our first definition of fractional dimension d_f (or *Hausdorf-Besicovitch* dimension D_H) is based on our knowledge that a line has dimension 1; a triangle, dimension 2, and a cube, dimension 3. It seems perfectly reasonable to ask if there is some mathematical formula, that agrees with our experience for regular objects, for determining the fractal dimension of an object. For simplicity, let us consider objects that have the same length L on each side, as do equilateral triangles and squares. We postulate that the dimension of an object is determined by the dependence of its mass upon its length:

$$M(L) \stackrel{\text{def}}{=} \mathcal{A}L^{d_f}. \tag{24.1}$$

Here $\mathcal{A}$ is a constant and the power d_f is the *fractal dimension.* As you may verify, this rule works with the regular figures of our experience, so it must make some sense. Yet we will see that when we apply it to some unusual objects, this rule produces fractional values for d_f.

24.3 PROBLEM 1: THE SIERPIŃSKI GASKET

We generate our first fractal, shown in Fig. 24.1, by playing a game in which you pick points and place dots on them. Here are the rules (which you may try out in the margins of this book):

1. Draw an equilateral triangle with vertices and coordinates:

$$\text{Vertex 1}: \quad (a_1, b_1), \tag{24.2}$$
$$\text{Vertex 2}: \quad (a_2, b_2), \tag{24.3}$$
$$\text{Vertex 3}: \quad (a_3, b_3). \tag{24.4}$$

2. Place a dot at an arbitrary point $P = (x_0, y_0)$ within the triangle.

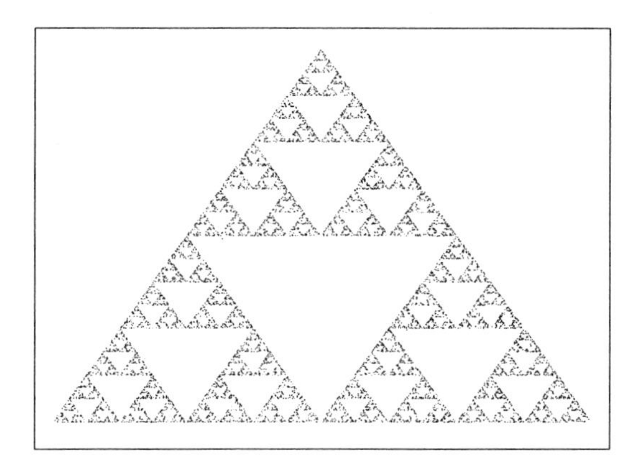

Fig. 24.1 A Sierpiński gasket containing 15,000 points. This pattern has been found in the tile work of medieval monasteries and on the shells of sea animals. Each filled part of this figure is self-similar.

3. Find the next point by selecting randomly the integer 1, 2, or 3:

   (a) If 1, place a dot halfway between P and vertex 1.

   (b) If 2, place a dot halfway between P and vertex 2.

   (c) If 3, place a dot halfway between P and vertex 3.

4. Keep repeating the process, using the last dot as the new P.

Mathematically, the coordinates of successive points are given by the formula

$$(x_{k+1}, y_{k+1}) = \frac{(x_k, y_k) + (a_n, b_n)}{2}, \tag{24.5}$$

$$n = \text{Integer}(1 + 3r_i), \tag{24.6}$$

where r_i is a random number between 0 and 1, and where the *Integer* function outputs the closest integer smaller than, or equal to, the argument. After 15,000 points, you should obtain a collection of dots like Fig. 24.1.

24.4 IMPLEMENTATION: SIERPIN.C

Write a program to produce a Sierpiński gasket. Determine empirically the fractal dimension of your figure. Assume that each dot has mass 1 and that $\rho = CL^{\alpha}$. You can have the computer do the counting by defining an array *box* of all 0 values and then change a 0 to a 1 when a dot is placed there.

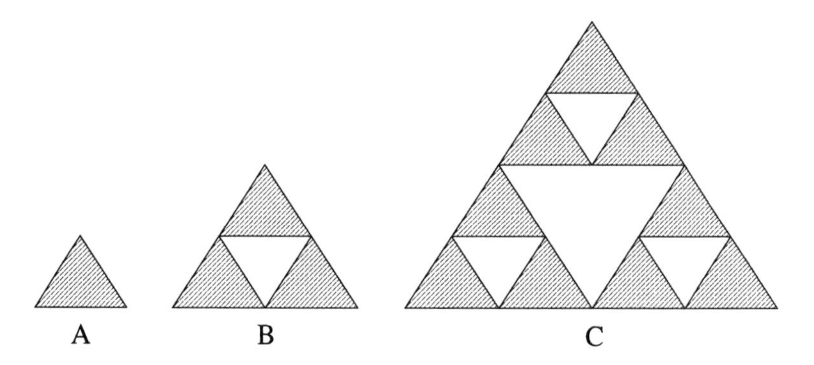

Fig. 24.2 Another construction of the Sierpinski gasket. Here an inverted equilateral triangle is repeatedly removed from the center of all filled equilateral triangles, and then the figure is scaled up so each filled triangle has side r.

24.5 ASSESSMENT: DETERMINING A FRACTAL DIMENSION

The topology of Fig. 24.1 was first analyzed by the Polish mathematician Sierpiński. Observe that there is the the same structure in a small region as there is in the entire figure. In other words, if the figure were infinitely dense, any part of the figure could be scaled up in size and will be similar to the whole. This property is called *self-similarity.*

As shown in Fig. 24.2, we construct a similar object by removing an inverted equilateral triangle from the center of all filled equilateral triangles to create the next figure. We then repeat the process ad infinitum, scaling up the triangles so each one has side $r = 1$ after each step.

To see what is unusual about this type of object, we look at how its density (mass/area) changes with size. Assume that each triangle has mass m, and assign unit density to the single triangle:

$$\rho(L = r) = \frac{M = m}{r^2} \stackrel{\text{def}}{=} 1 \qquad \text{(Fig. 24.2A).} \tag{24.7}$$

Next, for the equilateral triangle with side $L = 2$, the density is

$$\rho(L = 2r) = \frac{(M = 3m)}{(2r)^2} = \frac{3}{4}\frac{m}{r^2} = \frac{3}{4} \qquad \text{(Fig. 24.2B).} \tag{24.8}$$

We see that the extra white space in Fig. 24.2B leads to a density that is $\frac{3}{4}$ that of the previous stage. For the structure in Fig. 24.2C, we obtain

$$\rho(L = 4r) = \frac{(M = 9m)}{(4r)^2} = \left(\frac{3}{4}\right)^2 \frac{m}{r^2} = \left(\frac{3}{4}\right)^2 \qquad \text{(Fig. 24.2C).} \tag{24.9}$$

We see that as we continue the construction process, the density of each new structure is $\frac{3}{4}$ that of the previous one. This is unusual. For ordinary objects the density is an *intensive* quantity independent of the size of an object. For this strange object, there is a power-law dependence of the density on the size of the object

$$\rho = CL^{\alpha}, \tag{24.10}$$

where C is some constant. To determine the power α, we plot the logarithm of the density ρ versus the logarithm of the length L for successive structures. We obtain a straight line with slope:

$$\alpha = \frac{\Delta \log \rho(L)}{\Delta \log L} = \frac{\log 1 - \log(3/4)}{\log 1 - \log 2} = \frac{\log 3}{\log 2} - 2 \simeq -0.41504. \tag{24.11}$$

This means that

$$\rho \propto \frac{1}{L^{0.415}}. \tag{24.12}$$

As is evident in Fig. 24.2, as the gasket gets larger and larger, it contains more and more open space. So even though its mass approaches infinity, its density approaches zero! And since a two-dimensional figure like a solid triangle has a constant density as its length increases, a 2-D figure would have $\alpha = 0$. Since the Sierpiński gasket has a nonzero α value, it is not a 2-D figure!

The *fractal dimension* d_f is defined by (24.1):

$$M(L) \stackrel{\text{def}}{=} AL^{d_f} = \rho(L)L^2 = AL^{2+\alpha}. \tag{24.13}$$

We know α from (24.11), and so:

$$d_f = \alpha + 2 = \frac{\log 3}{\log 2} \simeq 1.58496. \tag{24.14}$$

We see that the Sierpiński gasket has a dimension between that of a 1-D line and a 2-D triangle; that is, it has a fractional dimension.

24.6 PROBLEM 2: HOW TO GROW BEAUTIFUL PLANTS

It seems paradoxical that natural processes that are subject to various chance influences can produce objects of high regularity. For example, it is hard to believe that something as beautiful and symmetric as the fern in Fig. 24.3 has random elements in it. Nonetheless, there is a clue here in that much of the fern's beauty arises from the similarity of each part to the whole. In addition, even though each fern in nature is somewhat different from all other ferns, they are all quite similar. Clearly, we should think fractal. Your **problem** is to be discover if a simple, yet random, algorithm can draw ferns. If the algorithm produces objects that resemble ferns, then, presumably, you have

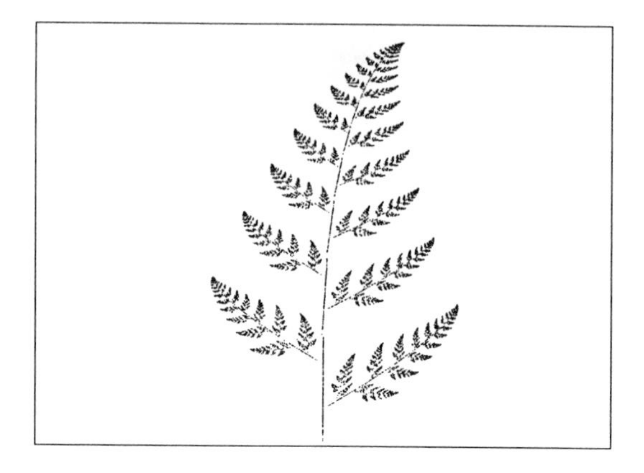

Fig. 24.3 A fern after 30,000 iterations of the algorithm (24.18). If you enlarge this, you will see that each frond has similar structure.

uncovered the mathematics of the shape of ferns.

24.7 THEORY: SELF-AFFINE CONNECTIONS

In equation (24.5), which defines mathematically how a Sierpiński gasket is constructed, a factor of $\frac{1}{2}$ multiplies the x_n. This factor is called a *scaling factor*. We now consider a more general transformation of a point $P = (x, y)$ into another point $P' = (x', y')$:

$$(x', y') = s(x, y). \tag{24.15}$$

This is called *scaling*. If the scale factor $s > 0$, an amplification is obtained, while if $s < 0$, a reduction is obtained. In our definition of the Sierpiński gasket (24.5), we also added in a constant a_n. This is a *translation operation*. In general, it has the form

$$(x', y') = (x, y) + (a_x, a_y). \tag{24.16}$$

Another operation, which was not used in the Sierpiński gasket, is a *rotation* by angle θ:

$$x' = x\cos\theta - y\sin\theta, \qquad y' = x\sin\theta + y\cos\theta. \tag{24.17}$$

The entire set of transformations, scalings, rotations, and translations, define an *affine transformation* ("affine" denotes a close relation between successive points). The transformation is still considered affine even if it is a more general linear transformation with the coefficients not all related to one θ (in that case, we can have contractions and reflections). What is important is that the object created with these rules turn out to be self-similar; each step leads to new parts of the object that bear the same relation to the ancestor parts as did the ancestors to theirs. This is what makes the object look similar at all scales.

24.8 IMPLEMENTATION: BARNSLEY'S FERN, FERN.C

We apply an affine transformation to obtain Barnsley's Fern. We extend the dots game by selecting new points using rules given by chance:

$$(x, y)_{n+1} = \begin{cases} (0.5, 0.27y_n), & \text{with } 2\% \text{ probability}, \\ (-0.139x_n + 0.263y_n + 0.57 \\ \quad 0.246x_n + 0.224y_n - 0.036), & \text{with } 15\% \text{ probability}, \\ (0.17x_n - 0.215y_n + 0.408 \\ \quad 0.222x_n + 0.176y_n + 0.0893), & \text{with } 13\% \text{ probability}, \\ (0.781x_n + 0.034y_n + 0.1075 \\ \quad -0.032x_n + 0.739y_n + 0.27), & \text{with } 70\% \text{ probability}. \end{cases} \tag{24.18}$$

To select a transformation with probability $\mathcal{P}$, we select a uniform random number r in the interval $[0, 1]$ and perform the transformation if r is in a range proportional to $\mathcal{P}$:

$$\mathcal{P} = \begin{cases} 2\%, & \text{if } r < 0.02, \\ 15\%, & \text{if } 0.02 \leq r \leq 0.17, \\ 13\%, & \text{if } 0.17 < r \leq 0.3, \\ 70\%, & \text{if } 0.3 < r < 1. \end{cases} \tag{24.19}$$

The rules (24.18) and (24.19) can be combined into one:

$$(x, y)_{n+1} = \begin{cases} (0.5, 0.27y_n), & \text{for } r < 0.02, \\ (-0.139x_n + 0.263y_n + 0.57 \\ \quad 0.246x_n + 0.224y_n - 0.036), & \text{for } 0.02 \leq r \leq 0.17, \\ (0.17x_n - 0.215y_n + 0.408 \\ \quad 0.222x_n + 0.176y_n + 0.0893), & \text{for } 0.17 < r \leq 0.3, \\ (0.781x_n + 0.034y_n + 0.1075, \\ \quad -0.032x_n + 0.739y_n + 0.27), & \text{for } 0.3 < r < 1. \end{cases} \tag{24.20}$$

Although the form (24.18) makes the basic idea clearer, the form (24.20) is easier to program.

The initial point in Barnsley's fern in Fig. 24.3 is taken as $(x_1, y_1) = (0.5, 0.0)$, and the points generated by repeated iterations are plotted. One difference from the Sierpiński gasket is that the entire fern is not self-similar, as you can see by noting how different are the stems and the fronds. Nevertheless, the stem can be viewed as a compressed copy of a frond, and the fractal obtained with (24.18) is still *self-affine* (yet with a fractal dimension

Fig. 24.4 A fractal tree created with the simple algorithm (24.21).

that varies for different parts of the figure).

24.9 EXPLORATION: SELF-AFFINITY IN TREES

Now that you know how to grow ferns, look up and notice the regularity in trees (such as Fig. 24.4). Can it be that this, also, arises from a self-affine structure?

Write a program similar to the one for the fern, with the following self-affine transformation:

$$(x_{n+1}, y_{n+1}) = \begin{cases} (0.05x_n, 0.6y_n), & 10\% \text{ probability}, \\ (0.05x_n, -0.5y_n + 1.0), & 10\% \text{ probability}, \\ (0.46x_n - 0.15y_n, 0.39x_n + 0.38y_n + 0.6), & 20\% \text{ probability}, \\ (0.47x_n - 0.15y_n, 0.17x_n + 0.42y_n + 1.1), & 20\% \text{ probability}, \\ (0.43x_n + 0.28y_n, -0.25x_n + 0.45y_n + 1.0), & 20\% \text{ probability}, \\ (0.42x_n + 0.26y_n, -0.35x_n + 0.31y_n + 0.7), & 20\% \text{ probability}. \end{cases} \tag{24.21}$$

Start your program at $(x_1, y_1) = (0.5, 0.0)$.

24.10 IMPLEMENTATION: NICE TREES, TREE.C

24.11 PROBLEM 3: BALLISTIC DEPOSITION

There are a number of physical processes in which particles are deposited on a surface to form a film. Because the particles are evaporated thermally from a hot filament, there is randomness in the emission process. But the films produced by deposition regularly turn out to have well-defined structures. Your **problem** is to simulate this growth process on the computer.

24.12 METHOD

The idea of simulating random depositions began with [Vold 59], who used tables of random numbers to simulate the sedimentation of moist spheres in hydrocarbons. We shall examine a method of simulation [Fer 90] which results in the deposition shown in Fig. 24.5.

Consider particles falling onto and sticking to a horizontal line of length L. For simplicity, we assume that there are 200 sites for particles on the line. All particles start from the same height, but to simulate their different velocities, we assume that they start at random distances from the left side of the line.

The simulation consists of generating uniform random sites between 0 and L, and having the particle stick to the site on which it lands. Because a realistic situation may have columns of aggregates of different heights, the particle may be stopped before it makes it to the line, or may bounce around until it falls into a hole. We therefore assume that if the column height at which the particle lands is greater than that of both its neighbors, it will add to that height. If the particle lands in a hole, or if there is an adjacent hole, it will fill up the hole. We speed up the simulation by setting the height of this hole equal to the maximum of its neighbors:

1. Choose a random site r.

2. Let the array h_r be the height of the column at site r.

3. Make the decision:

$$h_r = \begin{cases} h_r + 1, & \text{if } h_r \geq h_{r-1}, h_r > h_{r+1}, \\ \max[h_{r-1}, h_{r+1}], & \text{if } h_r < h_{r-1}, h_r < h_{r+1}. \end{cases} \tag{24.22}$$

The results of a simulation of this sort are shown in Fig. 24.5. We can see that there are several empty regions scattered throughout the line. This is an indication of the statistical nature of the process while the film is growing. Simulations by Fereydoon found that the average height increases linearly with time and that the surface is a fractal.

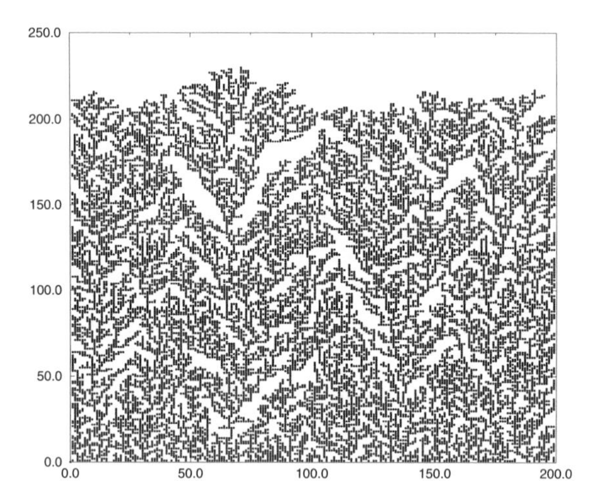

Fig. 24.5 A simulation of the ballistic deposition of 20,000 particles on a substrate. The top is the final surface, and the open areas arise from the method of printout.

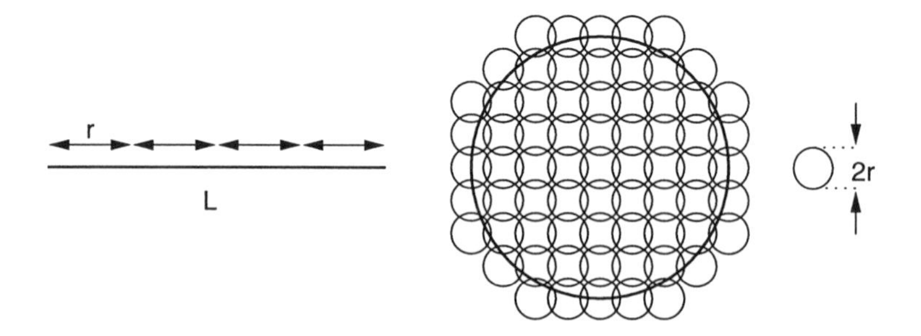

Fig. 24.6 Examples of "box" counting for a 1-D and 2-D figure.

24.13 IMPLEMENTATION: BALLISTIC DEPOSITION, FILM.C

24.14 PROBLEM 4: LENGTH OF THE COASTLINE OF BRITAIN

Mandelbrot asked the classic question "What is the length of the coastline of Britain?" While this is an interesting mathematical question, it is even more interesting that he was able to answer it [Mand 82]. The answer lends another dimension to our study of fractals, namely, using box counting to determine fractal dimension. As illustrated on the left in Fig. 24.6, consider a line of length L broken up into segments of length r. The number of segments or "boxes" needed to cover the line is related to the size r of the box by

$$N = \frac{L}{r} \propto \frac{1}{r}. \tag{24.23}$$

Another definition of fractional dimension is the power of r in this expression as $r \to 0$. In our example, it tells us that the line has dimension $d_f = 1$.

If, as illustrated on the right of Fig. 24.6, we now ask how many little circles of radius r it takes to *cover* or fill a circle of area A, we would find

$$N = \lim_{r \to 0} \frac{A}{\pi r^2} \quad \Rightarrow \quad d_f = 2, \tag{24.24}$$

as expected. Likewise, counting the number of little spheres or cubes that can be packed within a large sphere tells us that a sphere has dimension $d_f = 3$. In general, if it takes N "little" spheres or cubes of side $r \to 0$ to cover some object, then the fractal dimension d_f can be deduced from

$$N \propto \left(\frac{1}{r}\right)^{d_f} \propto s^{d_f} \quad (r \to 0), \tag{24.25}$$

$$d_f = \lim_{r \to 0} \frac{\log N(r)}{\log s}. \tag{24.26}$$

Here $s \propto 1/r$ is called the *scale*, and as we make the boxes smaller and smaller, we make the scale higher and higher.

24.15 MODEL: THE COAST AS A FRACTAL

The length of the coastline of an island is the perimeter of that island. While the concept of perimeter is clear for regular geometric figures, defining the perimeter for an irregular object such as an island is more challenging. We assume that the perimeter of an object is proportional to the number of boxes needed to cover the object. For our model, we assume that the coastline is a fractal (some fractals do in fact look very much like coastlines). We determine

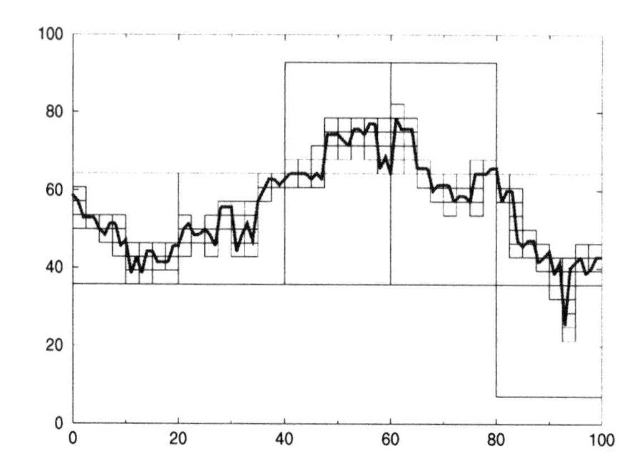

Fig. 24.7 Example of box counting for a coastline. Notice how the coastline is being covered by boxes of two different sizes (scales).

the fractal dimension d_f of the coastline; then for any scale s at which we made the determination, the perimeter P is given by

$$P \propto s^{d_f(s)}. \tag{24.27}$$

We then get our final answer by taking the limit as $s \to \infty$.

24.16 METHOD: BOX COUNTING

Rather than ruin your eyes with a geographic map, we use a mathematical one. Specifically, with a little imagination you will see that the top portion of the graph in Fig. 24.5 looks like a natural coastline. Determine d_f by covering this figure, or one you have generated, with a semitransparent piece of graph paper,[1] and counting the number of boxes containing any part of the coastline. We illustrate this in Fig. 24.7.

1. Print out your coastline graph with the same physical dimensions for the vertical and horizontal scales.

2. The vertical height in our printout was 17 cm. This sets the scale of the graph as 1:17, or $s = 17$.

[1] Yes, we are suggesting a painfully analog technique based on the theory that trauma leaves a lasting impression. If you prefer, you can store your output as a matrix of 1 and 0 values and let the computer do the counting.

3. The largest boxes on our graph paper were 1 cm × 1 cm. We found that the coastline passed through $N = 24$ of these large boxes (i.e., 24 large boxes covered the coastline for $s = 17$).

4. The next smaller boxes on our graph paper were 0.5 cm × 0.5 cm. We found that 51 smaller boxes covered the coastline for a scale of $s = 34$.

5. The smallest boxes on our graph paper were 1 mm × 1 mm. We found that 406 smallest boxes covered the coastline for a scale of $s = 170$.

6. Equation (24.26) tells us that as the box sizes get smaller and smaller, we should have

$$\log N \simeq \log A + d_f \log s, \tag{24.28}$$

$$\Rightarrow \quad d_f \simeq \frac{\Delta \log N}{\Delta \log s} = \frac{\log N_2 - \log N_1}{\log s_2 - \log s_1} = \frac{\log(N_2/N_1)}{\log(s_2/s_1)}. \tag{24.29}$$

Clearly, only the relative scales matter because the proportionality constants cancel out in the ratio. When you plot $\log N$ versus $\log s$, you should obtain a straight line (the third point verifies that it is a line). In our example, we found a slope $d_f = 1.23$.

As is obvious, the perimeter of the coastline is proportional to the number of boxes we need to cover it, and so

$$P \propto N \propto s^{1.23}. \tag{24.30}$$

If we keep making the boxes smaller and smaller so that we are looking at the coastline at higher and higher scale, *and* if the coastline is a fractal with self-similarity at all levels, then the scale s keeps getting larger and larger with no limits (or at least until we get down to some quantum limits). This mean

$$P \propto \lim_{s\to\infty} s^{1.23} = \infty. \tag{24.31}$$

We (well, actually Mandelbrot) conclude that, in spite of being only a small island, the coastline of Britain is, indeed, infinite.

24.17 PROBLEM 5: CORRELATED GROWTH IN FORESTS AND FILMS

It is an empirical fact that in nature there is an increased likelihood for a plant to grow if there is another one nearby. (This is also valid for the "growing" of surface films.) We see an example of this *correlation* in Fig. 24.8. Your **problem** is to include correlations in a simulation.

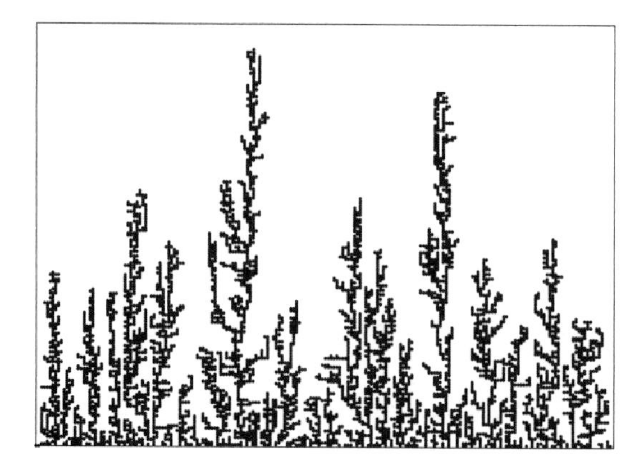

Fig. 24.8 A scene as might be seen in the undergrowth of a forest or in a correlated ballistic deposition.

24.18 METHOD: CORRELATED BALLISTIC DEPOSITION

A variation of the ballistic deposition algorithm, known as *correlated ballistic deposition*, simulates mineral deposition onto substrates in which dendrites form [Tait 90]. We extend this algorithm to include the likelihood that a freshly deposited particle will attract another particle. As illustrated in Fig. 24.9, we assume that the probability of sticking $\mathcal{P}$ depends on the distance d that the added particle is from the last one:

$$\mathcal{P} = cd^{\eta}. \tag{24.32}$$

Here η is a parameter and c is a constant that sets the probability scale.[2] For our implementation we choose $\eta = -2$, which means that there is an inverse square attraction between the particles (less probable as they get farther apart).

As in our study of uncorrelated deposition, a uniform random number in the interval $[0, L]$ determines the column in which the particle is deposited. We use the same rules about the heights as before, but now a second random number is used in conjunction with (24.32) to decide if the sticks. For example, if the computed probability is 0.6 and if $r < 0.6$, the particle is accepted (sticks); if $r > 0.6$, the particle is rejected.

[2]The absolute probability, of course, must be less than one, but it is nice to choose c so that the relative probabilities produce a graph with easily seen variations.

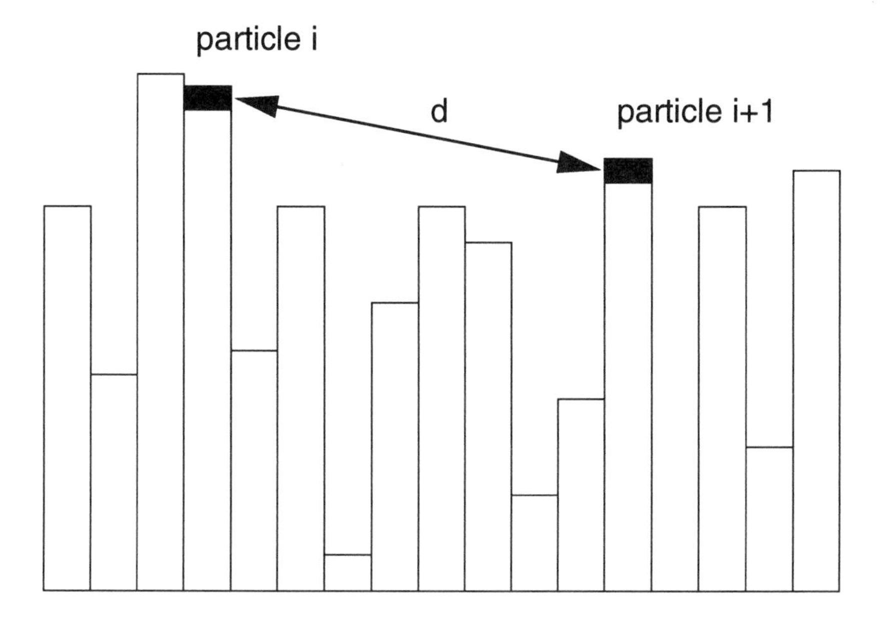

Fig. 24.9 The probability of particle $i + 1$ sticking in some column depends on the distance d from the previously deposited particle i.

24.19 IMPLEMENTATION: CORRELATED BALLISTIC DEPOSITION, COLUMN.C

24.20 PROBLEM 6: A GLOBULAR CLUSTER

Consider a bunch of grapes on an overhead vine. Your **problem** is to determine how its tantalizing shape arises. As a hint, you are told that these shapes, as well as others such as dendrites, colloids, and thin-film structure, appear to arise from an aggregation process that is limited by diffusion.

24.21 MODEL: DIFFUSION-LIMITED AGGREGATION

A model of diffusion-limited aggregation (DLA) has successfully explained the relation between a cluster's perimeter and mass [W&S 83]. We start with a 2-D lattice containing a seed particle in the middle. We draw a circle around the particle and place another particle on the circumference of the circle at some random angle. We then release the second particle and have it execute a random walk, much like the one we studied in Chapter 6, *Deterministic Randomness*, but restricted to vertical or horizontal jumps between lattice sites.

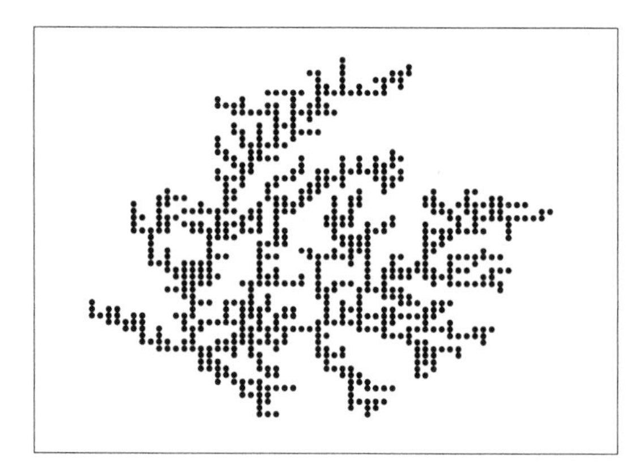

Fig. 24.10 A globular cluster of particles of the type that might occur in a colloid.

This is a type of *Brownian motion* that simulates the diffusion process. To make the model more realistic, we let the length of each step vary according to a random Gaussian distribution. If at some point during its random walk, the particle finds another particle within one lattice spacing, they stick together and the walk terminates. If the particle passes outside the circle from which it was released, it is lost forever. The process is repeated as often as desired. A typical cluster growth is illustrated in Fig. 24.10.

24.22 METHOD

1. Write a subroutine that generates random numbers with a Gaussian distribution.[3]

2. Define a lattice of points *grid* with all elements initially zero.

3. Place the seed at the center of the lattice, that is, let $grid(200, 200) = 1$.

4. Imagine a circle of radius 180 lattice constants centered at site $(200, 200)$. This is the circle from which we release particles.

5. Determine the angular position of the new particle on the circle's circumference by generating a uniform random angle between 0 and 2π.

6. Compute the x and y positions of the new particle on the circle.

[3]We indicated how to do this in §7.18.4 and provide a sample code on the diskette and the Web.

7. Determine whether the particle moves horizontally or vertically by generating a uniform random number $0 < r_{xy} < 1$ and applying the rule

$$\text{if } r_{xy} \begin{cases} < 0.5, & \text{the motion is vertical,} \\ > 0.5, & \text{the motion is horizontal.} \end{cases} \tag{24.33}$$

8. Generate a Gaussian-weighted random number in the interval $[-\infty, \infty]$. This is the size of the step, with the sign indicating positive or negative direction.
9. We know now the total distance and direction in which the particle will move. It will "jump" one lattice spacing at a time until this total distance is covered.
10. Before a jump, check whether a nearest-neighbor site is occupied.
    (a) If occupied, the particle stays at its present position and the walk is over.
    (b) If unoccupied, the particle jumps one lattice spacing.
    (c) Continue the checking and jumping until the total distance is covered, until the particle sticks, or until it leaves the circle.
11. Once one random walk is over, another particle can be released and the process repeated.

Because many particles "get lost," you may have to generate hundreds of thousands of particles to form a cluster of several hundred particles. This may take some time.

24.23 IMPLEMENTATION: DIFFUSION-LIMITED AGGREGATION, DLA.C

24.24 ASSESSMENT: FRACTAL ANALYSIS OF THE DLA GRAPH

A cluster generated with the DLA technique is shown in Fig. 24.10. We wish to analyze it to see if the structure is a fractal, and, if so, to determine its dimension. The analysis is a variation of the one used to determine the length of the coastline of Britain.

1. Draw a square of length L, small relative to the size of the cluster, around the seed particle.[4] ("Small" might be seven lattice spacings to a side.)

[4]Being old fashioned, we actually did this by hand. It would be rather straightforward to write a simple program to do this by counting elements in a matrix.

2. Count the number of particles within the square.

3. Compute the density ρ by dividing the number of particles by the number of sites available in the box (49 in our example).

4. Repeat the procedure using larger and larger squares centered about the seed.

5. Stop when the cluster is covered.

6. The (box-counting) fractal dimension d_f is estimated from a log–log plot of the density ρ versus L. If the cluster is a fractal, then (24.13) tells us that $\rho \propto L^{d_f-2}$, and the graph should be a straight line of slope $d_f - 2$.

The graph we generated had a slope of -0.36, which corresponds to a fractal dimension of 1.66. Because random numbers are involved, the graph you generate will be different, but the fractal dimension may be close.[5]

24.25 PROBLEM 7: FRACTAL STRUCTURES IN BIFURCATION GRAPH

Recall the project on the logistics map where we plotted the values of the stable population numbers versus growth parameter μ. Take one of the bifurcation graphs you produced, or use one of the figures in Chapter 13, *Unusual Dynamics of Nonlinear Systems*, and determine the fractal dimension of different parts of the graph by using the same technique that was applied to the coastline of Britain.

[5] Actually, the structure is a multifractal, and if you want to analyze it in detail, a more sophisticated theory and simulation should be used.

Part III

PARTIAL DIFFERENTIAL EQUATIONS

25

Electrostatic Potentials

25.1 INTRODUCTION: TYPES OF PDE'S

Natural quantities such as temperature and pressure vary continuously in both space and time. Such being our world, the function or *field* $U = U(x, y, z, t)$ used to describe these quantities must contain space and time coordinates as independent variables. The independence of each variable means that the derivatives in the equations must be partial derivatives. The equations are then *partial differential equations* (PDEs) in contrast to ordinary differential equations (ODEs).

Solving PDEs differs from solving ODEs in a number of ways. In particular, the *initial condition* (the $t = 0$ solution) must be known not just at one point, but throughout all of space. In addition, we must constrain the solution for all times by requiring it to have a specified form in some region of space [the *boundary conditions*, e.g., $U(x = a, y, z, t) = 12$]. This makes the algorithms for the solutions of PDEs more complicated than those for ODEs.

As time evolves, the changes in the field $U(x, y, z, t)$ at any one position affects the field at neighboring points. While the time evolution is similar to *ordinary differential equations*, there are now couplings to simultaneous variations in the space dimensions, and so our algorithms will make *finite-difference* steps in both time and space. For a realistic problem requiring a reliable level of precision, this may lead to an incredibly large number of coupled equations. In the next few chapters we will investigate various problems leading to different types of partial differential equations and correspondingly different approaches to solving them.

The most general form of a 2-D time-independent PDE is

$$A(x,y)\frac{\partial^2 U}{\partial x^2} + 2B(x,y)\frac{\partial^2 U}{\partial x \partial y} + C(x,y)\frac{\partial^2 U}{\partial y^2} = F(x,y,U,\frac{\partial U}{\partial x},\frac{\partial U}{\partial y}), \quad (25.1)$$

where A, B, C, and F are general functions. For the special case where

$$B^2(x,y) = A(x,y)C(x,y) \quad (25.2)$$

for all x and y, the equation is called *parabolic*. An example with $B = C = 0$, is the 1-D heat equation:

$$\frac{\partial U(x,t)}{\partial t} = \frac{k}{c\rho}\frac{\partial^2 U(x,t)}{\partial x^2}. \quad (25.3)$$

When $B^2 > AC$ for all x and y, the equation is called *hyperbolic*. An example with $B = 0$ and $AC < 0$ is the 2-D wave equation:

$$\frac{\partial^2 \psi(x,y,t)}{\partial x^2} + \frac{\partial^2 \psi(x,y,t)}{\partial y^2} = \frac{1}{c^2}\frac{\partial^2 \psi(x,y,t)}{\partial t^2}. \quad (25.4)$$

Here the y of (25.1) has become the time variable. When $AC > B^2$ for all x and y, the equation is *elliptic*. An example is Laplace's equation (25.7).

25.2 PROBLEM: DETERMINING AN ELECTROSTATIC POTENTIAL

Your **problem** is to find the electric potential for all points *inside* the charge-free square shown in Fig. 25.1. The bottom and sides of the region are made up of wires that are "grounded" (kept at 0 V). The top has a different wire running across it, connected to a battery that keeps it at a constant 100 V. Once the electric potential is known, you should also determine the nature of the electric field $\mathbf{E}$.

25.3 THEORY: LAPLACE'S EQUATION (ELLIPTIC PDE)

It is known from classical electrodynamics [Jack 75] that the electric potential $U(\mathbf{x})$ satisfies Poisson's PDE:

$$\nabla^2 U(\mathbf{x}) = -4\pi\rho(\mathbf{x}). \quad (25.5)$$

Here $\rho(\mathbf{x})$ is the charge density at the spatial location $\mathbf{x}$, and we leave off its time dependence because we are dealing with an electrostatics problem. In charge-free regions of space, that is, regions where $\rho(\mathbf{x}) = 0$, the scalar

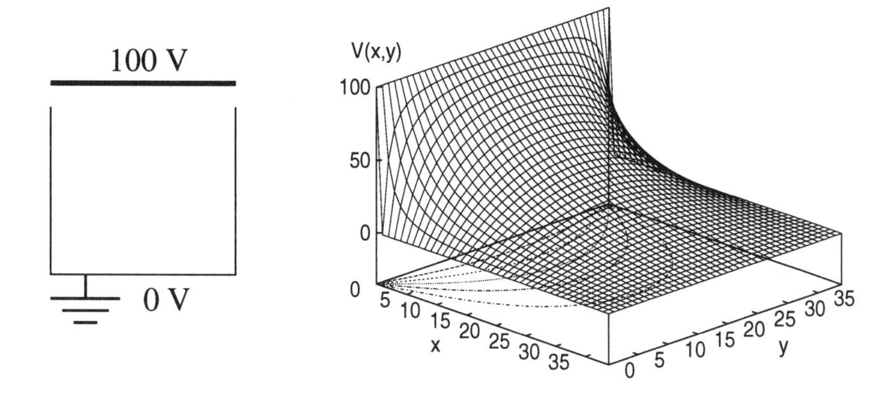

Fig. 25.1 (*Left*) The region of space within a square in which we want to determine the electric potential. There is a wire at the top kept at a constant 100 V and a grounded wire at the sides and bottom. (*Right*) The electric potential V ($\equiv U$) for this geometry. The projection onto the xy plane gives the equipotential lines.

potential satisfies *Laplace's equation*:

$$\nabla^2 U(\mathbf{x}) = 0. \tag{25.6}$$

This equation is an elliptic PDE with physical applications beyond electrostatics. In 2-D rectangular coordinates it takes the form

$$\boxed{\frac{\partial^2 U(x,y)}{\partial x^2} + \frac{\partial^2 U(x,y)}{\partial y^2} = 0,} \tag{25.7}$$

which shows that the potential depends simultaneously on x and y. The mathematical problem is to find $U(x,y)$ within a boundary, given its values along the boundary.

25.4 METHOD, NUMERICAL: FINITE DIFFERENCE

Mathematical theorems tell us that if the potential $U(x,y)$ is known along a specific boundary, solutions must exist inside the boundary.[1]

[1]The solution outside the boundary cannot always be found analytically, although its existence is physically obvious. It can be found numerically. Of course, we view infinity as our "boundary," and make the potential zero there. Then being outside the square is equivalent to being "between" two conductors, which is a problem we can solve.

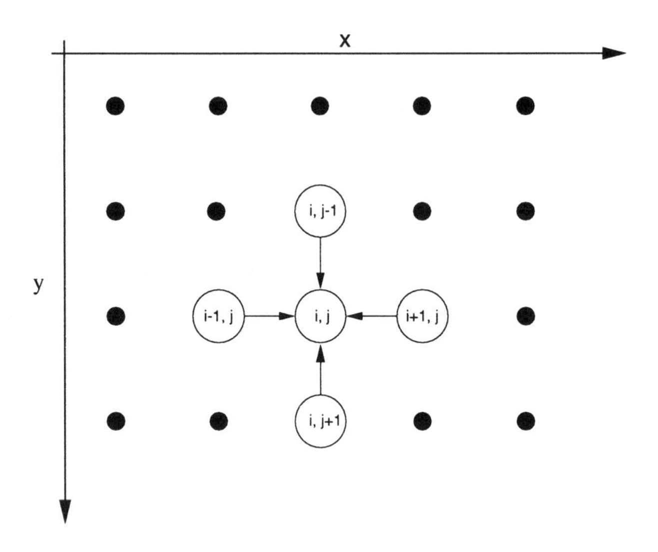

Fig. 25.2 The algorithm for Laplace's equation in which the potential at the point $(x, y) = (i, j)\Delta$ equals the average of the potential values at the four nearest-neighbor points. The boundary conditions are imposed by having the potentials along the perimeter remain at fixed values.

To deduce an algorithm (illustrated in Fig. 25.2) for the numeric solution of (25.7), we follow similar steps to those used in §8.1 to derive the forward-difference algorithm for differentiation. We expand the potential at the points $(x \pm \Delta x, y)$ as a Taylor series in the x variable:

$$U(x + \Delta x, y) = U(x, y) + \frac{\partial U}{\partial x}\Delta x + \frac{1}{2}\frac{\partial^2 U}{\partial x^2}(\Delta x)^2 + \cdots, \tag{25.8}$$

$$U(x - \Delta x, y) = U(x, y) - \frac{\partial U}{\partial x}\Delta x + \frac{1}{2}\frac{\partial^2 U}{\partial x^2}(\Delta x)^2 + \cdots. \tag{25.9}$$

When we add these equations and keep terms of order $(\Delta x)^2$, the linear terms in Δx cancel and we obtain a central-difference approximation to the second derivative for the partial derivative $\partial^2 U(x, y)/\partial x^2$:

$$\frac{\partial^2 U(x, y)}{\partial x^2} \simeq \frac{U(x + \Delta x, y) + U(x - \Delta x, y) - 2U(x, y)}{(\Delta x)^2}. \tag{25.10}$$

Likewise, we expand $U(x, y \pm \Delta y)$ as a Taylor series in y to obtain

$$\frac{\partial^2 U(x, y)}{\partial y^2} \simeq \frac{U(x, y + \Delta y) + U(x, y - \Delta y) - 2U(x, y)}{(\Delta y)^2}. \tag{25.11}$$

We expect the error in these finite-difference equations to be proportional to Δ^4. Substituting these approximations into (25.7) leads to a finite-difference approximation (Fig. 25.2) to Laplace's PDE:

$$\frac{U(x+\Delta x,y)+U(x-\Delta x,y)-2U(x,y)}{(\Delta x)^2} + \frac{U(x,y+\Delta y)+U(x,y-\Delta y)-2U(x,y)}{(\Delta y)^2} \simeq 0. \tag{25.12}$$

The reader will notice that (25.12) is a relation among the solutions at five points in space. To utilize it, we divide space up into a lattice, and solve for U at each site on the lattice. This results in a set of linear algebraic equations. One approach is to solve these linear equations explicitly as a (big) matrix problem, using the computer to do the matrix algebra. This is attractive as a direct solution, but it requires a great deal of computing time and memory.

The approach we follow is a simple one. We assume equal grid spacings, $\Delta x = \Delta y = \Delta$ in (25.12) to obtain

$$\boxed{U(x,y) \simeq \tfrac{1}{4}\left[U(x+\Delta,y)+U(x-\Delta,y)+U(x,y+\Delta)+U(x,y-\Delta)\right].} \tag{25.13}$$

Equation (25.13) is our basic algorithm for Laplace's equation in two dimensions. It approximates the potential at point (x, y) as the average of the potential values at the four nearest neighbors, as shown in Fig. 25.2. It is applied as part of an iterative scheme in which we start with the solution along the boundaries and an initial guess for the rest of the solution, use (25.13) to obtain an improved solution, and keep repeating the algorithm until stability is attained.

As is often true in the numerical solution of PDEs, this algorithm is not of as high an order as those used to solve ordinary differential equations (e.g., *rk4*), and so for a given step size Δ, is not as accurate. We will see the effect of this crudeness in the large number of iterations needed to obtain convergence. For serious calculations that use large amounts of computer time, a more accurate, "industrial-strength" algorithm is probably worth the programming complexity.

25.5 METHOD, ANALYTIC: POLYNOMIAL EXPANSIONS

We want the analytic solution of Laplace's equation

$$\frac{\partial^2 U(x,y)}{\partial x^2} + \frac{\partial^2 U(x,y)}{\partial y^2} = 0, \tag{25.14}$$

with the boundary conditions given along a square of side L. We assume that the potential is the product of independent functions of x and y:

$$U(x,y) = X(x)Y(y), \tag{25.15}$$

and substitute this product into (25.14). After dividing the resulting equation by $X(x)Y(y)$ we obtain

$$\frac{d^2X(x)/dx^2}{X(x)} + \frac{d^2Y(y)/dy^2}{Y(y)} = 0. \tag{25.16}$$

Because $X(x)$ is a function of only x and $Y(y)$ of only y, the derivatives in (25.16) are *ordinary* as opposed to *partial* derivatives. Since $X(x)$ and $Y(y)$ are assumed to be independent, the only way (25.16) can be valid for *all* values of x and y is for each term in (25.16) to be equal to a constant:

$$\frac{d^2Y(y)/dy^2}{Y(y)} = -\frac{d^2X(x)/dx^2}{X(x)} = k^2. \tag{25.17}$$

We now have two, noncoupled ordinary differential equations:

$$\frac{d^2X(x)}{dx^2} + k^2X(x) = 0, \tag{25.18}$$

$$\frac{d^2Y(y)}{dy^2} - k^2Y(y) = 0. \tag{25.19}$$

We shall see that this choice of sign for the constant matches the boundary conditions and gives us periodic behavior in x. The other choice of sign would give periodic behavior in y, and that would not work. The solutions for $X(x)$ are periodic and those for $Y(y)$ are exponential:

$$X(x) = A\sin kx + B\cos kx, \tag{25.20}$$

$$Y(y) = Ce^{ky} + De^{-ky}. \tag{25.21}$$

The $x = 0$ boundary condition, $U(x = 0, y) = 0$, can be met only if $B = 0$. The $x = L$ boundary condition, $U(x = L, y) = 0$, can be met only for values of k for which

$$kL = n\pi, \quad n = 1, 2, \ldots. \tag{25.22}$$

Accordingly, for each value of n there is a solution for X that we label as $X_n(x)$:

$$X_n(x) = A_n \sin\left(\frac{n\pi}{L}x\right). \tag{25.23}$$

For each value of k_n which satisfies the x boundary conditions, the y solution $Y(y)$ must satisfy the boundary condition $U(x, y = 0) = 0$. This requires

$D = -C$ in (25.21), and so

$$Y_n(y) = C(e^{k_n y} - e^{-k_n y}) \equiv 2C \sinh\left(\frac{n\pi}{L}y\right). \tag{25.24}$$

Because we are solving linear equations, the principle of linear superposition holds and this means that the most general solution is the sum of the products $X_n(x)Y_n(y)$:

$$U(x,y) = \sum_{n=1}^{\infty} E_n \sin\left(\frac{n\pi}{L}x\right) \sinh\left(\frac{n\pi}{L}y\right). \tag{25.25}$$

The E_n values are arbitrary constants and are fixed by requiring the solution to satisfy the remaining boundary condition at $y = L$. In general, this boundary condition could be that the potential has some specific functional dependence. For our **problem** the boundary condition is $U(x, L) = 100$ V, and so

$$\sum_{n=1}^{\infty} E_n \sin\frac{n\pi}{L}x \sinh n\pi = 100 \text{ V}. \tag{25.26}$$

We determine the constants E_n by projection. We multiply both sides of the equation by $\sin m\pi/Lx$, with m an integer, and integrate from 0 to L:

$$\sum_{n}^{\infty} E_n \sinh n\pi \int_0^L dx \sin\frac{n\pi}{L}x \sin\frac{m\pi}{L}x = \int_0^L dx 100 \sin\frac{m\pi}{L}x. \tag{25.27}$$

Yet the integral on the LHS is nonzero only for $n = m$, in which case we can solve for E_n:

$$E_n = \begin{cases} 0, & \text{for } n \text{ even}, \\ \frac{4(100)}{n\pi \sinh n\pi}, & \text{for } n \text{ odd}. \end{cases} \tag{25.28}$$

Finally, we obtain the potential at any point (x, y),

$$U(x,y) = \sum_{n=1,3,5,\ldots}^{\infty} \frac{400}{n\pi} \sin\left(\frac{n\pi x}{L}\right) \frac{\sinh(n\pi y/L)}{\sinh(n\pi)}. \tag{25.29}$$

At this point it is interesting to observe that the solution via the numerical algorithm (25.13) starts with the values of the potential on the boundaries and then propagates them through all space via many iterations. The numerical solution keeps repeating the algorithm until a stable solution results. In contrast, the analytic solution has the x and y dependence explicit via a double Fourier series, but must keep summing terms until a stable solution results.

It is important to notice when evaluating the analytic solution, that the sinh functions in (25.29) may overflow for large n. Some of these overflows can be avoided by expressing the quotient of the two hyperbolic sine functions in terms of exponentials:

$$\frac{\sinh(n\pi y/L)}{\sinh(n\pi)} = \frac{e^{n\pi(y/L-1)} - e^{-n\pi(y/L+1)}}{1 - e^{-2n\pi}}. \tag{25.30}$$

While the e^n/n term still gets large, the sum of term with alternating signs in (25.29) converges.

25.6 IMPLEMENTATION: SOLUTION ON LATTICE, LAPLACE.F

We divide the square into a lattice with equal spacing Δ in both in the x and y directions. The x and y variables are now discrete:

$$x = x_0 + i\Delta, \qquad y = y_0 + j\Delta, \qquad (i, j = 0, N_{\text{max}} = L/D). \tag{25.31}$$

We represent the potential by the array $U(N_{\text{max}}, N_{\text{max}})$. The finite-difference algorithm (25.13) is now

$$U(i,j) = \tfrac{1}{4}\left[U(i+1,j) + U(i-1,j) + U(i,j+1) + U(i,j-1)\right], \tag{25.32}$$

with the following boundary conditions:

$$\begin{array}{lcll} U(i, N_{\text{max}}) & = & 100, & \text{(top)}, \\ U(1,j) & = & 0, & \text{(left)}, \\ U(N_{\text{max}}, j) & = & 0, & \text{(right)}, \\ U(i,1) & = & 0, & \text{(bottom)}. \end{array} \tag{25.33}$$

The average (25.32) is used to find the potential at each position (i, j), starting from an edge and working inwards. Successive iterations are obtained by using (25.32) on the output of the previous iteration, the iterations ending when the changes in the potential are insignificant.

1. Write a program or modify the one on the diskette or Web to find the solution of Laplace's equation within a square of side L.

2. Increase the number of iterations until you see there is no significant change in the potential throughout the entire region. Because many computations are involved, do the calculation in double precision.

3. Impose the boundary conditions on the four sides of the square. There is an ambiguity in the top corners because they can be 100 V or 0.

4. Repeat the iteration several hundred times and observe the potential, row by row. Check whether the boundary conditions are always met.

5. After you have the numerical solution debugged and stable, compare with the analytic solution (25.29). Do not be surprised if you need to

sum thousands of terms before the "analytic" solution converges!

25.7 ASSESSMENT AND VISUALIZATION

1. Observe how the solution converges with successive iterations.

2. Observe how your choice of potential values for the corners leads to inaccuracies near the corners.

3. Repeat the process for different step sizes and judge if it is stable and convergent.

4. You will need hundreds or possibly thousands of iterations to obtain stable and accurate answers (we warned you that you pay a price for simplicity of algorithm).

5. Which solution is more precise, the analytic or numerical?

6. Use a plotting program to draw lines of constant U (potential). These are equipotential surfaces.

7. Either by hand or in some clever way, draw curves orthogonal to the equipotential lines, beginning and ending on the boundaries (where charges lie). The regions of high line density are regions of high electric force.

25.8 EXPLORATION

The numerical solution to the PDE can be used for any boundary conditions. The computer just calculates different numbers on the boundary and then iterates as before. Once you have a running program, have some fun and try out other boundary conditions. Two boundary conditions to try out are triangular

$$U(x,a) = \begin{cases} 200\frac{x}{L}, & \text{for } x \leq L/2, \\ 100(1-\frac{x}{L}), & \text{for } x \geq L/2, \end{cases} \tag{25.34}$$

and sinusoidal

$$U(x,a) = 100 \sin\left(\frac{2\pi x}{L}\right). \tag{25.35}$$

Unfortunately, you will have to recalculate the Fourier coefficients for these boundary conditions, or have confidence in your numerical solution and live without an analytic comparison.

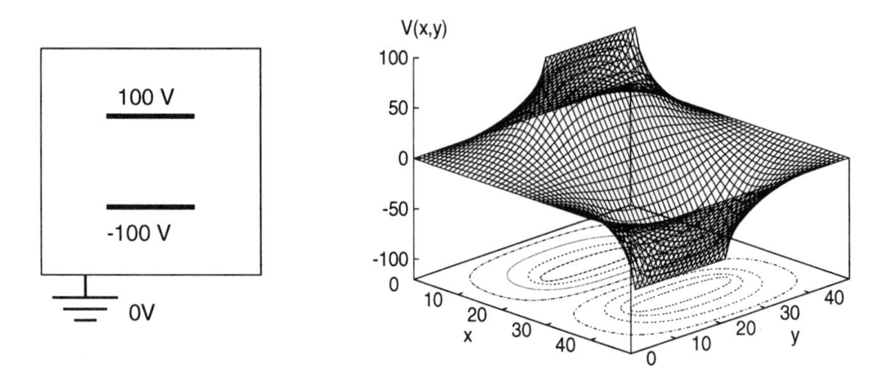

Fig. 25.3 (*Left*) the geometry of a parallel-plate capacitor within a box. A realistic capacitor would have the plates closer together in order to condense the field. (*Right*) The electric potential for this geometry. The projection on the xy plane gives the equipotential lines.

25.9 EXPLORATION: PARALLEL-PLATE CAPACITOR

The standard solution for a capacitor's field is for the region between two infinite plates. We want to see the edge effects and the exterior field when a finite capacitor is placed in a grounded box, as shown in Fig. 25.3. Modify the given program to satisfy these boundary conditions.

Plot the potential and equipotential surfaces. Sketch in the electric field lines (always orthogonal to the equipotential surfaces and beginning and ending on charges). Where is the electric field most intense, and how does it differ from that for an infinite capacitor? Results of our simulation are shown in Fig. 25.3.

25.10 EXPLORATION: FIELD BETWEEN SQUARE CONDUCTORS

You have designed a piece of equipment that is essentially a small metal box at 100 V within a larger, grounded one. You find that sparking occurs inside it, which indicates too large an electric field. You need to determine where the field is greatest so that you can change the geometry and eliminate the sparking.

Modify the program to satisfy these boundary conditions and to determine the field between the boxes (Gauss's law tells us that the electric field vanishes within the inner box because it contains no charge). Plot the potential and equipotential surfaces, and sketch in the electric field lines (always orthogonal to the equipotential surfaces and beginning and ending on charges). Deduce where the electric field is most intense and try redesigning the equipment to reduce the field.

26

Heat Flow

26.1 PROBLEM: HEAT FLOW IN A METAL BAR

In Fig. 26.1 we see a metal bar of length $L = 100$ cm and width w located on the x axis. It is insulated along its sides but not its ends. Initially the bar is at a uniform temperature of 100°C, and then both ends are placed in contact with ice water. Heat flows out of the noninsulated ends only. Your **problem** is to determine how the temperature will vary as we move along the bar at any instant of time, and how this variation changes with time.

26.2 MODEL: THE HEAT (PARABOLIC) PDE

A basic fact of nature is that heat flows from hot to cold; that is, from regions of high temperature to regions of low temperature. The rate of heat flow $\mathbf{H}$ through some material is proportional to the gradient of the temperature T within the material:

$$\mathbf{H} = -K\boldsymbol{\nabla}T(\mathbf{x}, t), \tag{26.1}$$

where K is the thermal conductivity of the material. The total amount of heat energy $Q(t)$ in the material at any one time is proportional to the integral of the temperature over the volume of the material:

$$Q(t) = \int d\mathbf{x} C\rho(\mathbf{x})T(\mathbf{x}, t), \tag{26.2}$$

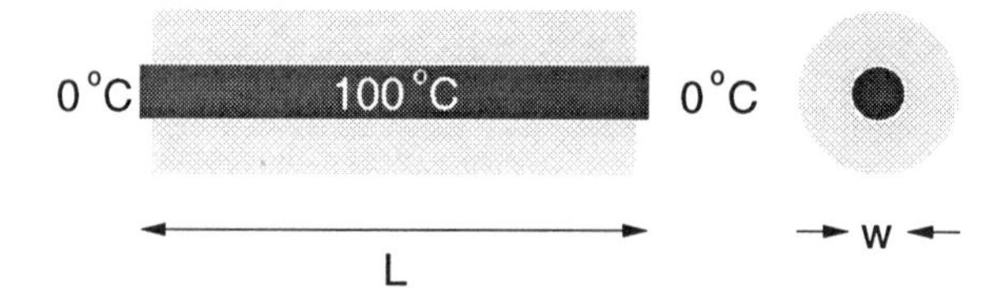

Fig. 26.1 A metallic bar insulated along its length with its ends in contact with heat reservoirs.

where C is the specific heat and ρ the density of the material. Because energy is conserved, the rate of decrease of Q with time must equal the amount of heat flowing out of the material. When this energy balance is struck and the divergence theorem applied, the *heat equation* is the result:

$$\frac{\partial T(\mathbf{x},t)}{\partial t} = \frac{K}{C\rho}\nabla^2 T(\mathbf{x},t), \tag{26.3}$$

where we assume that a constant density ρ.

Equation (26.3) is a parabolic PDE with space and time as independent variables. The setup of this problem implies that there is no temperature variation in directions perpendicular to the bar, and so we have one spatial coordinate to consider in our PDE:

$$\boxed{\frac{\partial T(x,t)}{\partial t} = \frac{K}{C\rho}\frac{\partial^2 T(x,t)}{\partial x^2}.} \tag{26.4}$$

We are given the initial temperature of the bar

$$T(x, t=0) = 100^o\mathrm{C}, \tag{26.5}$$

and must solve equation (26.4) to determine how the temperature changes with time and along the bar. The solution is constrained by the boundary conditions

$$T(x=0,t) = T(x=L,t) = 0, \tag{26.6}$$

that the ends of the bar are in ice water.

26.3 METHOD, ANALYTIC: POLYNOMIAL EXPANSIONS

The analytic approach is again based on the assumption that a solution exists in which the time and space dependences occur as separate functions:

$$T(x,t) = X(x)\mathcal{T}(t). \tag{26.7}$$

When (26.7) is substituted into (26.4), and the resulting equation is divided by the product $X(x)\mathcal{T}(t)$, there results two, noncoupled ODEs:

$$\frac{d^2X(x)}{dx^2} + \lambda^2 X(x) = 0, \quad (26.8)$$

$$\frac{d\mathcal{T}(t)}{dt} + \lambda^2 \frac{K}{C\rho}\mathcal{T}(t) = 0, \quad (26.9)$$

where λ is a constant to be determined. The boundary condition that the temperature equals zero at $x = 0$ demands the sine function:

$$X(x) = A \sin \lambda x. \quad (26.10)$$

The requirement that the temperature vanish at $x = L$ determines the possible values for the constant λ:

$$\sin \lambda L = 0 \Rightarrow \lambda = \lambda_n = \frac{n\pi}{L}, \quad n = 1, 2, \ldots, \quad (26.11)$$

$$\mathcal{T}(t) = e^{-\lambda_n^2 t/C\rho}. \quad (26.12)$$

The analytic solution is in this way

$$T(x,t) = A_n \sin\left(\frac{n\pi x}{L}\right) e^{-\lambda_n^2 t/C\rho}, \quad (26.13)$$

where n can be any odd integer and A_n is an arbitrary constant.

Because the principle of linear superposition holds, the most general solution to (26.4) can be written as a linear superposition of (26.13) using all values of n:

$$T(x,t) = \sum_{n=1}^{\infty} A_n e^{-\lambda_n^2 t/C\rho} \sin(\lambda_n x). \quad (26.14)$$

The Fourier expansion coefficients A_n are determined by the initial condition that at time $t = 0$ the entire bar has a temperature of $T = 100^\circ$C:

$$T(x, t=0) = T_0 \quad \Rightarrow \quad \sum_{n=1}^{\infty} A_n \sin(\lambda_n x) = T_0. \quad (26.15)$$

As before, we use the orthogonality of different sine functions to determine A_n by projection:

$$A_n = \frac{4T_0}{n\pi}, \quad n = 1, 3, 5, \ldots, \quad (26.16)$$

The full solution is consequently the infinite series

$$\boxed{T(x,t) = \sum_{n=1,3,..}^{\infty} \frac{4T_0}{n\pi} e^{-n^2\pi^2 Kt/(L^2 C\rho)} \sin\left(\frac{n\pi x}{L}\right).} \quad (26.17)$$

26.4 METHOD, NUMERICAL: FINITE DIFFERENCE

As with Laplace's equation, the numerical solution is based on converting a differential equation into an approximate finite-difference one. A custom algorithm is then used to generate the solution. The algorithm is derived by expanding $T(x, t + \Delta t)$ and $T(x + \Delta x, t)$ in Taylor series and keeping terms of lowest order in Δ:

$$
\begin{aligned}
T(x, t+\Delta t) &\simeq T(x,t) + \frac{\partial T(x,t)}{\partial t}\Delta t, && (26.18)\\
T(x+\Delta x, t) &\simeq T(x,t) + \frac{\partial T}{\partial x}\Delta x, && (26.19)\\
\Rightarrow \quad \frac{\partial T(x,t)}{\partial t} &\simeq \frac{T(x,t+\Delta t) - T(x,t)}{\Delta t}, && (26.20)\\
\frac{\partial^2 T(x,t)}{\partial x^2} &\simeq \frac{T(x+\Delta x,t) + T(x-\Delta x,t) - 2T(x,t)}{(\Delta x)^2}. && (26.21)
\end{aligned}
$$

The PDE (26.4) becomes the finite-difference equation:

$$
\begin{aligned}
\frac{T(x,t+\Delta t) - T(x,t)}{\Delta t} &\simeq \frac{K}{C\rho}\frac{T(x+\Delta x,t) + T(x-\Delta x,t) - 2T(x,t)}{(\Delta x)^2},\\
\Rightarrow \quad T(x,t+\Delta t) &\simeq T(x,t) + \frac{\Delta t}{(\Delta x)^2}\frac{K}{C\rho} && (26.22)\\
&\quad \times \left[T(x+\Delta x,t) + T(x-\Delta x,t) - 2T(x,t)\right].
\end{aligned}
$$

In discrete form, this is

$$
\boxed{T(i, j+1) = T(i,j) + \frac{K\Delta t\,[T(i+1,j) + T(i-1,j) - 2T(i,j)]}{C\rho(\Delta x)^2},} \qquad (26.23)
$$

where $x = i\Delta x$ and $t = j\Delta t$.

The algorithm described by (26.23) is pictured in Fig. 26.2. We see that the temperature at the point $[x = i\Delta x, t = (j+1)\Delta t]$ [the LHS of (26.23)] is computed from the temperature values at three points of an earlier time [the RHS of (26.23)]. The boundary conditions are imposed by having the temperature along the perimeter fixed. The initial conditions are imposed by having the initial temperature distribution used to generate the temperature at time Δt. Once the distribution at time Δt is known, it is used to generate the temperature distribution at time $2\Delta t$, and so forth.

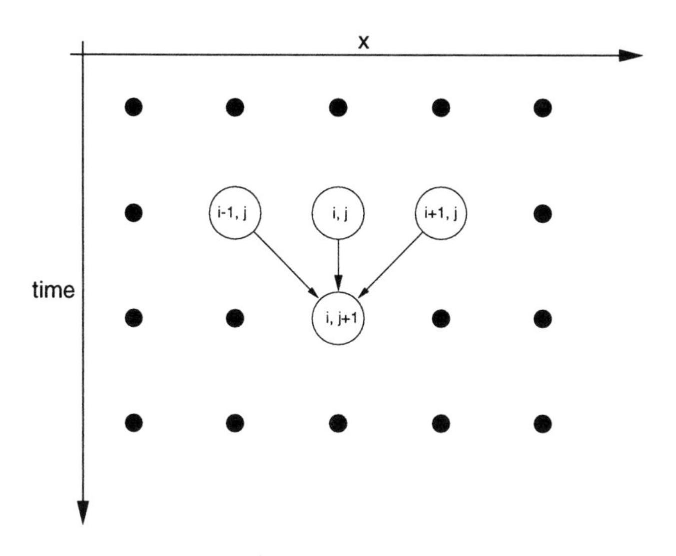

Fig. 26.2 The algorithm for the heat equation in which the temperature at the point $[x = i\Delta x, t = (j+1)\Delta t]$ is computed from the temperature values at three points of an earlier time.

26.5 ANALYTIC ASSESSMENT: ALGORITHM

Generally, PDEs are solved by converting them into finite-difference equations and then using the computer to find a numerical solution to the finite-difference equations. Sometimes we are able to find analytic solutions to the PDEs and test the numerics by comparing these to the analytic solution.[1] The heat equation is an exceptional casé in which there *also* exits an analytic solution to the finite-difference equation (26.23) [C&P 88]:

$$T(i,j) = A\left[1 - 4\frac{K}{C\rho}\frac{\Delta t}{(\Delta x)^2}\sin^2\left(\frac{i\pi\Delta x}{2l}\right)\right]^j \sin\left(\frac{i\pi\Delta x}{l}\right). \qquad (26.24)$$

While this analytic solution of the finite difference equation is not a valid solution to the PDE, it is helpful to provide understanding of the algorithm.

Results of the numerical simulation are shown in Fig. 26.3. If we compare the analytic solution of the PDE (26.17) with (26.24), we see that the valid solution decays exponentially with time, but this will not be true[2] for the

[1]In more realistic problems, or for more complicated boundary conditions, finding analytic solutions may be difficult or impossible. In those cases you may be able to find a "test case" similar enough to your problem to permit a check of your numerics.

[2]The argument is based on $\lim_{n\to\infty}(1 + x/n)^n = e^x$, and is a little tricky. For example, a

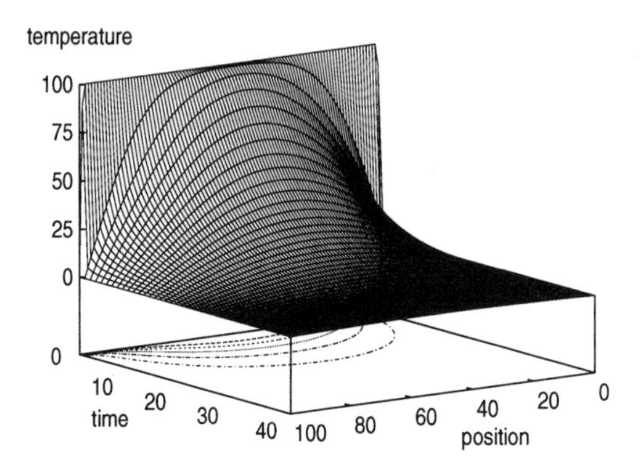

Fig. 26.3 The temperature versus position along the bar and versus time. The initial temperature is constant throughout the bar. The projected contours give the isotherms.

numeric solution unless

$$\frac{K}{C\rho} \frac{\Delta t}{(\Delta x)^2} \leq \tfrac{1}{4}. \tag{26.25}$$

If this condition is not met, the numeric solution will not decay in time and so, clearly, will be wrong. Equation (26.25) tells us that if we make the time step smaller we improve convergence, but if we decrease the space step without a simultaneous quadratic increase in the time step, we worsen convergence. (The lack of symmetry arises because the heat equation is second-order in space but only first-order in time.) Of course, for most problems you would not know the analytic solution of the finite-difference equation, and you would have to try out different combinations of Δx and Δt until a stable and reasonable solution is obtained. You may expect, nonetheless, that there may be choices for Δx and Δt for which the numeric solution fails, and that a separate decrease of Δx or Δt may not improve the solution.

26.6 IMPLEMENTATION, HEAT EQUATION, EQHEAT.F (.C)

Recall, we want to solve for the temperature distribution within an iron bar of length $L = 50$ cm with the boundary conditions

$$T(x = 0, t) = T(x = L, t) = 0, \tag{26.26}$$

$\frac{1}{2}$ on the RHS of (26.25) will give a stable solution, but it will not decay with time.

and initial conditions

$$T(x, t = 0) = 100^o\text{C}. \tag{26.27}$$

The constants appropriate to iron are

$$C = 0.113 \text{ cal/(g}^o\text{C)}, \quad K = 0.12 \text{ cal/(sg}^o\text{C)}, \quad \rho = 7.8 \text{ g/cc}. \tag{26.28}$$

1. Write or modify the program given on the diskette and Web to solve the heat equation.

2. Define a 2-D array `T(101,2)` for the temperature as a function of space and time. The first index is for the 100 space divisions of the bar, and the second index for present and past times (because thousands of time steps may be made, we save memory by saving only two times).

3. For time $t = 0$, (j=*1*), initialize `T` so that all points on the bar except the end points are at 100°C. Set the temperatures of the ends to 0°C.

4. Apply equation (26.21) to obtain the temperature at the next time.

5. Assign the present-time values of the temperature to the past values:

$$T(i, 1) = T(i, 2), \quad i = 1, \ldots, 101. \tag{26.29}$$

6. Start running with 50 time steps. Once you are confident the program is running properly, use thousands of steps to see the bar cool with time. For every ~500 time steps, print the time and temperature along the bar.

26.7 ASSESSMENT: CONTINUITY, NUMERIC VERSUS ANALYTIC

Extend your program to evaluate the analytic solution at the times and points used for in the numerical solution. You may have to sum the analytic solution over thousands of terms for stability and precision.

1. Make sure your program gives a temperature distribution that varies smoothly along the bar and which agrees with the boundary conditions.

2. Make sure your program gives a temperature distribution that varies smoothly with time and attains equilibrium. You may have to vary the time and space steps to obtain well behaved solutions.

3. Compare the analytic and numeric solutions (and the times needed to compute them). If the solutions differ, suspect the one which does not appear smooth and continuous.

26.8 ASSESSMENT: VISUALIZATION

1. Make 2-D plots of the temperature versus position along the bar.

2. Better yet, make a 3-D plot of the temperature versus position versus time.

3. Make a plot of the contours of constant temperature, the *isotherms.*

26.9 EXPLORATION

Stability test: Check that the temperature diverges with time if the constant C in equation (26.25) is made larger than 0.5.

Material dependence: Repeat the calculation for aluminum, $C = 0.217$ cal/(g°C), $K = 0.49$ cal/(g°C), $\rho = 2.7$ g/cc. Take note that the stability condition requires you to change the size of the time step.

Scaling: The shape of the temperature versus time curve may be the same for different materials, but not the scale. Which of the two bars cools faster?

Sinusoidal initial distribution: $\sin(\pi x/L)$. (This may seem somewhat artificial, but it leads to attractive graphs.) Use the same constants as in the example, and go out to 3000 time steps, printing out results every 150 steps to watch the bar cool. In this case we can compare to the analytic solution,

$$T(x,t) = \sin\frac{\pi x}{L} e^{-\pi^2 Rt/L^2}, \quad R = \frac{k}{C\rho}. \tag{26.30}$$

Two bars in contact: Assume that there are two identical bars, each 25 cm long, as shown in Fig. 26.4. One bar is kept in a heat bath at 100°C, and the other at 50°C. They are put in contact along one of their ends with their other ends kept at 0°C. Determine how the temperature varies with time and location.

Radiating bar (Newton's cooling): Imagine now, that instead of being insulated along its length, a single bar is in contact with an environment at a temperature T_e different from the initial temperature of the bar. In this case, Newton's law of cooling (radiation) says that the rate of temperature change due to radiation is

$$\frac{\partial T}{\partial t} = -h(T - T_e), \tag{26.31}$$

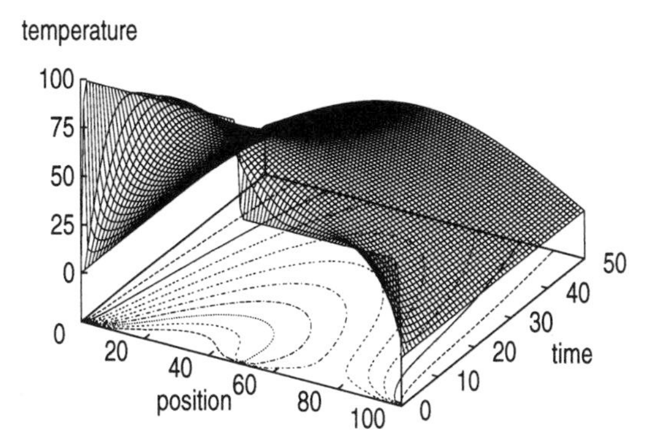

Fig. 26.4 The temperature versus position and time when two bars of differing temperature are placed in contact at $t = 0$. The contours projected onto the position-time plane give the isotherms.

where h is a positive constant. The heat equation is now modified to

$$\frac{\partial T(x,t)}{\partial t} = \frac{k}{C\rho}\frac{\partial^2 T}{\partial^2 x} - hT(x,t). \tag{26.32}$$

Modify the algorithm and program to include Newton's cooling along the length of the bar. Compare the cooling of this bar with that of the insulated bar.

27

Waves on a String

27.1 PROBLEM: A VIBRATING STRING

You pluck a string and some pattern of waves follow. Your **problem** is to predict this pattern when the pluck is 1 mm in height. You may have seen that if you pluck at one location and let go gently, a pulse or traveling wave is observed on the string. And if you shake the string just right, a standing-wave pattern, in which the nodes remain in the same place for all times, may result. Actually, we want to solve the problem for all of these possibilities.[1]

27.2 MODEL: THE WAVE EQUATION (HYPERBOLIC PDE)

As our model, we consider a string of length l, tied down at both ends as shown in Fig. 27.1. The string has a constant density per unit length ρ, a constant tension τ, and is subject to no frictional or gravitational forces. The vertical displacement of the string from its rest position is described by a function of two variables $y(x, t)$, where x is the horizontal location along the string and t the time. We assume that the displacement of the string y is only in the vertical direction.

To obtain a linear equation of motion (nonlinear PDEs are discussed in Chapters 28 and 29), we assume that the displacement and slope are small. If we isolate an infinitesimal section Δx of the string, we know from Newton's

[1]Some similar, but independent, studies can also be found in [Raw 96].

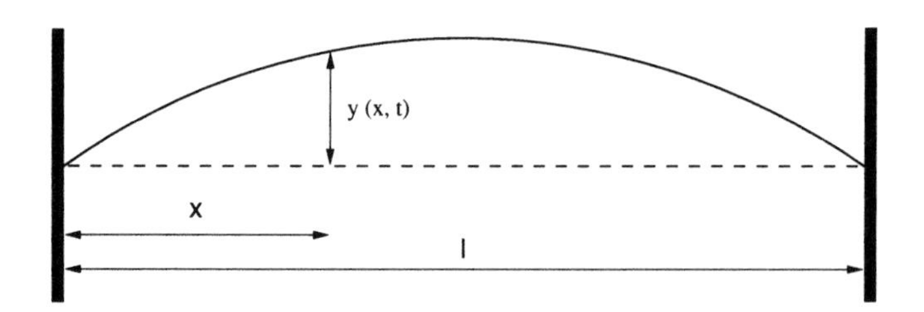

Fig. 27.1 A stretched string of length l tied down at both ends. The horizontal position along the string is given by x and the vertical disturbance of the string from its equilibrium position by $y(x,t)$.

second law of motion that the sum of the vertical forces on the string section must equal the mass times the vertical acceleration of the section:

$$\sum F_y = \rho \Delta x \frac{\partial^2 y}{\partial t^2}. \tag{27.1}$$

Here the forces are the components of the string's tension τ. The vertical components of the tension on each end of the segment change as the angle of the string changes, and we obtain those components by relating the slope of the string to $\partial y/\partial x$:

$$\sum F_y = \tau \left[\left(\frac{\partial y}{\partial x} \right)_{x+\Delta x} - \left(\frac{\partial y}{\partial x} \right)_x \right] = \tau \frac{\partial^2 y}{\partial x^2}, \tag{27.2}$$

$$\Rightarrow \quad \frac{\partial^2 y(x,t)}{\partial x^2} = \frac{1}{c^2} \frac{\partial^2 y(x,t)}{\partial t^2}. \tag{27.3}$$

Here the propagation speed is denoted by

$$c = \sqrt{\tau/\rho}. \tag{27.4}$$

Observe in these equations that y, the height of the string, is the dependent variable, and that the position along the string x and the time t are *both* independent variables. The existence of two independent variables makes this a PDE.

Because both ends of the string are tied down, the *boundary conditions* are that the displacements must vanish for all times at the end of the string:

$$y(0,t) = y(l,t) \equiv 0, \quad \text{(boundary conditions)}. \tag{27.5}$$

As stated in the **problem** specification, the *initial condition* is that at $t = 0$ the right side of the string is "plucked," that is, the string is lifted 1 mm at

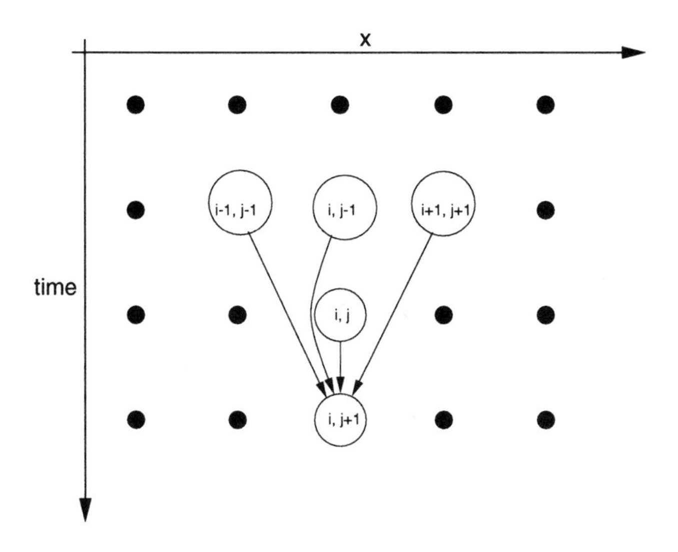

Fig. 27.2 Time steps in the algorithm for the vibrating string.

$x = 0.8l$. We model the "pluck" with the mathematical function:

$$y(x, t = 0) = \begin{cases} 1.25x/l, & \text{for } x \leq 0.8l, \\ 5.0(1 - x/l), & \text{for } x > 0.8l, \end{cases} \quad \text{(initial condition 1)}. \tag{27.6}$$

Because (27.3) is a second-order equation in time, a second initial condition (beyond initial displacement) is needed to determine the solution. We take that second condition to be that the plucked string is released from *rest*:

$$\frac{\partial y}{\partial t}(x, t = 0) = 0, \quad \text{(initial condition 2)}. \tag{27.7}$$

27.3 METHOD, NUMERICAL: TIME STEPPING

As was done with Laplace's equation in Chapter 25, *Electrostatic Potentials*, we look for a solution to our PDE on the 2-D grid shown in Fig. 27.2. In the present case the horizontal axis (first index) represents the position x along the string and the vertical axis (second index) represents time. We assign discrete variables to x and t:

$$x = i\Delta x, \qquad t = j\Delta t, \tag{27.8}$$

and represent y as $y(i, j)$. We convert the wave equation (27.3) into a finite-difference equation by expressing the second derivatives in terms of finite differences:

$$\frac{\partial^2 y(x,t)}{\partial t^2} \simeq \frac{y(i,j+1)+y(i,j-1)-2y(i,j)}{(\Delta t)^2}, \tag{27.9}$$

$$\frac{\partial^2 y(x,t)}{\partial x^2} \simeq \frac{y(i+1,j)+y(i-1,j)-2y(i,j)}{(\Delta x)^2}. \tag{27.10}$$

After substituting (27.9) and (27.10), we obtain final, discrete equation:

$$\boxed{y(i,j+1) = 2y(i,j)-y(i,j-1)+\frac{c^2}{c'^2}\left[y(i+1,j)+y(i-1,j)-2y(i,j)\right],} \tag{27.11}$$

where $c' = \Delta t/\Delta x$, is just a combination of numerical parameters with the dimension of velocity.

As shown in Fig. 27.2, (27.11) is a recurrence relation that propagates the wave from the two earlier times, j and $j-1$, and three nearby positions, $i-1$, i, and $i+1$, to a later time $j+1$ and a single position i (we are using Δx and Δt units). First starting the recurrence relation is a bit tricky because we need to know displacements from two earlier times, whereas the initial conditions are for only one time. To alleviate that difficulty, we convert the initial conditions (27.6) and (27.7) to finite-difference form, and use that to step backward in time! Explicitly, the *central-difference* approximation gives

$$\frac{\partial y}{\partial t}(x,t=0) = 0 \Rightarrow \frac{y(x,\Delta t)-y(x,-\Delta t)}{2\Delta t} = 0, \tag{27.12}$$

$$\Rightarrow y(i,-1) = y(i,1). \tag{27.13}$$

Imposing this condition onto (27.11) for the initial time yields

$$y(i,2) = y(i,1)+\frac{1}{2}\left(\frac{\Delta t}{\Delta x}\right)^2 c^2\left[y(i+1,1)+y(i-1,1)-2y(i,1)\right]. \tag{27.14}$$

We see that (27.14) takes the solution throughout all of space at the initial time $t = 0$ ($j = 1$), and propagates it forward to time Δt. Subsequent advances in time are produced by (27.11).

As we have seen in the finite-difference method applied to heat conduction, the success of the numerical method depends on the relative sizes of the time and space steps. For the present problem there is a similar stability criterion that tells us that the finite-difference solution will be stable if [Cour 28]

$$c \leq c' = \frac{\Delta x}{\Delta t}. \tag{27.15}$$

This means that the solution gets better with smaller *time* steps, but gets worse for smaller space steps. This appears somewhat surprising because the wave equation (27.3) is symmetric in x and t. Yet the symmetry is broken

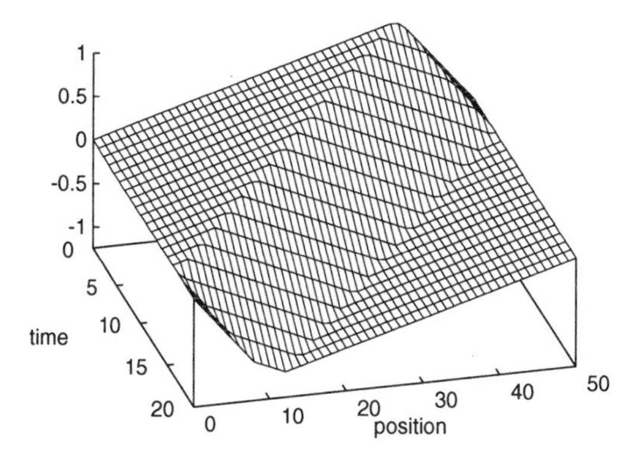

Fig. 27.3 The vertical displacement as a function of position and time of a string initially plucked near its right end. Observe how the initial pulse divides into waves traveling to the right and to the left, and how each traveling wave inverts after its reflection from a fixed end. Notice, too, that the traveling wave moving to the right hits the end first.

by the nonsymmetric way we specify the initial and boundary conditions. A typical numerical solution is shown in Fig. 27.3.

27.4 METHOD, ANALYTIC: NORMAL MODES

The analytic solution to (27.3) is obtained via the familiar separation-of-variables technique. We assume that a solution to the wave equation exists that is the product of a function of space times a function of time:

$$y(x,t) = X(x)T(t). \tag{27.16}$$

We substitute (27.16) into (27.3), divide the resulting equation by $y(x,t)$, and are left with an equation that has a solution only if there exists solutions to the two ODEs:

$$\frac{d^2T(t)}{dt^2} + \omega^2 T(t) = 0, \tag{27.17}$$

$$\frac{d^2X(x)}{dt^2} + k^2 X(x) = 0, \tag{27.18}$$

$$k \stackrel{\text{def}}{=} \frac{\omega}{c}. \tag{27.19}$$

Here the angular frequency ω and the wave vector k are determined by demanding that the solutions satisfy the boundary conditions.

The solution for $X(x)$ is required to satisfy the *boundary conditions* that the string is attached at both ends:

$$\begin{aligned} X(x=0,t) &= X(x=l,t)=0 && (27.20)\\ \Rightarrow \quad X_n(x) &= A_n \sin k_n x, && (27.21)\\ k_n &= \frac{2\pi(n+1)}{l}, \qquad n=0,1,\ldots. && (27.22) \end{aligned}$$

The corresponding solution for the time equation is

$$\begin{aligned} T_n(t) &= C_n \sin \omega_n t + D_n \cos \omega_n t, && (27.23)\\ \omega_n &= n\omega_0, \qquad \omega_0 \stackrel{\text{def}}{=} ck_0 = \frac{2\pi c}{l}. && (27.24) \end{aligned}$$

Solutions of the form (27.16) and (27.23) are the nth *normal modes*, where, by definition, each mode oscillates at a single frequency.

The *initial condition* (27.7) requires the C_n values in (27.23) to be zero. Putting the pieces together, this means that for a string with its ends fixed and initially at rest, there are solutions of the wave equation of the form

$$y_n(x,t) = \sin k_n x \cos \omega_n t, \quad (n=0,1,\ldots). \tag{27.25}$$

Since (27.3) is a linear equation in y, the principle of linear superposition holds and the most general solution can be written as the sum

$$y(x,t) = \sum_{n=0}^{\infty} B_n \sin k_n x \cos \omega_n t. \tag{27.26}$$

The Fourier coefficients B_n are determined by using the second initial condition (27.7), which describes how the wave is plucked. We start with

$$y(x,t=0) = \sum_{n}^{\infty} B_n \sin nk_0 x, \tag{27.27}$$

multiply both sides by $\sin mk_0x$, substitute the value of $y(x,0)$ from (27.7), and integrate from 0 to l to obtain

$$B_m = -0.0125 \frac{\sin 0.8\pi}{m^2\pi^2}. \tag{27.28}$$

We will compare (27.26) to our numerical solution. While it is in the nature of the approximation that the precision of the numerical solution depends on the choice of step sizes, it is also revealing to realize that the precision of the "analytic" solution depends on summing an infinite number of terms, which in real life can be done only approximately.

27.5 IMPLEMENTATION, WAVE EQUATION, EQSTRING.F (.C)

Modify the program given on the diskette or on the Web to solve for the behavior of the plucked string with ends fixed. Assume that the string has length $l = 1$ m, linear density $\rho = 0.01$ kg/m, and tension $\tau = 40$ N. The initial conditions are given by (27.6) and (27.7). You should get a solution that looks like the one in Fig. 27.3.

1. The program uses a two-dimensional array `y(101,3)`. The first index labels the x position along the string and the second, the three time values used in the recurrence relation (27.14). Time 1 is past, time 2 is present, and time 3 is the future.
2. Choose the space step $\Delta x = 0.01$, that is, 1 cm. Choose the time step Δt such that the stability condition (27.15) predicts a stable solution.
3. Use the initial condition (27.6) to assign `y(i,1)`.
4. Before running the program, print out the values of `y(i,1)` to check that you have assigned the initial displacements correctly.
5. Find the string's displacement for time *dt* by using the algorithm (27.14) to find `y(i,2)` for all i.
6. Check that the ends of the string remain fixed for all times, that is, `y(1,j)` $=$ `y(101,j)` $\equiv$ `0`.
7. Program up a loop that steps the time forward using the solutions from two earlier times, (27.11). Keep repeating this iteration, reassigning the last two columns of `y` to the first two:

```
y(i, 1) = y(i, 2)             Present time becomes past.
y(i, 2) = y(i, 3)           Future time becomes present.
```

8. For every five time increments, write out the displacement of the string.

27.6 ASSESSMENT: VISUALIZATION

Compare the analytic and numerical solutions, keeping at least 200 terms in the "analytic" solution.

1. Use your favorite graphics program to make a 3-D plot of displacement y versus position x versus time t. This is what we did in Fig. 27.3 with *gnuplot*.
2. Observe the motion of the peak as a function of time and use your graph to estimate the peak's propagation velocity c. Compare your deduced c to (27.4).

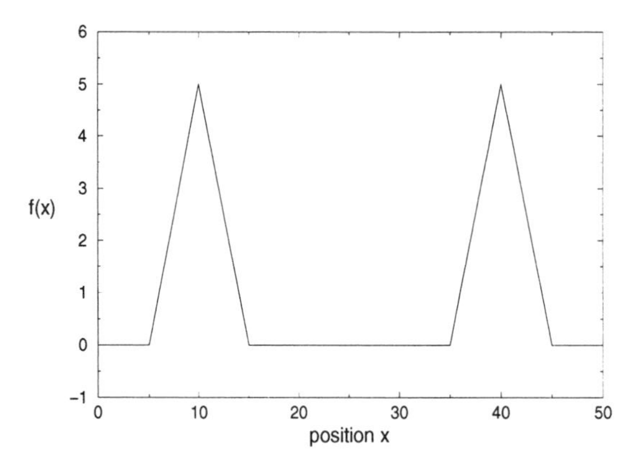

Fig. 27.4 The initial configuration of a string plucked in two places simultaneously. The resulting wavefront as a function of space and time is shown in Fig. 27.5.

27.7 EXPLORATION

Stability: Explore the use of different values for the steps Δx and Δt. Determine at which values the numerical solution becomes unstable. Does your determination agree with the stability condition (27.15)?

Two pulses: Consider a one-meter string with its ends tied down. It is under a tension of 40 N and it has a mass density of 10 g/m. Initially the string is plucked 5 mm at two points, as illustrated in Fig. 27.4. This initial condition is

$$\frac{y(x,t=0)}{0.005} = \begin{cases} 0, & 0.0 \le x \le 0.1, \\ 10x-1, & 0.1 \le x \le 0.2, \\ -10x+3, & 0.2 \le x \le 0.3, \\ 0, & 0.3 \le x \le 0.7, \\ 10x-7, & 0.7 \le x \le 0.8, \\ -10x+9, & 0.8 \le x \le 0.9, \\ 0, & 0.9 \le x \le 1.0. \end{cases} \tag{27.29}$$

Solve for the ensuing motion. In particular, observe whether the pulses move or just oscillate up and down. Our solution is given in Fig. 27.5.

Symmetric vibrating string: You may have seen that when a string is plucked near its end, a pulse reflects off the ends and bounces back and forth. We now want to see what happens if the string is plucked in its middle. Change the initial conditions of the model program to

$$\frac{y(x,t=0)}{0.01} = \begin{cases} x/l, & 0 \le x \le 0.5, \\ -x/l, & 0.5 \le x \le 1. \end{cases} \tag{27.30}$$

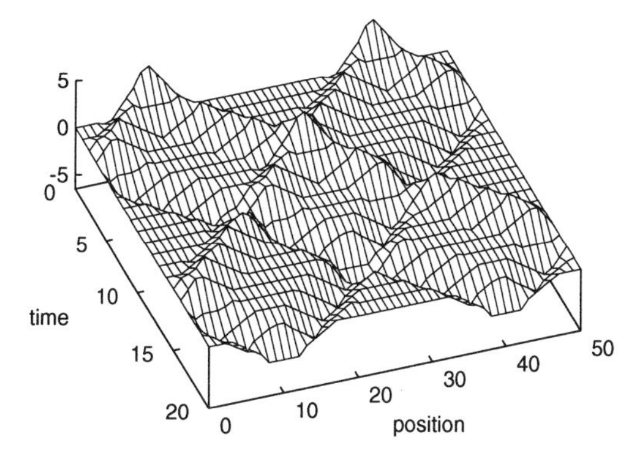

Fig. 27.5 The vertical displacement as a function of position and time of a string initially plucked simultaneously at two points, as shown in Fig. 27.4. Note that each initial peak breaks up into waves traveling to the right and to the left. The traveling waves invert on reflection from the fixed end. As a consequence of these inversions, the $t = 15$ wave is an inverted $t = 0$ wave.

Including friction: We have so far assumed that the string feels no resistance, clearly an idealization because we know that the notes on a guitar fade away rather quickly. The effect of friction on the motion of an element of string between x and $x + dx$ is to oppose the motion of that element. As a model, we assume that the force of friction is proportional to the vertical velocity $\partial y/\partial t$ of the string's element. This changes the wave equation to

$$\frac{\partial^2 y}{\partial t^2} + 2\kappa \frac{\partial y}{\partial t} = c^2 \frac{\partial^2 y}{\partial x^2}. \tag{27.31}$$

The constant κ is proportional to the viscosity of the medium in which the string is vibrating, and is inversely proportional to the density of the string.

Generalize the algorithm for the wave equation to include friction and observe the change in wave behavior. Start off with $T = 40$ N, $\rho = 10$ g/m, and $\kappa = 5$. A solution might look something like that in Fig. 27.6, where damping of the wave is evident. As a check, reverse the sign of κ and see if the wave grows in time (this would eventually violate our assumption of small oscillations).

Normal modes: We know from the analytic solution that there are *normal-mode* solutions to the wave equation that vibrate with one frequency. An example is given in Fig. 27.7. Explore what happens if a frictionless

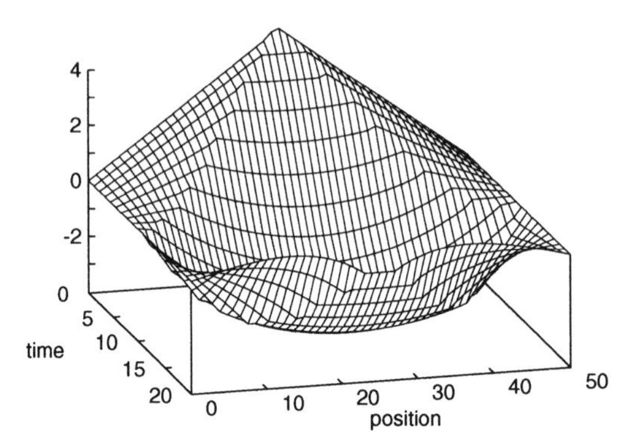

Fig. 27.6 The vertical displacement as a function of position and time of a string with friction initially plucked at its middle. Here, too, the initial pluck breaks up into waves traveling to the right and left that get reflected and inverted by the walls. Those parts of the wave with the greatest transverse (y) velocity experience the greatest friction; this effect distorts the wave and tends to smooth and dampen it.

string is initially placed in a normal mode, for example

$$y(x, t = 0) = 0.001 \sin 2\pi x. \tag{27.32}$$

Try other modes. See if the sum of two modes gives *beating.*

Variable density and tension: If the string has a variable density, then its tension will no longer be a constant, and waves propagate as shown in Fig. 27.8. In this case, the wave equation becomes [F&W 80]:

$$\frac{\partial}{\partial x}\left[T(x)\frac{\partial y(x,t)}{\partial x}\right] = \rho(x)\frac{\partial^2 y(x,t)}{\partial t^2}, \tag{27.33}$$

$$\frac{\partial T(x)}{\partial x}\frac{\partial y(x,t)}{\partial x} + T(x)\frac{\partial^2 y(x,t)}{\partial x^2} = \rho(x)\frac{\partial^2 y(x,t)}{\partial t^2}. \tag{27.34}$$

For constant tension and density, the propagation velocity $c = \sqrt{T/\rho}$. This no longer will be valid for variable densities, and you should observe whether the wave moves faster or slower in regions of high density. To be specific, assume that the density ans tension are proportional:

$$\rho(x) = \rho_0 e^{\alpha x}, \tag{27.35}$$

$$T(x) = T_0 e^{\alpha x}. \tag{27.36}$$

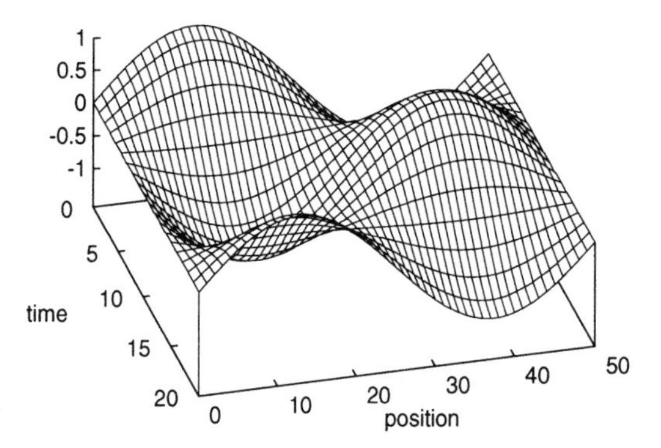

Fig. 27.7 The vertical displacement as a function of position and time of a string initially placed in a normal mode. Notice how the standing wave moves up and down with time.

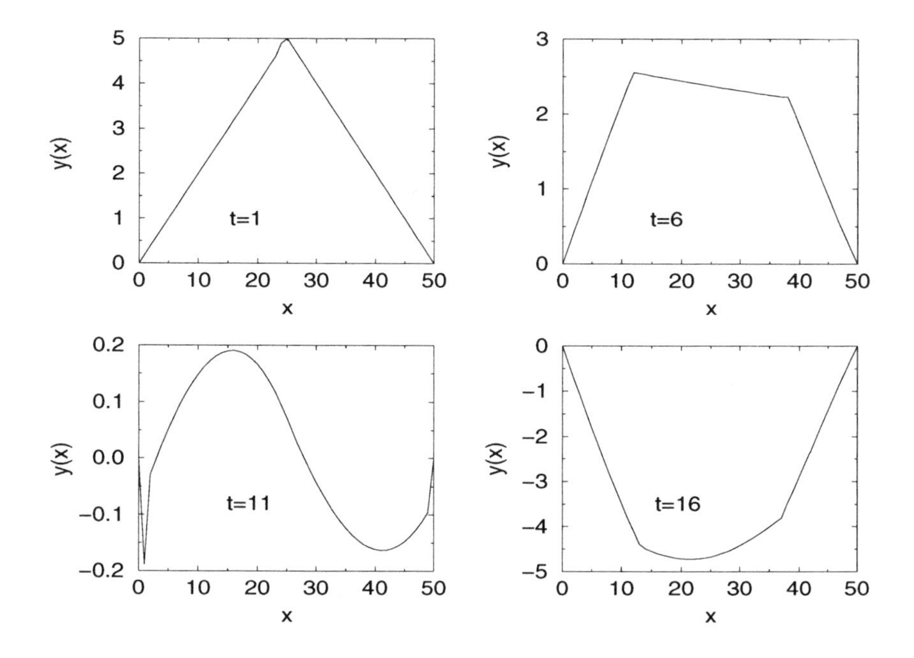

Fig. 27.8 Disturbance versus position for a string with variable density that is initially plucked at its center. The disturbances at four times are given. At $t = 6$ we see that the wave moves faster in the denser region to the right, but that its amplitude decreases because the string is heavier there.

Substitution of these relations into (27.34) yields the new wave equation:

$$\frac{\partial^2 y(x,t)}{\partial x^2} + \alpha\frac{T_0}{\rho_0}\frac{\partial y(x,t)}{\partial x} = \frac{T_0}{\rho_0}\frac{\partial^2 y(x,t)}{\partial t^2}. \tag{27.37}$$

This equation is similar to the wave equation with friction, only now the first derivative is with respect to x. The corresponding difference equations are

$$\begin{aligned} y(i,2) &= y(i,1) + \left(\frac{\Delta t}{\Delta x}\right)^2 \frac{T_0}{2\rho_0}[y(i+1,1) + y(i-1,1) - 2y(i,1)] \\ &\quad + \frac{1}{2}\frac{\alpha(\Delta t)^2 T_0}{\rho_0 \Delta x}[y(i+1,1) - y(i,1)], \qquad (27.38) \\ y(i,j+1) &= 2y(i,j) - y(i,j-1) + \frac{\alpha(\Delta t)^2 T_0}{\rho_0 \Delta x}[y(i+1,j) - y(i,j)] \\ &\quad + \left(\frac{\Delta t}{\Delta x}\right)^2 \frac{T_0}{\rho_0}[y(i+1,j) + y(i-1,j) - 2y(i,j)]. \qquad (27.39) \end{aligned}$$

Modify your program to handle this algorithm with $\alpha = 0.5$, $T_0 = 40$ N, and $\rho_0 = 0.01$ kg/m. Explain in words how the wave motion dampens. The behavior you obtain may look something like that shown in Fig. 27.8.

Part IV

NONLINEAR PARTIAL DIFFERENTIAL EQUATIONS

28

Solitons, the KdeV Equation ⊙

In this and the next chapter we look at soliton solutions of nonlinear wave equations. We have marked these chapters as optional because the material is more advanced and the techniques rather subtle. Nevertheless, we recommend that everyone at least read through these chapters because the material is fascinating and because the computer has been absolutely essential in the discovery and understanding of solitons. In recognition of the possible newness of this material to many students, we give additional background and explanatory materials.

28.1 INTRODUCTION

Up until now we have been dealing with *linear* partial differential equations. As is valid for ordinary differential equations, finding analytic solutions of linear equations is simplified by the principle of linear superposition, which tells us that the sum of two solutions is also a solution. When the description of a physical system is made more realistic by including higher-order effects, there frequently result nonlinear effects that may produce some unusual properties (as we are about to see). In most cases the nonlinear equations are no more difficult to solve numerically than linear equations, which is in contrast to trying to solve nonlinear equations analytically.

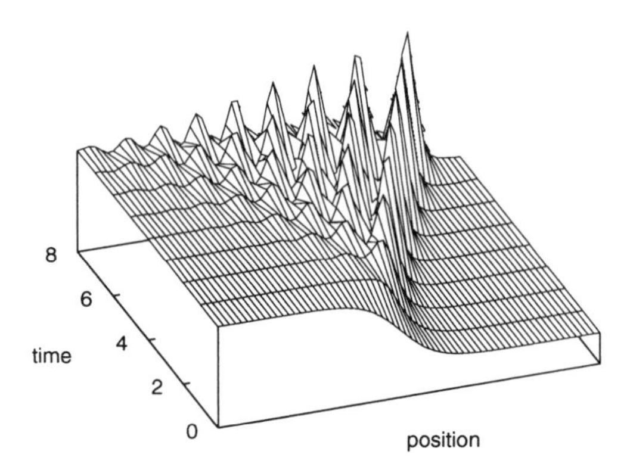

Fig. 28.1 A solution of the KdeV equation describing the behavior of shallow water waves. The single two-level waveform at time zero progressively breaks up into eight solitons as time increases. The taller solitons are then seen to be narrower and faster in their motion to the right.

28.2 PROBLEM: SOLITONS

Your **problem** is to discover whether nonlinear and dispersive systems can support waves with particle-like properties. While a logical response is that systems with dispersion have solutions that broaden in time and thereby lose their identity, consider Fig. 28.1 and the following experimental observation as the **problem** you need to explain. In 1834, J. Scott Russell observed a phenomenon on the Edinburgh–Glasgow canal [Russ 44]:

> *I was observing the motion of a boat which was rapidly drawn along a narrow channel by a pair of horses, when the boat suddenly stopped—not so the mass of water in the channel which it had put in motion; it accumulated round the prow of the vessel in a state of violent agitation, then suddenly leaving it behind, rolled forward with great velocity, assuming the form of a large solitary elevation, a rounded, smooth and well-defined heap of water, which continued its course along the channel apparently without change of form or diminution of speed. I followed it on horseback, and overtook it still rolling on at a rate of some eight or nine miles an hour, preserving its original figure some thirty feet long and a foot to a foot and a half in height. Its height gradually diminished, and after a chase of one or two miles I lost it in the windings of the channel. Such, in the month of August 1834, was my first chance interview with that singular and beautiful phenomenon ....*

Russell went on to produce these solitary waves in a laboratory and empirically deduced that their speed c is related to the depth h of the water in the canal and to the amplitude A of the wave by

$$c^2 = g(h + A), \tag{28.1}$$

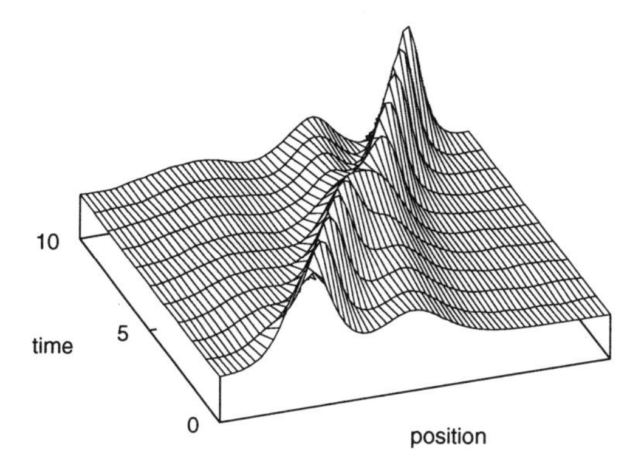

Fig. 28.2 Two shallow-water solitary waves crossing each other. The taller soliton, on the left at $t = 0$, catches up with the shorter one and overtakes it at $t \simeq 5$. The taller soliton is narrower because a soliton's height times its squared width is approximately constant. The taller soliton is faster because a soliton's height times its squared speed is approximately constant.

where g is the acceleration due to the gravity. Equation (28.1) implies an effect not found for linear systems, namely, that the waves with greater amplitudes travel faster than those with smaller amplitudes. Notice that this is different from *dispersion* in which waves of different wavelengths have different velocities. The former effect is illustrated in Fig. 28.2, where we see a tall soliton catching up with and passing through a short one. Russell also noticed that an initial, arbitrary waveform set in motion in the channel evolves into two or more waves that move at different velocities and progressively move apart until they form individual solitary waves. This effect is illustrated in Fig. 28.1, where we see a single step-like wave breaking up into approximately eight solitons (this shows why these eight solitons are considered the normal modes for this nonlinear systems).

28.3 THEORY: THE KORTEWEG–DE VRIES EQUATION

We want to understand these unusual water waves that occur in shallow, narrow channels such as canals [Abar 93, Tab 89]. The description of this "heap of water" was by [KdeV 95] in terms of the partial differential equation:

$$\frac{\partial u(x,t)}{\partial t} + \varepsilon u(x,t)\frac{\partial u(x,t)}{\partial x} + \mu\frac{\partial^3 u(x,t)}{\partial x^3} = 0. \qquad (28.2)$$

With a little hindsight it's possible to deduce the basic physics in (28.2) by inspection. There is a nonlinear term, $\varepsilon u \partial u/\partial t$, where the usual wave equation has a $c\partial u/\partial t$ term. This means that as long as u does not change too much, the waves propagate with a speed proportional to εu. In turn, this means that those parts of the wave that have a larger disturbance u move faster. This can lead to a sharpening of the wave and ultimately a *shock* wave. In contrast, the $\partial^3 u/\partial x^3$ term in (28.2) produces dispersive broadening that, for the proper conditions, can exactly compensate the narrowing caused by the nonlinear term.

Korteweg and deVries (KdeV) solved (28.2) and proved that the speed given by Russell, (28.1), is, in fact, correct. In more recent times, the KdeV equation was rediscovered 70 years later by [Z&K 65] who solved it numerically and discovered that a $\cos x/L$ initial condition broke up into eight solitary waves, similar to that shown in Fig. 28.1. They also found that those parts of the wave with larger amplitudes move faster than do those parts with smaller amplitudes, which is why the higher peaks tend to be on the right in Fig. 28.1. As if wonders never cease, Zabusky and Kruskal also observed from their numerical solution that the faster peaks actually passed through the slower one unscathed, as we can see in Fig. 28.2. Zabusky and Kruskal coined the name *soliton* for the solitary wave, and in the process launched a new branch of mathematics.

Before attempting to solve the KdeV equation, it's valuable to have some inkling of the physics lurking in its assorted parts. We can get that by looking at the equation in the limits in which different terms dominate. We start with the linear wave equation:

$$\frac{\partial u(x,t)}{\partial t} + c\frac{\partial u(x,t)}{\partial x} = 0. \tag{28.3}$$

Here we assign positive propagation velocity c to a wave traveling from left to right. An equation of this sort supports traveling-plane-wave solutions of the form

$$u(x,t) = e^{\pm i(kx-\omega t)}. \tag{28.4}$$

When this $u(x,t)$ is substituted into (28.3), we obtain a *dispersion relation*; that is, a relation between the frequency ω and the wave vector k:

$$\omega = \pm ck, \quad \text{or} \quad \omega^2 = c^2k^2. \tag{28.5}$$

The second form of (28.5) is fine as long as we remember that $c < 0$ implies a wave traveling from right to left. When a wave obeys a dispersion law such as (28.5), the wave velocity $c = \omega/k$ is independent of frequency ω and we call such wave propagation *dispersionless*.

Let us deduce a wave equation that describes waves traveling with a small amount of *dispersion*; that is, with a frequency that decreases slightly as the

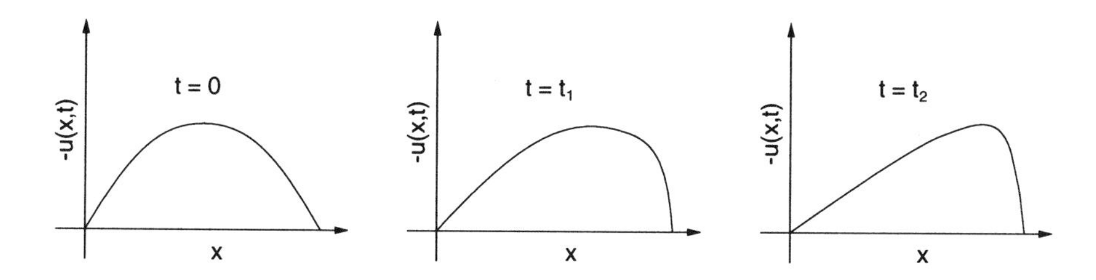

Fig. 28.3 A sketch of a shock wave accumulating as time progresses.

wave number k increases. We posulate

$$\omega^2 \simeq c^2k^2 - \Gamma k^4, \tag{28.6}$$

where only even powers of k occur to reflect the symmetry in the $\pm x$ direction. There are two forms for the solution:

$$\omega = \pm\sqrt{c^2k^2 - \Gamma k^4} \simeq \pm ck\left(1 - \frac{\Gamma k^2}{2c^2}\right), \tag{28.7}$$

$$\Rightarrow \quad \omega \simeq \pm ck \mp \beta k^3, \tag{28.8}$$

where β is a constant, and where the upper sign is for waves moving from left to right.

If plane-wave solutions like (28.4) arise from a wave equation, then we see that the ω term of the dispersion relation arises from a first-order time derivative, that the ck term arises from a first-order space derivative, and that the k^3 term arises from a third-order space derivative. As can be verified by substitution, the wave equation needed to produce the dispersion relation (28.8) is

$$\frac{\partial u(x,t)}{\partial t} + c\frac{\partial u(x,t)}{\partial x} + \beta\frac{\partial^3 u(x,t)}{\partial x^3} = 0. \tag{28.9}$$

Already we have a form close to the KdeV equation.

Next let us examine the small $\partial^3 u/\partial x^3$ limit of the the KdeV equation,

$$\frac{\partial u}{\partial t} + \epsilon u\frac{\partial u}{\partial x} \simeq 0. \tag{28.10}$$

Equation (28.10), which for small u is almost a linear equation, is an equation of the type that produces *shock waves* [Tab 89]. This is sketched in Fig. 28.3 for $\epsilon = -6$. This tells us that the nonlinear term in the KdeV equation introduces the possibility of shock waves into the solution.

28.4 METHOD, ANALYTIC

The trick in analytic approaches to these types of nonlinear equations is to look for steady-state solutions that have the form of a traveling wave; that is, with a specific dependence on x and t:

$$u(x,t) = f(\xi = x - ct). \tag{28.11}$$

The traveling-wave form (28.11) means that if we move with a constant speed c, we see a constant phase (yet the speed depends on the magnitude of u). There is no guarantee that this form of a solution exists, but if we are lucky, this substitution converts the partial differential equation into an ordinary differential equation that we can solve. If we take the KdeV equation and make the substitution (28.11), we obtain the ODE

$$\frac{d^3 f}{dt^3} - 6f\frac{df}{dt} - c\frac{df}{dt} = 0. \tag{28.12}$$

While solving this differential equation may still be somewhat of a challenge, mathematicians are good at that sort of thing and have come up with the inverse solution

$$\xi - \xi_0 = \int \frac{df}{\sqrt{2(f^3 + \frac{1}{2}cf^2 + df + e)}}, \tag{28.13}$$

where d and e are integration constants. If we demand as boundary conditions that f, f', and $f'' \to 0$ as $\xi \to \pm\infty$, we can invert (28.13) to obtain the solution

$$u(x,t) = \frac{-c}{2}\text{sech}^2\left[\frac{1}{2}\sqrt{c}(x - ct - \xi_0)\right]. \tag{28.14}$$

We see in (28.14) an amplitude that is proportional to the wave speed c, and a sech^2 function which gives a single lump-like wave. This is a typical mathematical form for a soliton.

28.5 METHOD, NUMERIC: FINITE DIFFERENCE

The KdeV equation is solved numerically using a centered, finite-difference scheme. The time derivative is expressed as a difference centered at t:

$$\frac{\partial u(x,t)}{\partial t} \simeq \frac{u(x,t+\Delta t) - u(x,t-\Delta t)}{2\Delta t}, \tag{28.15}$$

and we solve the equation on a spacetime grid:

$$x = i\Delta x, \qquad t = j\Delta t. \tag{28.16}$$

In terms of the discrete variables, the expansions of $u(x, t+\Delta t)$ and $u(x, t-\Delta t)$ in Taylor series gives

$$\frac{\partial u}{\partial t}(i,j) \simeq \frac{u(i,j+1) - u(i,j-1)}{2\Delta t}, \tag{28.17}$$

where terms of order $\mathcal{O}(\Delta t)^5$ are neglected. Similarly, the x derivative is

$$\frac{\partial u}{\partial x}(i,j) \simeq \frac{u(i+1,j) - u(i-1,j)}{2\Delta x}. \tag{28.18}$$

To approximate $\partial^3 u(x,t)/\partial x^3$, we expand $u(x,t)$ to $\mathcal{O}(\Delta t)^5$ about the four points $u(x \pm 2\Delta x, t)$ and $u(x \pm \Delta x, t)$. A typical expansion is

$$u(x \pm \Delta x, t) \simeq u(x,t) \pm (\Delta x)\frac{\partial u}{\partial x} + \frac{(\Delta x)^2}{2!}\frac{\partial^2 u}{\partial^2 x} \pm \frac{(\Delta x)^3}{3!}\frac{\partial^3 u}{\partial x^3}, \tag{28.19}$$

which we can solve for $\partial^3 u(x,t)/\partial x^3$. Finally, the factor $u(x,t)$ in the second term in (28.2) is taken as the average of the three values centered at (i,j) in a row (for time t or index j):

$$u(i,j) \simeq \frac{u(i+1,j) + u(i,j) + u(i-1,j)}{3}. \tag{28.20}$$

After substituting all of these expansions, we obtain the finite-difference form of the KdeV equation:

$$\boxed{\begin{aligned} u(i,j+1) &\simeq u(i,j-1) - \tfrac{\epsilon}{3}\tfrac{\Delta t}{\Delta x}\left[u(i+1,j) + u(i,j) + u(i-1,j)\right] \\ &\times \left[u(i+1,j) - u(i-1,j)\right] - \mu\tfrac{\Delta t}{(\Delta x)^3} \\ &\times \left[u(i+2,j) + 2u(i-1,j) - 2u(i+1,j) - u(i-2,j)\right]. \end{aligned}} \tag{28.21}$$

To apply this algorithm, we need to know $u(x,t)$ at present and past times to predict future times. We note that the solution for initial time $u(i,1)$ is known for all positions i because the initial condition on $u(x, t=0)$ is some known function of x. To find $u(i,2)$, we use a noncentered scheme in which we expand $u(x,t)$ keeping only two terms for the time derivative:

$$\begin{aligned} u(i,2) &\simeq u(i,1) \\ &-\frac{\epsilon}{6}\frac{\Delta t}{\Delta x}\left[u(i+1,1) + u(i,1) + u(i-1,1)\right]\left[u(i+1,j) - u(i-1,j)\right] \\ &-\frac{\mu}{2}\frac{\Delta t}{(\Delta x)^3}\left[u(i+2,1) + 2u(i-1,1) - 2u(i+1,1) - u(i-2,1)\right]. \end{aligned} \tag{28.22}$$

The keen observer will note that there are still some undefined columns of points, namely, $u(1,j)$, $u(2,j)$, $u(N_{\text{max}}-1,j)$, and $u(N_{\text{max}},j)$, where N_{max} is the total number of grid points. A simple technique for determining their values is to assume that $u(1,2)=1$ and $u(N_{\text{max}},2)=0$. To obtain $u(2,2)$ and $u(N_{\text{max}}-1,2)$, assume that $u(i+2,2)=u(i+1,2)$ and $u(i-2,2)=u(i-1,2)$ [avoid $u(i+2,2)$ for $i=N_{\text{max}}-1$, and $u(i-2,2)$ for $i=2$]. To carry out these steps, approximate (28.22) so that

$$u(i+2,2)+2u(i-1,2)-2u(i+1,1)-u(i-2,2)\rightarrow u(i-1,2)-u(i+1,2). \quad (28.23)$$

The stability condition for this method of solution is

$$\frac{\Delta t}{\Delta x}\left[\epsilon|u|+4\frac{\mu}{(\Delta x)^2}\right]\leq 1 \qquad \text{(stability)}. \quad (28.24)$$

The truncation error is:

$$\mathcal{E}(u)=\mathcal{O}[(\Delta t)^3]+\mathcal{O}[\Delta t(\Delta x)^2]. \quad (28.25)$$

These last two equations are illuminating. They show that smaller time and space steps do lead to smaller truncation error, but, as discussed in Chapter 3, *Errors and Uncertainties in Computations*, not necessarily smaller *total* error because roundoff error increases with more steps. We also see that a progressive decrease of the space steps, or even of both space and time steps, will ultimately lead to instability. Care and experimentation are clearly required.

28.6 IMPLEMENTATION: KDEV SOLITONS, SOLITON.F (.C)

Modify or run the program given on the diskette and Web that solves the KdeV equation (28.2) for the initial condition:

$$u(x,t=0)=\frac{1}{2}\left[1-\tanh\left(\frac{x-25}{5}\right)\right], \quad (28.26)$$

with parameters $\epsilon=0.2$ and $\mu=0.1$. Start with $\Delta x=0.4$ and $\Delta t=0.1$. These constants are chosen to satisfy (28.24) with $|u|=1$.

1. Define a 2-D array `u(131,3)` with the the first index corresponding to the position x and the second to the time t. With our choice of parameters, the maximum value for x is $130\times 0.4=52$.

2. Initialize the time to $t=0$ and assign values to `u(i,1)` using (28.26).

3. Assign values to `u(i,2)`, i=3, 4, ..., 129 corresponding to the next time interval. Use (28.22) to advance the time, but note that you cannot start at $i=1$ nor end at $i=131$ because (28.22) would include `u(132,2)`

and `(u,132)`, which are beyond the limits of the array.

4. Increment the time and assume that `u(1,2)`=1 and `u(131,2)`=0. To obtain `u(2,2)` and `u(130,2)`, assume that `u(i+2,2)`=`u(i+1,2)` and `u(i-2,2)`=`u(i-1,2)`. Avoid `u(i+2,2)` for `i=130`, and `u(i-2,2)` for `i=2`. To do this, approximate (28.22) so that (28.23) is satisfied.

5. Increment time and compute `u(i,j)` for `j=3` and for `i=3`, 4, ..., 129, using equation (28.21). Again follow the same procedures to obtain the missing array elements `u(2,j)` and `u(130,j)` [set `u(1,j)`=1.0 and `u(131,j)`=0]. As you print out the numbers during the iterations, you will be convinced that it was a good choice.

6. Set `u(i,1)`=`u(i,2)` and `u(i,2)`=`u(i,3)` for all i. In this way you are ready to find the next `u(i,j)` in terms of the previous two rows.

7. Repeat the previous two steps some 2000 times. Write your solution out to a file after every ~250 iterations.

28.7 ASSESSMENT: VISUALIZATION

1. Use your favorite graphics tool to plot your results as a 3-D graph of disturbance u versus position *and* versus time.

2. Observe the wave profile as a function of time and try to confirm Russell's observation that a taller soliton travels faster than a smaller one.

28.8 EXPLORATION: TWO SOLITONS CROSSING

Explore what happens when a tall soliton collides with a short one. Do they bounce off each other? Do they go through each other? Do they interfere? Do they destroy each other? Does the tall soliton still move faster than the short one after collision? (One result we obtained is shown in Fig. 28.2.) Start off by placing a tall soliton of height 0.8 at $x = 12$, and a smaller soliton in front of it at $x = 26$:

$$u(x, t = 0) = 0.8 \left[1 - \tanh^2\left(\frac{3x}{12} - 3\right)\right] + 0.3 \left[1 - \tanh^2\left(\frac{4.5x}{26} - 4.5\right)\right]. \tag{28.27}$$

The procedure is now similar to the example.

1. The KdeV equation (28.2) now has the constants $\mu = 0.1$, $\epsilon = 0.2$, $\Delta = 0.029$, and the numerical algorithm has step sizes $\Delta x = 0.4$, $\Delta t = 0.1$.

2. Establish the initial condition (28.27).

3. Try 4000 time steps with printouts every 400 iterations.

28.9 EXPLORATION: PHASE-SPACE BEHAVIOR

Construct phase-space plots of the KdeV equation for various parameter values. Note that only very specific sets of parameters produce solitons. In particular, by correlating the behavior of the solutions with your phase-space plots, show that the soliton solutions correspond to the *separatrix* solutions to the KdeV equation.

28.10 EXPLORATION: SHOCK WAVES

Study the solutions of the simple PDE (28.10) and show that it produces shock waves.

29

Sine–Gordon Solitons ⊙

This chapter extends the optional study of solitons that began in the previous chapter. Here we concentrate on 2-D solitons or *pulsons* as might occur in field-theoretic models for elementary particles, or as disturbances in an nonlinear elastic medium. We base our implementations on the work of a number of sources [C&L 81, C&O 78, Argy 91].

29.1 PROBLEM 1: ELEMENTARY PARTICLES FROM FIELD EQUATIONS

For field theory to produce what we normally think of as particles, we need solutions to the field equations that are confined to a region of space for a long time and that do not radiate away their energy. Your **problem** is to see if something as singular and individual as an elementary particle can occur as the solution of a wave equation.

29.2 THEORY: PULSON BEHAVIOR OF CIRCULAR RING SOLITONS

There arises in field theories the wave equation for mesons:

$$\nabla^2 u(\mathbf{x},t) - \frac{\partial^2 u(\mathbf{x},t)}{\partial t^2} = m^2 u(\mathbf{x},t) + g^2 u^3(\mathbf{x},t). \tag{29.1}$$

While (29.1) is similar to the relativistic Klein–Gordon equation [L 96], it also contains a nonlinear u^3 term that leads to solitons like those studied in Chapter 28, *Solitons, the KdeV Equation.* Sine–Gordon solitons are a useful model for elementary particles.

Equation (29.1) can be obtained as an approximate form of the 2-D sine–Gordon equation,

$$\frac{\partial^2 u}{\partial x^2} + \frac{\partial^2 u}{\partial y^2} - \frac{\partial^2 u}{\partial t^2} = \sin u, \tag{29.2}$$

when the $\sin u$ on the RHS of (29.2) is expanded. Because both equations have similar solutions, and because the sine–Gordon equation (29.2) is more common, we shall study it. Before doing that, we look at another, somewhat more concrete system that has the same equation of motion.

29.3 PROBLEM 2: DISPERSIONLESS DISPERSIVE CHAINS

In 1955, Fermi, Ulam, and Pastu were investigating how a 1-D chain of coupled oscillators disperses waves. Since different harmonics traveled through the chain with different speeds [as we will show in (29.9)], a pulse, which is the sum of many harmonics, has each of its components traveling with a different speed and correspondingly broadens as time progresses. As the pulse travels down the chain, the very large number of k values (large Δk) needed initially to create a sharp pulse (small Δx), decreases because the chain cuts off the higher and lower k values.[1] A smaller Δk naturally leads to a larger Δx; that is, to a broadening pulse. When the oscillators were made more realistic by introducing a nonlinear terms into Hooke's law

$$F(x) \simeq -k(x + \alpha x^2), \tag{29.3}$$

it was found that for the right conditions, a sharp pulse could survive indefinitely.

Your **problem** is to explain this discovery.

29.4 THEORY: COUPLED PENDULA

We studied earlier how a nonlinearity introduces higher harmonics to a system. If entering just right, the additional harmonics tend to sharpen the wave to the same extent that the dispersion tends to broaden it, and a stable pulse results. This is now recognized as a soliton, and is what we wish to study.

[1] The understanding of this phenomenon follows from basic Fourier analysis, which tells us that the space uncertainty and wave vector uncertainty are related by $\Delta x \Delta k \geq \sqrt{2\pi}$.

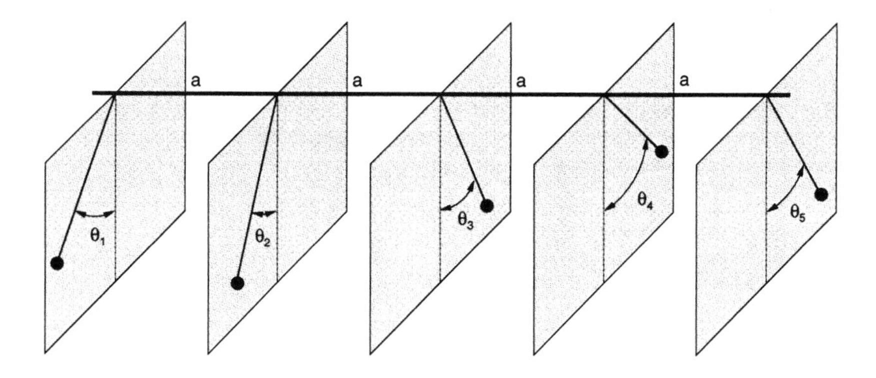

Fig. 29.1 A 1-D chain of pendula coupled with a torsion bar on top. The pendula swing in planes perpendicular to the length of the bar.

The sine–Gordon equation (SGE) arises naturally in classical mechanics.[2] As a specific model, consider a 1-D chain of identical pendula connected by a torsion bar, as illustrated in Fig. 29.1. For each pendulum, θ measures its displacement from equilibrium. While gravity causes each pendulum to return to its equilibrium position, the torques from the twisting of the bar causes the motion of one pendulum to influence its neighbors. The equation of motion for pendulum j follows from Newton's law for rotational motion in terms of torques τ:

$$\sum_{j\neq i} \tau_{ji} = I\alpha_j \tag{29.4}$$

$$-\kappa(\theta_j - \theta_{j-1}) - \kappa(\theta_j - \theta_{j+1}) - mg\sin\theta_j = I\frac{d^2\theta_j(t)}{dt^2}, \tag{29.5}$$

$$\kappa(\theta_{j+1} - 2\theta_j + \theta_{j-1}) - mg\sin\theta_j = I\frac{d^2\theta_j(t)}{dt^2}, \tag{29.6}$$

where I is the moment of inertia of each pendulum, $\alpha_j = d^2\theta_j(t)/dt^2$ is the angular acceleration of pendulum j, and κ is the torque constant for each bar.

The nonlinearity in (29.5) arises from the gravitational $\sin\theta$ term; similar solutions would result if we also included nonlinear terms from the bar. As it stands, (29.5) is a set of coupled nonlinear equations, with the number of equations equal to the number of oscillators. If all the pendula are set off swinging together, $\theta_i \equiv \theta_j$, the coupling terms would vanish and we would have our old friend, the equation for a realistic (albeit, very thick) pendulum.

[2]The name "sine–Gordon" is either a reminder that the SGE is like the Klein–Gordon equation of relativistic quantum mechanics with a $\sin u$ added to the RHS, or a reminder of how clever one can be in thinking up names.

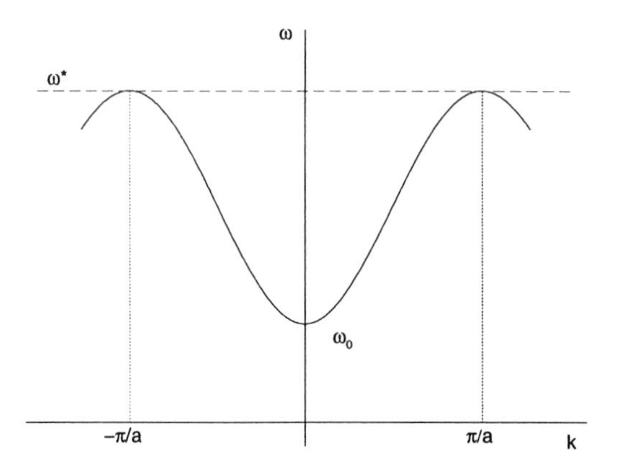

Fig. 29.2 The dispersion relation for a linearized chain of pendula.

29.4.1 Dispersion in Linear Chain

To understand the dispersion aspect of (29.5), we look at its linear version

$$\frac{d^2\theta_j(t)}{dt^2} + \omega_0^2\theta_j(t) = \frac{\kappa}{I}(\theta_{j+1} - 2\theta_j + \theta_{j-1}), \tag{29.7}$$

where $\omega_0 = mg/I$ is the natural frequency for any one pendulum. Because we are interesting in seeing if waves of a single frequency can propagate on a chain such as this, we *guess* a single-frequency, traveling-wave solution:

$$\theta_j(t) = Ae^{i(\omega t - ka_j)} + \text{cc}, \tag{29.8}$$

where it is understood that the physical solution we want is the real one obtained by adding in the complex conjugate (cc). Substitution of (29.8) into the wave equation (29.7) produces the *dispersion relation:*

$$\omega^2 = \omega_0^2 + \frac{2\kappa}{I}(1 - \cos ka), \quad \text{(nonlinear dispersion)}. \tag{29.9}$$

This is shown in Fig. 29.2.

In dispersionless propagation we have

$$\lambda = \frac{2\pi}{\omega} \quad \Rightarrow \quad \omega = ck, \quad \text{(dispersionless)}. \tag{29.10}$$

This means that waves of all frequencies can propagate and that the frequency is proportional to the wave vector. In contrast, in order to have solutions of

(29.9) with real k; that is, to have solutions that propagate, we must have

$$\omega_0 \leq \omega \leq \omega^*, \quad \text{(waves propagate)}. \tag{29.11}$$

The minimum frequency ω_0 and the maximum frequency ω^* are related through the limits of $\cos ka$ in (29.9),

$$(\omega^*)^2 = \omega_0^2 + \frac{4\kappa}{I}. \tag{29.12}$$

Waves with frequency less than ω_0 do not propagate, and we shall see that it is unphysical to have waves with wavelength $\lambda < 2a$ and $\omega > \omega^*$.

We see from the dispersion relation that for $k = 0$, which corresponds to all torsion bars untwisted, ω equals the natural frequency of a free pendulum ω_0. For nonzero values of k, the bars add an additional restoring torque to the gravitational one, and this leads to an increased ω (shorter period). Yet when we get to $k = \pi/a$ ($\omega = \omega^*$), we have a wavelength $\lambda = 2a$, which is the shortest wavelength a chain of repeat distance a can support. A wavelength of a, which is half of the shortest wavelength, would produce exactly the same configuration and therefore be physically indistinguishable.

29.4.2 Continuum Limit, the Sine–Gordon Equation

If the wavelengths in a pulse are much longer than the repeat distance a; that is, if $ka \ll 1$, the chain can be approximated as a continuous medium. In this case a becomes the continuous variable x, and the system of coupled ordinary differential equations becomes the single, partial differential equation:

$$\theta_{j+1} \simeq \theta_j + \frac{\partial \theta}{\partial x}\Delta x, \tag{29.13}$$

$$\Rightarrow \quad (\theta_{j+1} - 2\theta_j + \theta_{j-1}) \simeq \frac{\partial^2 \theta}{\partial x^2}\Delta x^2 = \frac{\partial^2 \theta}{\partial x^2}a^2, \tag{29.14}$$

$$\Rightarrow \quad \frac{\partial^2 \theta}{\partial t^2} - \frac{\kappa a}{I}\frac{\partial^2 \theta}{\partial x^2} = \frac{mg}{I}\sin\theta. \tag{29.15}$$

To obtain the standard form of the sine–Gordon equation

$$\frac{\partial^2 \theta}{\partial t^2} - \frac{\partial^2 \theta}{\partial x^2} = \sin\theta, \tag{29.16}$$

we measure time in units of $\sqrt{I/mg}$ and distances in units of $\sqrt{\kappa a/(mg)}$.

The 3-D generalization of (29.16) extends the spatial derivatives:

$$\boxed{\frac{\partial^2 \theta}{\partial t^2} - \nabla^2 \theta = \sin\theta.} \tag{29.17}$$

In either case, the $\sin\theta$ on the RHS makes the equation nonlinear for all but the smallest θ values.

29.5 SOLUTION: ANALYTIC

The nonlinearity of the sine–Gordon equation (29.16) makes it hard to solve analytically. While there are no overwhelming obstacles to solving it numerically, the solutions appear complicated because there occurs a continuous spectrum of coupled propagation modes. As done in Chapter 28, *Solitons, the KdeV Equation*, the trick is to guess a functional form of a traveling wave as the solution and hope that this converts the PDE into an ODE:

$$\theta(x,t) \stackrel{?}{=} \theta(\xi = t \pm x/v), \tag{29.18}$$

$$\Rightarrow \frac{d^2\theta}{d\xi^2} = \frac{v^2}{v^2-1}\sin\theta. \tag{29.19}$$

After a little inspection and retrospection, you may recall that (29.19) is our old friend, the equation of motion for the realistic pendulum with no driving force and no friction. The constant v is a velocity in natural units, and separates different regimes of the motion:

$$\begin{array}{lll} v<1: & \text{all pendula initially pointing down} \ \downarrow\downarrow\downarrow\downarrow\downarrow & \text{(stable)},\\ v>1: & \text{all pendula initially pointing up} \ \uparrow\uparrow\uparrow\uparrow\uparrow & \text{(unstable)}. \end{array} \tag{29.20}$$

Because (29.19) is the equation of motion for a pendulum, we know from classical mechanics and Chapter 9, *Differential Equations and Oscillations*, that there is an integral of the motion (usually associated with the total energy) and we can use that to obtain an integral solution for θ. While this does not yield a general solution, for motion along the separatrix ($E = \pm 1$) we obtain the characteristic soliton form,

$$\theta(x-vt) = \begin{cases} 4\tan^{-1}\left(\exp\left[+\frac{x-vt}{\sqrt{1-v^2}}\right]\right), & \text{for } E=1,\\ 4\tan^{-1}\left(\exp\left[-\frac{x-vt}{\sqrt{1-v^2}}\right]\right)+\pi, & \text{for } E=-1. \end{cases} \tag{29.21}$$

This soliton corresponds to a solitary *kink* traveling with $v = -1$ that flips the pendula around by 2π as it moves down the chain. There is also an *antikink* in which the initial $\theta = \pi$ values are flipped to final $\theta = -\pi$ along the chain.

29.6 SOLUTION: NUMERIC

We solve the SGE in a finite region of 2-D space and for positive times:

$$-x_0 < x < x_0, \qquad -y_0 < y < y_0, \qquad 0 \le t. \tag{29.22}$$

We take $x_0 = y_0 = 7$ and impose the *boundary conditions* that the derivative of the displacement vanishes at the ends of the region:

$$\frac{\partial u}{\partial x}(-x_0, y, t) = \frac{\partial u}{\partial x}(x_0, y, t) = \frac{\partial u}{\partial y}(x, -y_0, t) = \frac{\partial u}{\partial y}(x, y_0, t) = 0. \tag{29.23}$$

We also impose the *initial condition* that at time $t = 0$ the waveform is that of a droplet (see Fig. 29.3) with its surface at rest:

$$u(x, y, t = 0) = 4 \tan^{-1}(e^{3-\sqrt{x^2+y^2}}), \tag{29.24}$$

$$\frac{\partial u}{\partial t}(x, y, t = 0) = 0. \tag{29.25}$$

We use a discretization process similar to the ones used previously with time step Δt and space steps $\Delta x = \Delta y$, and thereby transform from continuous variables (x, y, t) to discrete variables (m, l, n):

$$x = m\Delta x, \qquad y = l\Delta x, \qquad t = n\Delta t. \tag{29.26}$$

We adopt a notation in which the time index is used as a superscript and the space indices as subscripts:

$$\boxed{u^n_{m,l} \stackrel{\text{def}}{=} u(m\Delta x, l\Delta x, n\Delta t).} \tag{29.27}$$

After making all the usual expansions, the finite-difference SGE is obtained:

$$\begin{aligned} u^{n+1}_{m,l} \simeq & -u^{n-1}_{m,l} + 2\left[1 - 2\left(\frac{\Delta t}{\Delta x}\right)^2\right] u^n_{m,l} \\ & + \left(\frac{\Delta t}{\Delta x}\right)^2 \left(u^n_{m+1,l} + u^n_{m-1,l} + u^n_{m,l+1} + u^n_{m,l-1}\right) \\ & -\Delta t^2 \sin\left[\tfrac{1}{4}\left(u^n_{m+1,l} + u^n_{m-1,l} + u^n_{m,l+1} + u^n_{m,l-1}\right)\right]. \end{aligned} \tag{29.28}$$

The discrete form for the initial condition of vanishing velocity (29.25), becomes

$$\partial u(x, y, 0)/\partial t = 0 \quad \Rightarrow \quad u^2_{m,l} = u^0_{m,l}. \tag{29.29}$$

This will be useful in getting the time propagation started.

To make the algorithm simpler and ensure stability, we assume that the

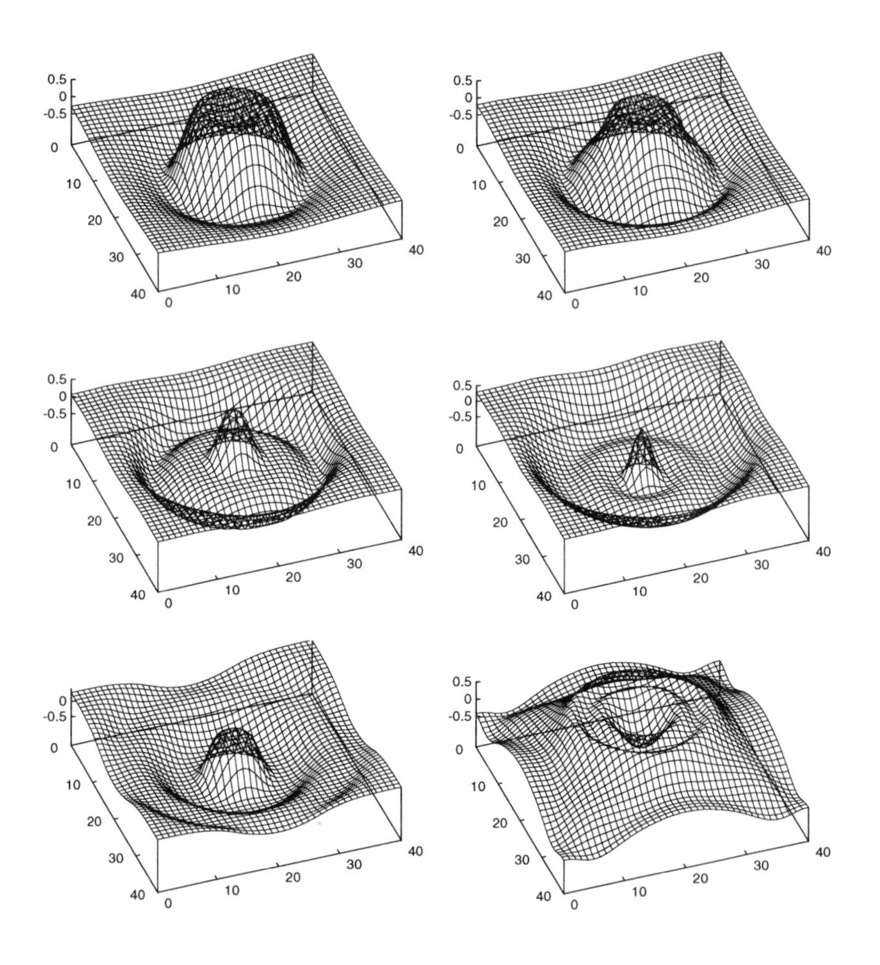

Fig. 29.3 A circular ring soliton viewed as a model for an elementary particle. The six frames correspond to times 8, 20, 40, 60, 80, and 120.

time and space steps to be proportional:

$$\Delta t = \frac{\Delta x}{\sqrt{2}}. \tag{29.30}$$

In this way all the $u^n_{m,l}$ terms drop out and the SGE becomes

$$\begin{aligned} u^2_{m,l} &\simeq \tfrac{1}{2}\left(u^1_{m+1,l} + u^1_{m-1,l} + u^1_{m,l+1} + u^1_{m,l-1}\right) \\ &- \frac{\Delta t^2}{2} \sin\left[\tfrac{1}{4}\left(u^1_{m+1,l} + u^1_{m-1,l} + u^1_{m,l+1} + u^1_{m,l-1}\right)\right]. \end{aligned} \tag{29.31}$$

It is not necessary to know $u^0_{m,l}$ to solve these equations. There are, nonetheless, some lattice points that cannot be obtained from these equations, specifically, the edges and corners. They are obtained by applying the boundary conditions (29.23):

$$\frac{\partial u}{\partial z}(x_0, y, t) = \frac{u(x+\Delta x, y, t) - u(x, y, t)}{\Delta x} = 0, \tag{29.32}$$

$$\Rightarrow \quad u^n_{1,l} = u^n_{2,l}. \tag{29.33}$$

Similarly, the other derivatives in (29.23) give

$$u^n_{N_{\max},l} = u^n_{N_{\max}-1,l} \tag{29.34}$$

$$u^n_{m,2} = u^n_{m,1} \tag{29.35}$$

$$u^n_{m,N_{\max}} = u^n_{m,N_{\max}-1}, \tag{29.36}$$

where $N_{\max}$ is the total number of grid points used in any one space dimension.

29.7 IMPLEMENTATION, 2-D SOLITONS, TWODSOL.F (.C)

1. Define an array $u(N_{\max}, N_{\max}, 3)$ with $N_{\max} = 201$ for the space slots and 3 for the time slots.

2. The solution (29.24) for the initial time $t = 0$ is placed in $u(m, l, 1)$.

3. The solution for the second time Δt is placed in $u(m, l, 2)$, and the solution for the next time, $2\Delta t$, is placed in $u(m, l, 3)$.

4. Assign values to the constants, $\Delta x = \Delta y = \frac{7}{100}$, $\Delta t = \Delta x/\sqrt{2}$, $y_0 = x_0 = 7$.

5. Start off at $t = 0$ with the initial conditions and impose the boundary conditions to this initial solution. This is the solution for the first time step, defined over the entire 201×201 grid.

6. For the second time step, increase time by Δt and use (29.31) for all

points in the plane. Do not include the edge points.

7. At the edges, for $i = 1, 2, \ldots, 200$, set

$$
\begin{aligned}
u(i,1,2) &= u(i,2,2), & (29.37)\\
u(i,N_{\max},2) &= u(i,N_{\max}-1,2) & (29.38)\\
u(1,i,2) &= u(2,i,2), & (29.39)\\
u(N_{\max},i,2) &= u(N_{\max}-1,i,2). & (29.40)
\end{aligned}
$$

8. To get values for the four points in the corners for the second time step, again use initial condition (29.31):

$$
\begin{aligned}
u(1,1,2) &= u(2,1,2), & (29.41)\\
u(N_{\max},1,2) &= u(N_{\max}-1,1,2), & (29.42)\\
u(1,1,N_{\max}) &= u(2,N_{\max},2), & (29.43)\\
u(N_{\max},N_{\max},2) &= u(N_{\max}-1,N_{\max}-1,2). & (29.44)
\end{aligned}
$$

9. For the third time step (the future), use (29.31).

10. Continue the propagation forward in time, reassigning the future to the present, and so forth. In this way the solutions for only three time steps need to be stored.

29.8 VISUALIZATION

We see in Fig. 29.3 the time evolution of a circular ring soliton for the stated initial conditions (these results are not critically dependent on the initial conditions). We note that the ring at first shrinks in size, then expands, and then shrinks back into another (but not identical) ring soliton. A small amount of the particle does radiate away, and in the last frame we can notice some interference between the radiation and the boundary conditions. An animation of this sequence can be found on the Web.

30

Confined Electronic Wave Packets ⊙

30.1 PROBLEM: A CONFINED ELECTRON

An electron is initially confined to a one-dimensional region of space the size of an atom. Your **problem** is to determine how long in time the electron remains confined. This is different from the problem of a particle confined to a box considered in Chapter 10, *Quantum Eigenvalues; Zero-Finding and Matching*. There we had a time-independent situation in which we had to solve for the spatial wave function; here we have a time-dependent problem in which we know the wave function at time zero, and even though that state is not in an eigenstate or stationary state of the Hamiltonian, we wish to determine it for all future times.

30.2 MODEL: TIME-DEPENDENT SCHRÖDINGER EQUATION

We use a wave function (or wave packet) $\psi(x,t)$ that is a function of the position x and time t to describe a localized electron. We assume that the electron is initially localized around $x = 5$, and model this by a Gaussian wave function multiplying a plane wave:

$$\psi(x, t=0) = \exp\left[-\frac{1}{2}\left(\frac{x-5.0}{\sigma_0}\right)^2\right] e^{ik_0 x}. \tag{30.1}$$

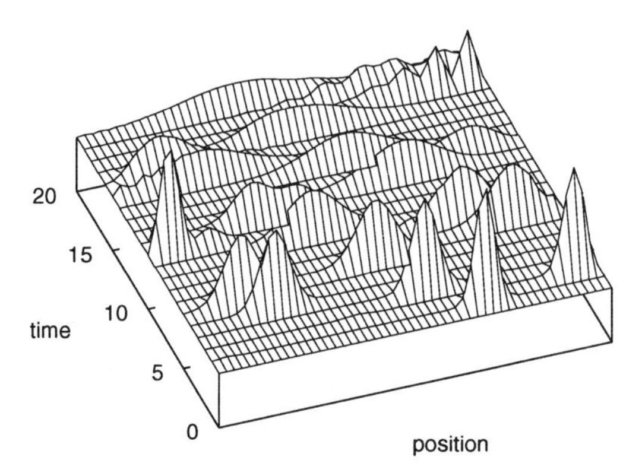

Fig. 30.1 The position as a function of time of a localized electron confined to a square well. The electron is initially on the right with a Gaussian wave packet. In time, the wave packet spreads out and collides with the walls.

The behavior of this wave packet as a function of time, when placed in a square well, is shown in Fig. 30.1. The behavior, when placed in an harmonic oscillator potential, is shown in Fig. 30.2.

As you may verify by applying the momentum operator $\tilde{p} = id/dx$, the wave packet (30.1) does not correspond to an electron with a definite momentum (that is, it is not an eigenstate of $\tilde{p}$).[1] However, if the width σ_0 of the Gaussian is made very large, the electron gets spread over a sufficiently large region of space to consider the wave packet as a plane wave of momentum k_0 with a slowly varying amplitude.

The time and space evolution of a quantum particle is described by the time-dependent Schrödinger equation (here in one dimension):

$$i\frac{\partial\psi(x,t)}{\partial t} = \tilde{H}\psi(x,t). \tag{30.2}$$

Here $\tilde{H}$ is the Hamiltonian operator:

$$\tilde{H} = -\frac{1}{2m}\frac{\partial^2}{\partial x^2} + V(x), \tag{30.3}$$

where we have set $2m = 1$ to keep the equations simple, and use a partial x derivative because ψ is also a function of t.

An important aspect of quantum mechanics is that the wave function is

[1] We use natural units in which $\hbar = 1$, so there is no difference between momentum and wave numbers [L 96, Appendix A.1].

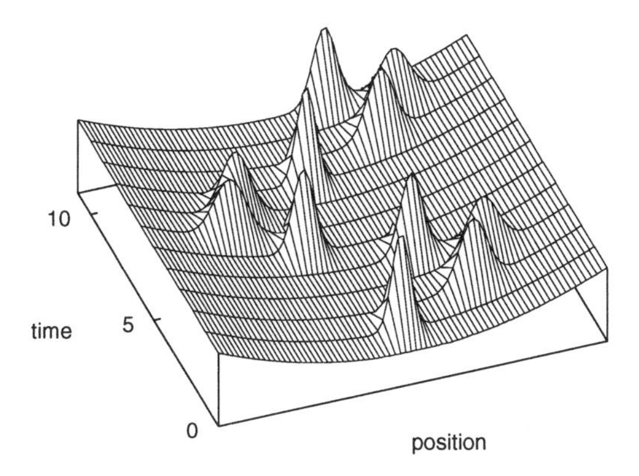

Fig. 30.2 The probability density as a function of time for an electron confined to a 1-D harmonic oscillator potential well. The electron's initial localization is described by the Gaussian wave packet (30.1). Because the wave packet is an eigenfunction of the potential, it does not break up on collision but instead returns to its original form.

complex (because it is not directly observable, this is not a problem). Even though many computer languages can handle complex functions, we will find it advantageous to decompose the wave function into its real and imaginary parts:

$$\psi(x,t) = R(x,t) + iI(x,t). \tag{30.4}$$

Substitution of (30.4) into Schrödinger equation (30.2) produces the coupled PDEs:

$$\frac{\partial R(x,t)}{\partial t} = +\tilde{H}I(x,t) = -\frac{1}{2m}\frac{\partial^2 I(x,t)}{\partial x^2} + V(x)I(x,t), \tag{30.5}$$

$$\frac{\partial I(x,t)}{\partial t} = -\tilde{H}R(x,t) = +\frac{1}{2m}\frac{\partial^2 R(x,t)}{\partial x^2} + V(x)R(x,t), \tag{30.6}$$

where the Hamiltonian operator $\tilde{H}$ is assumed real.

30.3 METHOD, NUMERIC: FINITE DIFFERENCE

The time-dependent Schrödinger equation can be solved with both implicit and explicit methods. An *implicit* method [Gold 67] converts the PDEs into a very large set of simultaneous linear equations involving the wave function evaluated at each grid point, and then solves these linear equations by matrix inversion for each time. This can cause problems when the matrices get very large. For our project, we modify the *explicit* method described by [Ask 77] and [Viss 91]. This is an iterative scheme that avoids the inversion of large matrices.

There are some challenges in solving the Schrödinger equation. First, we need an algorithm that converges and is stable, the usual concerns. Second, we need an algorithm that ensures, at least to some order, that probability is conserved with time; otherwise the electron will fade away right before our eyes. A good solution to the probability problem is to determine the real and imaginary parts of the wave function at slightly different or "staggered" times. Explicitly, the real part R is determined at times $0, \Delta t, \ldots$, and the imaginary part I at $\frac{1}{2}\Delta t, \frac{3}{2}\Delta t$, and so forth. The algorithm is based on (what else) the Taylor expansions of R and T:

$$R(x, t + \tfrac{1}{2}\Delta t) \simeq R(x, t - \tfrac{1}{2}\Delta t) \tag{30.7}$$
$$-2\left\{\alpha\left[I(x + \Delta x, t) + I(x - \Delta x, t)\right] - 2\left[\alpha + V(x)\Delta t\right] I(x, t)\right\},$$
$$I(x, t + \tfrac{1}{2}\Delta t) \simeq I(x, t - \tfrac{1}{2}\Delta t) \tag{30.8}$$
$$+2\left\{\alpha\left[R(x + \Delta x, t) + R(x - \Delta x, t)\right] - 2\left[\alpha + V(x)\Delta t\right] R(x, t)\right\},$$
$$\alpha = \frac{\Delta t}{2(\Delta x)^2}. \tag{30.9}$$

In discrete form, these equations become our algorithm:

$$R_i^{n+1} = R_i^n - 2\left\{\alpha\left[I_{i+1}^n + I_{i-1}^n\right] - 2\left[\alpha + V_i \Delta t\right] I_i^n\right\}, \tag{30.10}$$
$$I_i^{n+1} = I_i^n + 2\left\{\alpha\left[R_{i+1}^n + R_{i-1}^n\right] - 2\left[\alpha + V_i \Delta t\right] R_i^n\right\}, \tag{30.11}$$

where the superscript n indicates the time $t = n\Delta t$ and the subscript i, the position $x = i\Delta x$.

In order to conserve probability to a higher level of precision, the probability density ρ is defined in terms of the wave function evaluated at three different times:

$$\rho(t) = \begin{cases} R^2(t) + I(t + \frac{\Delta t}{2})I(t - \frac{\Delta t}{2}), & \text{for integer } t, \\ I^2(t) + R(t + \frac{\Delta t}{2})R(t - \frac{\Delta t}{2}), & \text{for half-integer } t. \end{cases} \tag{30.12}$$

While this definition of ρ may seem strange, it reduces to the usual one for $\Delta t \to 0$, and so can be viewed as part of the art of numerical analysis. You will verify, if you do as told, that with this definition, the integral of the probability over all space is approximately constant from one time to the next:

$$\sum_x \rho(x, t + \tfrac{1}{2}\Delta t) \simeq \sum_x \rho(x, t). \tag{30.13}$$

We refer the reader to [Koon 86] and [Viss 91] for details on the stability of the algorithm and on the behavior of the evolution matrix $\exp(iH\Delta t)$.

30.4 IMPLEMENTATION: WAVE PACKET IN WELL, SQWELL.F

On the diskette and the Web you will find a program that solves for the motion of the wave packet (30.1) inside the infinite potential well:

$$V(x) = \begin{cases} \infty, & \text{for } x < 0, \\ 0, & \text{for } 0 \le x \le 15, \\ \infty, & \text{for } x > 15. \end{cases} \tag{30.14}$$

1. Define arrays `R(751,2)` and `I(751,2)` for the real and imaginary parts of the wave function, and `Rho(751)` for the probability density. The first subscript refers to the x position on the grid and the second to the present and future times.

2. Use the values $\sigma_0 = 0.5$, $\Delta x = 0.02$, $k_0 = 17\pi$, and $\Delta t = \frac{1}{2}\Delta x^2$.

3. Use equation (30.1) for the initial wave packet to define `R(j,1)` for all j at $t = 0$, and `I(j,1)` at $t = \frac{1}{2}\Delta t$.

4. Set `Rho(1)=Rho(751)` $= 0.0$ because the wave function must vanish at the infinitely high well walls.

5. Increment time by $\frac{1}{2}\Delta t$. Use (30.10) to compute `R(j,2)` in terms of `R(j,1)`, and (30.11) to compute `I(j,2)` in terms of `I(j,1)`.

6. Repeat the steps through all of space; that is, for i=2–750.

7. Throughout all of space, replace the present wave packet (second index equal to 1) by the future wave packet (second index 2).

8. Repeat the time stepping many times, ultimately ~5000, but do not let your program run that long until you are sure it is working properly.

30.5 ASSESSMENT: VISUALIZATION, AND ANIMATION

1. Output the probability density `Rho` on a coarse grid, say, about every fifth grid point. For crude animation, output the entire space behavior after every 200 time steps.

2. Make a 3-D plot of probability versus position versus time. This should look like Fig. 30.1 or Fig. 30.2.

3. Make a movie showing the wave function as a function of time.

4. Check how well probability is conserved for early and late times. Determine the integral of the probability over all of space, $\int_0^\infty dx \rho(x)$, and see if it changes with time (its explicit value doesn't matter because that's just normalization).

5. What might be a good explanation of why collisions with the walls cause the wave packet to broaden and break up? (*Hint:* The collisions do not appear so disruptive when a Gaussian wave packet is confined within a harmonic oscillator potential well.)

30.6 EXPLORATION: 1-D HARMONIC OSCILLATOR

Modify the sample program to describe the motion of a Gaussian wave packet in the harmonic oscillator potential:

$$V(x) = \tfrac{1}{2}x^2 \quad (-\infty \le x \le \infty). \tag{30.15}$$

Take the initial momentum of the wave packet as $k_0 = 3\pi$ and the time and space steps as $\Delta x = 0.02$ and $\Delta t = \frac{1}{4}\Delta x^2$. Note that the wave packet appears to breathe, yet returns to it initial shape!

30.7 IMPLEMENTATION: WAVE PACKET IN HARMONIC WELL, HARMOS.F

30.8 PROBLEM: TWO-DIMENSIONAL CONFINEMENT

Consider now an electron moving in 2-D space as shown in Fig. 30.3. This adds another degree of freedom to the problem, which means that we must solve the 2-D time-dependent Schrödinger equation:

$$i\frac{\partial\psi(x,y,t)}{\partial t} = \tilde{H}\psi(x,y,t), \tag{30.16}$$

$$i\frac{\partial\psi(x,y,t)}{\partial t} = -\left(\frac{\partial^2\psi}{\partial x^2} + \frac{\partial^2\psi}{\partial y^2}\right) + V(x,y)\psi, \tag{30.17}$$

where we have chosen units in which $2m = \hbar = 1$. To be more specific, have the electron move in an infinitely long tube with a parabolic cross section:

$$V(x,y) = 0.9x^2, \quad (-9.0 \le x \le 9.0), \quad (0 \le y \le 18.0). \tag{30.18}$$

Assume that the electron's initial localization is described by a Gaussian wave packet in two dimensions:

$$\psi(x,y,t=0) = e^{ik_{0x}x}e^{ik_{0y}y}\exp\left[-\frac{(x-x_0)^2}{2\sigma_0^2}\right]\exp\left[-\frac{(y-y_0)^2}{2\sigma_0^2}\right]. \tag{30.19}$$

We show the tube potential and the wave packet in Fig. 30.3.

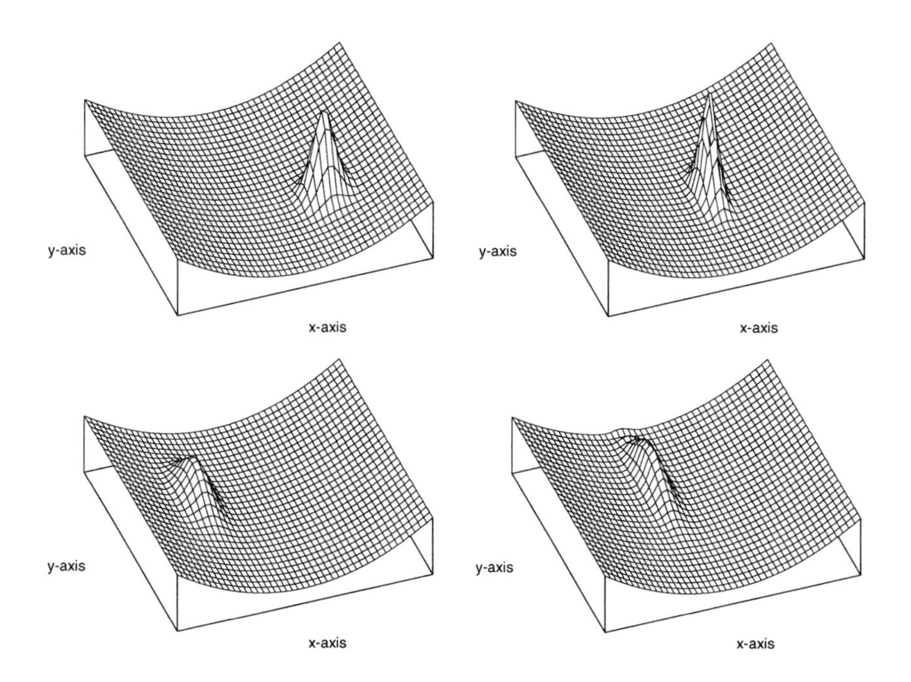

Fig. 30.3 The probability density as a function of x and y of an electron confined to a 2-D parabolic "tube." The electron's initial localization is described by a Gaussian wave packet in both the x and y directions. The times are 100, 300, 500, and 750.

30.9 METHOD: NUMERICAL

One way to develop an algorithm for solving the time-dependent Schrödinger equation in two dimensions is to extend the 1-D algorithm. Rather than do that, we apply quantum theory directly to obtain a more powerful algorithm. First we note that equation (30.17) can be integrated in a formal sense [L 96, p.4] to obtain the operator solution:

$$\psi(x,y,t) = U(t)\psi(x,y,t=0) = e^{-i\tilde{H}t}\psi(x,y,t=0). \tag{30.20}$$

From this formal solution we deduce that a wave packet can be moved ahead by a time Δt with the action of the time evolution operator:

$$\psi(t+\Delta t) = U(\Delta t)\psi(t), \tag{30.21}$$

$$U(\Delta t) = e^{-i\tilde{H}\Delta t}. \tag{30.22}$$

If the operator U were known exactly, it would provide the exact advance of the solution by one time step:

$$\psi_{i,j}^{n+1} = U(\Delta t)\psi_{i,j}^{n}, \tag{30.23}$$

where the superscripts denote time and the subscripts denote the two spatial variables,

$$\psi_{i,j}^{n} \stackrel{\text{def}}{=} \psi(i\Delta x, j\Delta y, n\Delta t). \tag{30.24}$$

Likewise, the inverse of the time evolution operator moves the solution back one time step:

$$\psi^{n-1} = U^{-1}(\Delta t)\psi^{n} = e^{+i\tilde{H}\Delta t}\psi^{n}. \tag{30.25}$$

While it would be nice to have an algorithm based on a direct application of (30.23), the references show that the resulting algorithm is not stable. That being so, we base our algorithm on an indirect application [Ask 77], namely, the relation between the difference in ψ^{n+1} (30.23) and ψ^{n-1} (30.25):

$$\psi^{n+1} = \psi^{n-1} + [e^{-i\tilde{H}\Delta t} - e^{i\tilde{H}\Delta t}]\psi^{n}, \tag{30.26}$$

where the difference in sign of the exponents is to be noted. The algorithm derives from combining the $O(\Delta x^2)$ expression for the second derivative obtained from the Taylor expansion,

$$\frac{\partial^2\psi}{\partial x^2} \simeq -\frac{1}{2}\left[\psi_{i+1,j}^{n} + \psi_{i-1,j}^{n} - 2\psi_{i,j}^{n}\right], \tag{30.27}$$

with the corresponding-order expansion of the evolution equation (30.26). When the resulting expression for the second derivative is substituted into the 2-D time-dependent Schrödinger equation, there results[2]

$$\begin{aligned}\psi_{i,j}^{n+1} &= \psi_{i,j}^{n-1} - 2i\left[\left(4\alpha + \Delta t V_{i,j}\right)\psi_{i,j}^{n}\right. \\ &\left. -\alpha\left(\psi_{i+1,j}^{n} + \psi_{i-1,j}^{n} + \psi_{i,j+1}^{n} + \psi_{i,j-1}^{n}\right)\right], \qquad (30.28) \\ \alpha &= \frac{\Delta t}{2(\Delta x)^2}. \qquad (30.29)\end{aligned}$$

We convert these complex equations into coupled real equations by substituting the real and imaginary parts of the wave function, $\psi = R + iI$, into (30.28):

$$R_{i,j}^{n+1} = R_{i,j}^{n-1} + 2\times \qquad (30.30)$$
$$\left[\left(4\alpha + \Delta t V_{i,j}\right) I_{i,j}^{n} - \alpha\left(I_{i+1,j}^{n} + I_{i-1,j}^{n} + I_{i,j+1}^{n} + I_{i,j-1}^{n}\right)\right],$$
$$I_{i,j}^{n+1} = I_{i,j}^{n-1} - 2\times \qquad (30.31)$$
$$\left[\left(4\alpha + \Delta t V_{i,j}\right) R_{i,j}^{n} + \alpha\left(R_{i+1,j}^{n} + R_{i-1,j}^{n} + R_{i,j+1}^{n} + R_{i,j-1}^{n}\right)\right].$$

This is the basic algorithm we use to integrate the 2-D Schrödinger equation. To determine the probability, we generalize [Viss 91] the expression used for one-dimension to:

$$\rho(t) = \begin{cases} R^2(t) + I(t + \frac{1}{2}\Delta t)I(t - \frac{1}{2}\Delta t), & \text{for integer time,} \\ R(t + \frac{1}{2}\Delta t)R(t - \frac{1}{2}\Delta t) + I^2(t), & \text{for half-integer time.} \end{cases} \qquad (30.32)$$

Although probability is not conserved exactly with this algorithm, the error is two orders higher than that in the wave function, and this is usually quite satisfactory. If it is not satisfactory, then we need to use smaller steps.

30.10 EXPLORATION: 2-D HARMONIC OSCILLATOR

Determine the motion of a 2-D Gaussian wave packet within the 2-D harmonic oscillator potential:

$$V(x, y) = 0.3(x^2 + y^2), \quad (-9.0 \leq x \leq 9.0), \quad (-9.0 \leq y \leq 9.0). \qquad (30.33)$$

Center the initial wave packet at $(x, y) = (3.0, -3)$ with momentum $(k_{0x}, k_{0y}) = (3.0, 1.5)$.

[2]For reference sake, note that the constants in the equation change as the dimension of the equation change; that is, there will be different constants for the 3-D equation, and therefore our constants are different from the references!

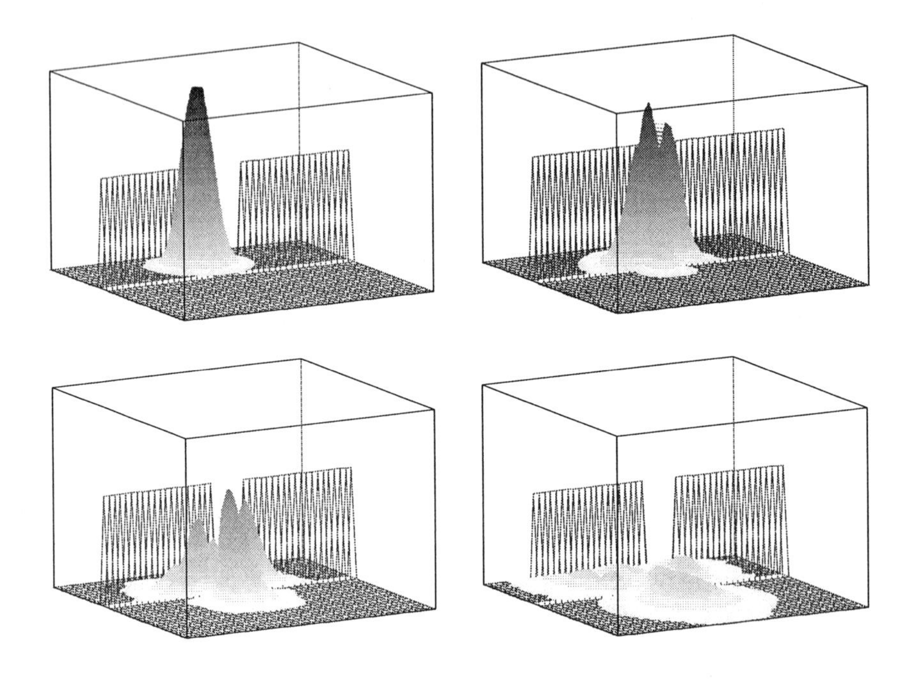

Fig. 30.4 The probability density as a function of position and time for an electron incident upon and passing through a slit.

30.11 EXPLORATION: SINGLE-SLIT DIFFRACTION, SLIT.F

Young's single-slit experiment has a wave passing through a small slit, which causes the emerging wavelets to interfere with each other. In quantum mechanics, where we represent a particle by a wave packet, this means that an interference pattern should be formed when a particle passes through a small slit. Consider a Gaussian wave packet of initial width 3 incident on a slit of width 5, as shown in Fig. 30.4.

Appendix A: Analogous Elements in Fortran and C

C	Fortran
Program Structure	
main ()	program f
double f(double x)	function f(x)
void f(x)	subroutine f(x)
Operations	
pow(x,y)	x**y
x = y;	x = y
for (k=1;k<=kmax;k++)	do 2 k=1,kmax,1
for (k=kmax;k>=1;k--)	do 2 k=kmax,1,-1
Data Type Declarations	
double x, y;	real*8 x, y
int i, j;	integer i, j
double y [100] [100];	real*8 y(100,100)
int ii [100] [100];	integer ii(100,100)
define min 0.0	data min/0.0/
char jan ;	character jan

C	Fortran[a]
Input and Output to Screen	
scanf("%lf" x1);	read(*,*) x1
printf("Enter radius ");	write(*,*) 'Enter radius'
printf("radius=%f ", radius);	write(*,*) 'radius= ', radius
Input and Output to Files	
FILE *fout;	
out = fopen("bound.dat", "w");	open(7,file='bound.dat',status='n
fprintf(fout, "%lf", Din);	write(10,f10.2) Din
fscanf(fin, "%lf", Din);	read(7, 100) Din
Control Structure: if	
if (den == 0)	if(den .eq. 0) then
{ printf("Trouble");	write(*,*) 'Trouble'
}	
Control Structure: if .. else	
if (den == 0)	if (den .eq. 0) then
{ x = 0;	x = 0;
else	else
{ x = x1/den; }	x = x1/den
Control Structure: for	
for (k=1;k<=kmax;k++)	do 2 k=1,kmax,1
{ term = r*term;	term = r*term
sum = sum+term; }	2 continue
Switch	
switch (choice)	goto (1,2), choice
{	1 series = first
case 1:	2 series = second
series = first;	
break;	
case2:	
series = second;	
break;	
}	

[a] This converter borrows freely from [Thom 92].

Appendix B: Programs on Floppy Diskette

The floppy diskette contains source codes in C and Fortran written by the authors, by Hans Kowallik, and by the students in PH465/565, Computational Physics, at Oregon State University. They are copyrighted by Rubin H. Landau, who makes no claim as to their precision or accuracy. If necessary, corrected versions of the programs can be loaded over the Web after accessing Landau's home page at `http://www.physics.orst.edu`.

In some cases, for example with the Monte-Carlo calculations, the user may have to substitute their own computer's random number generator. If only a subroutine is compiled, the compiler may complain (as well it should) that the `main` program is missing. Some of the Fortran programs contain `Implicit none` statements as an encouragement to safe programming; if your compiler does not support this statement, it should be commented out or removed.

The DOS format diskette contains the directories `C_PROGS`, containing all the C programs, and `F_PROGS`, containing all the Fortran programs. The diskette should be readable on PC, Macintosh, and Unix systems (the latter may require some DOS-emulation utilities as described in [L&F 93]). The user may want to create directories with these names on their computer and install our programs into them. The tabulations below are first in numerical order by section number, and then in alphabetical order by program name.

Section	Program	Description
2.4 (2.5)	**area.f (.c)**	Area of a circle, sample program
2.13	**over.f (.c)**	Determine overflow and underflow limits
2.15	**limit.f (.c)**	Determine machine precision
2.18	**complex.f (.c)**	Dealing with complex numbers on a computer
2.24	**exp-bad.f (.c)**	Bad algorithm for calculating exponential
2.23	**exp-good.f (.c)**	Good algorithm for calculating exponential
3.8	**bessel.f (.c)**	Spherical Bessel functions via up and down recursion
4.7	**integ.f (.c)**	Integration; trapezoid, Simpson, Gauss
4.7	**gauss.f (.c)**	Points and weights for Gaussian quadrature
5.4	**lagrange.f (.c)**	Lagrange interpolation of tabular data
5.8	**spline.f**	Spline fit with SLATEC's dbint4 and dbvalu
5.15	**fit.f (.c)**	Least-squares fit to decay spectrum
6.5	**call.f (.c)**	Pseudo-random numbers using drand48 or rand
6.11	**walk.f (.c)**	Random-walk simulation
7.6	**decay.f (.c)**	Spontaneous radioactive decay simulation
7.10	**pond.f (.c)**	Monte Carlo integration (stone throwing)
7.14	**int_10d.f (.c)**	10-D Monte Carlo integration
8.4	**diff.f (.c)**	Differentiation; forward, central, extrapolated
9.11	**rk4.f (.c)**	ODE, 4th-order Runge–Kutta, harmonic oscillatc
12.7	**fourier.f (.c)**	Discrete Fourier transformation (DFT)
12.7	**invfour.c**	Inverse discrete Fourier transformation
10.7	**numerov.c**	Numerov method for ODE eigenvalues, eigenfunctions, particle in potential well
13.7	**bugs.f (.c)**	Bifurcation diagram for logistic map
15.9.3	**lineq.c**	Solve matrix equation Ax=b using LAPACK routine sgesv
16.5	**bound.c (.f)**	Bound states via integral equation with LAPACK dgeev and Gauss quadrature
17.5	**scatt.f**	Scattering via integral equation, delta-shell with LUfactor, LUSolve (included)
19.3	**tune.f**	Matrix program to be tuned for performance
19.5	**tune1.f**	Matrix program with basic optimization

Section	Program	Description
19.6	**tune2.f**	Matrix program with vector tuning
19.7	**tune3.f**	Matrix program with modified vector tuning
19.8	**tune4.f**	Matrix program with RISC tuning
21.5	**unim1d.cpp**	Object-oriented, uniform 1-D motion
21.5.2	**unimot2d.cpp**	Object-oriented, uniform 2-D motion
21.5.4	**accm2d.cpp**	Objected-oriented, accelerated 2-D motion
21.6	**shms.cpp**	Object-oriented, simple harmonic motion
22.6	**ising.f (.c)**	Ising model of magnetic dipole string
23.6	**qmc.f (.c)**	Feynman path integral, quantum Monte Carlo
24.4	**sierpin.c**	Creates Sierpiński gasket (fractal)
24.8	**fern.c**	Create fractal fern-like pattern
24.10	**tree.c**	Creates a fractal tree
24.13	**film.c**	Ballistic deposition simulation (fractal)
24.19	**column.c**	Correlated ballistic deposition to form fractal
24.23	**dla.c**	Diffusion-limited aggregation fractal
25.6	**laplace.f**	Solution of Laplace equation
26.6	**eqheat.f (.c)**	Solution of heat equation
27.5	**eqstring.f (.c)**	Solution of wave equation
28.6	**soliton.f (.c)**	Solves KdeV equation for solitons
29.7	**twodsol.f (.c)**	Solves the sine–Gordon equation for 2-D soliton
30.4	**sqwell.f**	Solution of t-dependent Schrödinger equation Gaussian wave packet in infinite square well
30.7	**harmos.f**	Solution of t-dependent Schrödinger equation Gaussian wave packet in a quadratic potential
30.11	**slit.f**	Solution of t-dependent Schroedinger equation, 2-D wave packet passing through slit

Program	Description	Section
accm2d.cpp	Objected-oriented, accelerated 2-D motion	21.5.4
area.f (.c)	Area of a circle, sample program	2.4 (2.5)
bessel.f (.c)	Spherical Bessel functions via up and down recursion	3.8
bound.c (.f)	Bound states via integral equation with LAPACK dgeev and Gauss quadrature	16.5
bugs.f (.c)	Bifurcation diagram for logistic map	13.7
call.f (.c)	Pseudo-random numbers using drand48 or rand	6.5
column.c	Correlated ballistic deposition to form fractal	24.19
complex.f (.c)	Dealing with complex numbers on a computer	2.18
decay.f (.c)	Spontaneous radioactive decay simulation	7.6
diff.f (.c)	Differentiation; forward, central, extrapolated	8.4
dla.c	Diffusion-limited aggregation fractal	24.23
eqheat.f (.c)	Solution of heat equation	26.6
eqstring.f (.c)	Solution of wave equation	27.5
exp-bad.f (.c)	Bad algorithm for calculating exponential	2.24
exp-good.f (.c)	Good algorithm for calculating exponential	2.23
fern.c	Create fractal fern-like pattern	24.8
film.c	Ballistic deposition simulation (fractal)	24.13
fit.f (.c)	Least-squares fit to decay spectrum	5.15
fourier.f (.c)	Discrete Fourier transformation (DFT)	12.7
gauss.f (.c)	Points and weights for Gaussian quadrature	4.7
harmos.f	Solution of t-dependent Schrödinger equation Gaussian wave packet in a quadratic potential	30.7
int_10d.f (.c)	10-D Monte Carlo integration	7.14
integ.f (.c)	Integration; trapezoid, Simpson, Gauss	4.7
invfour.c	Inverse discrete fourier transformation	12.7
ising.f (.c)	Ising model of magnetic dipole string	22.6
lagrange.f (.c)	Lagrange interpolation of tabular data	5.4
laplace.f	Solution of Laplace equation	25.6
limit.f (.c)	Determine machine precision	2.15
lineq.c	Solve matrix equation Ax=b using LAPACK routine sgesv	15.9.3

Program	Description	Section
numerov.c	Numerov method for ODE eigenvalues, eigenfunctions, particle in potential well	10.7
over.f (.c)	Determine overflow and underflow limits	2.13
pond.f (.c)	Monte Carlo integration (stone throwing)	7.10
qmc.f (.c)	Feynman path integral, quantum Monte Carlo	23.6
random.f (.c)	Too simple random number generator	6.5
rk4.f (.c)	ODE, 4th-rder Runge–Kutta, harmonic oscillator	9.11
scatt.f	Scattering via integral equation, delta-shell with LUfactor, LUSolve (included)	17.5
shms.cpp	Object-oriented, simple harmonic motion	21.6
sierpin.c	Creates Sierpiński gasket (fractal)	24.4
slit.f	Solution of t-dependent Schrödinger equation, 2-D wave packet passing through slit	30.11
soliton.f (.c)	Solves KdeV equation for solitons	28.6
spline.f	Spline fit with SLATEC's dbint4 and dbvalu	5.8
sqwell.f	Solution of t-dependent Schrödinger equation Gaussian wave packet in infinite square well	30.4
tree.c	Creates a fractal tree	24.10
tune.f	Matrix program to be tuned for performace	19.3
tune1.f	Matrix program with basic optimization	19.5
tune2.f	Matrix program with vector tuning	19.6
tune3.f	Matrix program with modified vector tuning	19.7
tune4.f	Matrix program with RISC tuning	19.8
twodsol.f (.c)	Solves the sine–Gordon equation for 2-D soliton	29.7
unim1d.cpp	Object-oriented, uniform 1-D motion	21.5
unimot2d.cpp	Object-oriented, uniform 2-D motion	21.5.2
walk.f (.c)	Random-walk simulation	6.11

Appendix C: Listing of C Programs

```
/*
cccccccccccccccccccccccccccccccccccccccccccccccccccccccccccccccccccccccc
c   Objected Oriented Programming of accelerated motion in 2D          c
c                                                                      c
c  UNIX (DEC OSF, IBM AIX): cpp accm2d.cpp                             c
c                                                                      c
cccccccccccccccccccccccccccccccccccccccccccccccccccccccccccccccccccccccc
*/
#include<stdio.h>
#include <stdlib.h>
/*---------------- Um1D Class Definition  ----------------------------*/
class Um1D                                          /* class is created  */
{
   protected:              /* so children classes may access the data  */
      double delt;
      int steps;                        /* Time steps to write in file  */
      double x(double tt);                          /* makes x=xo+v*dt  */
   private:
      double x00, vx, time;
   public:
      Um1D(double x0, double dt, double vx0, double tott);
      ~Um1D(void);                 /* Class Constructor and Destructor  */
      void archive();                     /* send x vs t to disk file  */
};
/*---------------- Um1D Constructor and Destructor -------------------*/
Um1D::Um1D(double x0, double dt, double vx0, double tott)
{
//CONSTRUCTOR Um1D: initializes position, veloc., time, delta t
   x00    = 0;
   delt   = dt;
   vx     = vx0;
   time   = tott;
   steps  = (int)(tott/delt);
}
/*--------------------------------*/
Um1D::~Um1D(void)
{
//DESTRUCTOR: Um1D
   printf("Class Um1D destroyed\n");
}
/*--------------- Um1D Methods -----------------------------------*/
double Um1D::x(double tt)
{
// METHOD x: returns X=Xo+dt*v
   return (x00 + tt*vx);
}
/*--------------------------------*/
void Um1D::archive()
{
// METHOD archive: Produces disk file with X at several time intervals
   FILE  *pf;
   int i;
   double xx, tt;

   pf = fopen("unim1d.dat","w");
   tt = 0.0;

   for(i = 1 ; i <= steps ; i += 1)
   {
      xx = x(tt);          /* computes X=Xo+t*v , changing Xo   */
      fprintf(pf,"%f  %f\n", tt, xx);
      tt = tt + delt;
   }
   fclose(pf);
}
/*-------------------------------------*/
```

Fig. C.1 accm2d.cpp: Objected-oriented programming of accelerated motion in two dimensions.

```
/*---------------- Um2D Class Definition --------------------------*/
class Um2D : public Um1D
{
// CLASS Um2D: child class of UmD1, for 2 dimensional uniform motion
   protected:
      double y(double tt);
   private:
      double y00,vy;
   public:
      Um2D(double x0,double dt,double vx0,double tott,
           double y0,double vy0);   /* constructor or the Um2D class  */
      ~Um2D(void);                   /* destructor of the Um2D class  */
      void archive()      /* override Um1D definition for 2D output  */
};
/*---------------- Um2D Constructor and Destructor -------------------*/
Um2D::Um2D(double x0,double dt,double vx0,double tott,
           double y0,double vy0):
      Um1D(x0,dt,vx0,tott)
{
// CONSTRUCTOR Um2D: initializes y position and y velocity
   y00   = y0;
   vy    = vy0;
}
/*------------------------------------*/
Um2D::~Um2D(void)
{
// DESTRUCTOR Um2D
   printf("Class Um2D is destroyed\n");
}
/*---------------- Um2D Methods ------------------------------------*/
double Um2D::y(double tt)
{
// METHOD ypos: returns Y=Y0+dt*v
   return (y00 + tt*vy);
}
/*------------------------------------*/
void Um2D::archive()
{
// METHOD archive: override of Um1D.archive, for 2-dim uniform motion
   FILE  *pf;
   int i;
   double xx, yy, tt;

   pf = fopen("unimot2d.dat","w");
   tt = 0.;
   for(i = 1 ; i <= steps ; i += 1)
   {
      xx = x(tt);
      yy = y(tt);
      fprintf(pf,"%f  %f\n",yy,xx);            /* data is now x vs y  */
      tt = tt + delt;
   }
   fclose(pf);
}
/*---------------- Accm2D Class Definition ---------------------*/
class Accm2D : public Um2D
{
// CLASS Accm2D: child of Um2D to deal with accelerated motion
   protected:
      void xy(double *xxac, double *yyac, double tt);
   private:
      double ax, ay;
   public:
   Accm2D(double x0,double dt,double vx0,double tott,
          double y0,double vy0,double accx,double accy);
   ~Accm2D(void);
   void archive();
};
```

Fig. C.2 accm2d.cpp (*continued*).

```
/*---------------- Accm2D Constructor and Destructor ----------------*/
Accm2D::Accm2D(double x0,double dt,double vx0,double tott,
               double y0,double vy0,double accx,double accy):
          Um2D(x0, dt, vx0, tott, y0, vy0)
{
// CONSTRUCTOR Accm2D: initializes accelx, accely
   ax = accx;
   ay = accy;
}
/*-----------------------------------*/
Accm2D::~Accm2D(void)
{
// DESTRUCTOR Accm2D
   printf(" Class Accm2D destroyed\n");
}
/*---------------- Accm2D Methods -----------------------------------*/
void Accm2D::xy(double *xxac, double *yyac, double tt)
{
// METHOD xy: returns x=x(t)+0.5*t*t*ax, y=y(t)+0.5*t*t*ay
   double dt2;

   dt2   = 0.5*tt*tt;
   *xxac = x(tt) + ax*dt2;
   *yyac = y(tt) + ay*dt2;
}
/*-----------------------------------*/
void Accm2D::archive()
{
// METHOD archive: override Um2D.archive; record data accelerated motion
   FILE  *pf;
   int i;
   double tt, xxac, yyac;
   pf = fopen("accm2d.dat","w");
   tt = 0.;
   for(i = 1 ; i <= steps ; i += 1)
   {
      xy(&xxac, &yyac, tt);
      fprintf(pf,"%f  %f\n", xxac, yyac);
      tt = tt + delt;
   }
   fclose(pf);
}
/*-----------------------------------*/
/*---------------- Main Program ---------------------------------------*/
main()
{
   double inix, iniy, inivx, inivy, aclx, acly, dtim, ttotal;
   inix   = 0.;
   dtim   = 0.1;
   inivx  = 14.;
   ttotal = 4.;
   iniy   = 0.;
   inivy  = 14.;
   aclx   = 0.;
   acly   = -9.8;
   Accm2D acmo2d(inix, dtim, inivx, ttotal, iniy, inivy, aclx, acly);
   printf("  \n");
   printf("  \n");
   acmo2d.archive();
}
```

Fig. C.3 accm2d.cpp (*continued*).

```c
/*
*************************************************************************
*  area.c: Area of a circle, sample program                            *
*                                                                       *
*  UNIX (DEC OSF, IBM AIX): cc area.c -lm                              *
*                                                                       *
*************************************************************************
*/
#include <stdio.h>

#define PI 3.1415926535897932385E0

main()
{
   double radius, area;                               /* define variables */

   printf("Enter the radius of a circle \n");          /* ask for radius */
   scanf("%lf", &radius);                             /* read in radius  */

   area = radius * radius * PI;                           /* area formula */

   printf("radius=%f, area=%f\n", radius, area);     /* print results */
}
```

```c
/*
*************************************************************************
*  bound.c: Bound states in momentum space of delta shell potential    *
*             using the LAPACK dgeev and Gaussian integration           *
*                                                                       *
*  UNIX (IBM AIX): cc -c bound.c                                        *
*                  f77 bound.o -lblas -llapack                          *
*  UNIX (DEC OSF): rename sgesv to sgesv_                               *
*                  cc bound.c -ldxml                                    *
*                                                                       *
*  comment: LAPACK has to be installed on your system and the file     *
*           gauss.c has to be in the same directory                     *
*           The energy eigenvalue is printed to standard output and     *
*           the corresponding eigenvector to the file bound.dat         *
*************************************************************************
*/
#include <stdio.h>
#include <math.h>
#include "gauss.c"

#define min 0.0                               /* integration limits  */
#define max 200.0
#define size 64                               /* grid points */
#define lambda -1.0                           /* parameters for potential  */
#define u 0.5
#define b 2.0
#define PI 3.1415926535897932385E0

main()
{
   int i, j , c1, c2, c5, ok;
   char c3, c4;
   double A[size][size], AT[size*size];          /* hamiltonian */
   double V[size][size];                         /* potential */
   double WR[size], WI[size];                    /* eigenvalues */
   double VR[size][size], VL[1][1];              /* eigenvectors */
   double WORK[5*size];                          /* work space */
   double k[size], w[size];                      /* points, weights */
                                                 /* for integration */
```

Fig. C.4 area.c: Area of a circle; bessel.c: spherical Bessel functions by recursion.

```
   FILE *out;                          /* save data in bound.dat */
   out = fopen("bound.dat","w");

   gauss(size, 0, min, max, k, w);     /* call gauss integration */


   for(i=0; i<size; i++)               /* set up hamiltonian matrix */
   {
      for(j=0; j<size; j++)
      {
         VR[i][j] = (lambda*b*b/(2*u))*
                (sin(k[i]*b)/(k[i]*b))*(sin(k[j]*b)/(k[j]*b));
         if(i == j)
         {
            A[i][j] = k[i]*k[i]/(2*u) + (2/PI)*VR[i][j]*k[j]*k[j]*w[j];
         }
         else A[i][j]=(2/PI)*VR[i][j]*k[j]*k[j]*w[j];
      }
   }

   for(i=0; i<size; i++)               /* transform matrix */
   {                                   /* so we can pass it to */
      for(j=0; j<size; j++)            /* a FORTRAN routine */
      {
         AT[j+size*i] = A[j][i];
      }
   }
   c1 = size;                  /* we have to do this so we can */
   c3 = 'N';                   /* pass pointers to the lapack */
   c4 = 'V';                   /* routine */
   c5 = 5*size;
   c2=1;                       /* for unreferenced arrays */
   /* call for AIX  */
   dgeev(&c3,&c4,&c1,AT,&c1,WR,WI,VL,&c2,VR,&c1,WORK,&c5,&ok);
   /* call for DEC  */
/* dgeev_(&c3,&c4,&c1,AT,&c1,WR,WI,VL,&c2,VR,&c1,WORK,&c5,&ok); */

   if(ok == 0)                            /* look for bound state */
   {
      for (j=0; j<size; j++)
      {
         if (WR[j]<0)
         {
            printf("The eigenvalue of the bound state is\n");
            printf("\tlambda= %f\n", WR[j]);
            for (i=0; i<size; i++)
            {
               fprintf(out, "%d\t%e\n", i, VR[j][i]);
            }
            break;
         }
      }
   }
   printf("eigenvector saved in bound.dat\n");
   fclose(out);
}
```

Fig. C.5 bound.c (*continued*).

```c
/*
*************************************************************************
*  bugs.c: Bifurcation diagram for logistic map                        *
*                                                                      *
*  UNIX (DEC OSF, IBM AIX): cc bugs.c                                  *
*                                                                      *
*  comment: plot without conneting datapoints with lines               *
*************************************************************************
*/
#include<stdio.h>

#define m_min 0.0                      /* minimum for m  */
#define m_max 4.0                      /* maximum for m  */
#define step 0.01                      /* stepsize for m  */

main()
{
   double m, y;
   int i;

   FILE *output;                       /* save data in bugs.dat */
   output = fopen("bugs.dat","w");

   for (m=m_min; m<=m_max; m+=step)    /* loop for m  */
   {
      y  = 0.5;                        /* arbitrary starting value */
      for (i=1; i<=200; i++)           /* ignore transients */
      {
          y = m*y*(1-y);
      }
      for(i=201; i<=401; i++)          /* then record 200 points  */
      {
         y  = m*y*(1-y);
         fprintf(output, "%.4f\t%.4f\n", m, y);
      }
   }
   printf("data stored in bugs.dat.\n");
   fclose(output);
}
```

```c
/*
*************************************************************************
*  call.c: Creates pseudo-random numbers using drand48 or rand         *
*                                                                      *
*  UNIX (DEC OSF, IBM AIX): cc call.c -lm                              *
*                                                                      *
*  comment: If your compiler complains about drand48, srand48          *
*           uncomment the define statements further down.              *
*************************************************************************
*/
#include <stdio.h>
#include <stdlib.h>
#include <math.h>

/* if you don't have drand48 uncomment the following two lines */
/*    #define drand48 1.0/RAND_MAX*rand
      #define srand48 srand                */

main()
{
   int i, seed;
   double x;

   printf("enter seed\n");                          /* user plants seed */
   scanf("%i", &seed);

   srand48(seed);                                   /* seed drand 48 */

   for (i=1; i<=10; i++)
   {
      x = drand48();               /* random number between 0 and 1 */
      printf("Your random number is: %f\n",x);
   }
}
```

Fig. C.6 bugs.c: Bifurcation diagram for logistic map; pseudo-random numbers using drand48 or rand.

```
/*
************************************************************************
*  column.c: Correlated ballistic deposition to form fractal          *
*                                                                      *
*                                                                      *
*  UNIX (DEC OSF, IBM AIX): cc column.c                                *
*                                                                      *
*  comment: If your compiler complains about drand48, srand48          *
*           uncomment the define statements further down.              *
*           Plot data without connecting datapoints with lines.        *
************************************************************************
*/
#include <stdio.h>
#include <stdlib.h>

/* if you don't have drand48 uncomment the following two lines    */
/*    #define drand48 1.0/RAND_MAX*rand
      #define srand48 srand                                         */

#define max 100000                          /* number of iterations */
#define npoints 200                           /* no. of open spaces */
#define seed 68111                      /* seed for number generator */

main()
{
   int i, hit[200], dist, r, x, y, oldx, oldy;
   double  pp, prob;

   FILE *output;
   output = fopen("column.dat","w");

   srand48(seed);                    /* seed the random-number generator */

   for (i=0; i<npoints; i++) hit[i] = 0;           /* clear the array */

   oldx = 100;
   oldy = 0;

   for(i=1; i<=max; i++)
   {
      r = (int) (npoints*drand48());
      x = r-oldx;
      y = hit[r]-oldy;
      dist = x*x + y*y;
      if(dist == 0)                              /* probability of sticking */
      {                                           /* depends on distance to */
         prob = 1.0;                                   /* the last particle */
      }
      else prob = 9.0/(dist);                           /* nu=-2.0, c=0.9 */
      pp = drand48();
      if(pp < prob)
      {
         if((r>0) && (r<(npoints-1)))
         {
            if((hit[r] >= hit[r-1]) && (hit[r] >= hit[r+1]))
            {
               hit[r]++;
            }
            else if(hit[r-1] > hit[r+1])
            {
               hit[r] = hit[r-1];
            }
            else
            {
               hit[r] = hit[r+1];
            }
            oldx = r;
            oldy = hit[r];
            fprintf(output, "%d\t%d\n", r, hit[r]);
         }
      }
   }
   printf("data stored in column.dat\n");
   fclose(output);
}
```

Fig. C.7 column.c: Correlated ballistic deposition to form fractal.

```c
/*
************************************************************************
*  complex.c: Dealing with complex numbers on a computer              *
*                                                                      *
*  UNIX (DEC OSF, IBM AIX): cc complex -lm                            *
*                                                                      *
*  related information: definition of math functions in math.h file   *
************************************************************************
*/
#include <stdio.h>
#include <math.h>
#define PI 3.1415926535897932385E0
typedef struct {float re,im;} complx;          /* complex number type */

main()
{
   float phi;                               /* angle in complex plane */
   complx z;                                        /* complex number */
   complx c_sqrt(complx z);                 /* sqrt of complex number */
   complx c_log(complx z);                   /* log of complex number */
                                             /* print header of table */
   printf("\nPhi(n)\tx\ty\tsqrt(z)\t\tlog(z)\t\tatan\tatan2\n");

   for (phi=0; phi<=2.5; phi+=0.1)                  /* loop for angle */
   {
      z.re = cos(phi*PI);                    /* split into real, im. */
      z.im = sin(phi*PI);                           /* phi,real,im */
      printf("%.1f*pi\t%+.3f\t%+.3f\t",phi, z.re, z.im);
                                                    /* complex sqrt */
      printf("%.3f%+.3fi", c_sqrt(z).re, c_sqrt(z).im);
                                                     /* complex log */
      printf("\t%+.3f%+.3fi", c_log(z).re, c_log(z).im);
      printf("\t%+.3f",atan(z.im/z.re));                /* arctan y/x */
      printf("\t%.3f\n",atan2(z.im,z.re));              /* atan2(x,y) */
   }
}
/*------------------------end of main program -----------------------*/
```

```c
/* function calculates sqrt of complex number */
complx c_sqrt (complx z)
{
   float     magnitude,angle;
   complx   value;

   magnitude = pow(z.re*z.re + z.im*z.im,0.25);    /* calc polar rep  */
   angle     = 0.5*atan2(z.im,z.re);               /* and take sqrt   */
   value.re  = magnitude*cos(angle);               /* cartesian rep   */
   value.im  = magnitude*sin(angle);
   return (value);
}
/*--------------------------------------------------------------------*/

/* function calculates log of complex number */
complx c_log (complx z)
{
   float     magnitude,angle;
   complx   value;

   magnitude = sqrt(z.re*z.re + z.im*z.im);          /* calc polar rep */
   angle     = atan2(z.im,z.re);
   value.re  = log(magnitude);                             /* calc log */
   value.im  = angle;                                 /* cartesian rep */
   return (value);
}
```

Fig. C.8 complex.c: Dealing with complex numbers.

```
/*
************************************************************************
*  decay.c: Spontaneous radioactive decay simulation                   *
*                                                                      *
*  UNIX (DEC OSF, IBM AIX): cc decay.c                                 *
*                                                                      *
*  comment: If your compiler complains about drand48, srand48          *
*           uncomment the define statements further down               *
************************************************************************
*/

#include <stdio.h>
#include <stdlib.h>

/* if you don't have drand48 uncomment the following two lines */
/*    #define drand48 1.0/RAND_MAX*rand
      #define srand48 srand                */

#define lambda 0.01                        /* the decay constant */
#define max 1000                           /* number of atoms at t=0 */
#define time_max 500                       /* time range */
#define seed 68111                         /* seed for number generator */

main()
{
   int atom, time, number, nloop;
   double decay;

   FILE *output;                           /* save data in decay.dat */
   output = fopen("decay.dat","w");

   number = nloop = max;                   /* initial value */
   srand48(seed);                          /* seed number generator */

   for(time=0; time<=time_max; time++)          /* time loop */
   {
      for (atom=1; atom<=number; atom++)        /* atom loop */
      {
         decay = drand48();
         if(decay < lambda) nloop--;            /* an atom decays */
      }
      number = nloop;
      fprintf(output, "%d\t%f\n", time, (double)number/max);
   }
   printf("data stored in decay.dat\n");
   fclose(output);
}
```

Fig. C.9 decay.c: Spontaneous radioactive decay simulation.

```c
/* diff.c:  Differentiation; forward, central, extrapolated difference *
*                                                                      *
*  UNIX (DEC OSF, IBM AIX): cc diff.c -lm                              *
*                                                                      *
*  comment: results saved as x y1 y2 y3                                *
*/
#include <stdio.h>
#include <math.h>
#define h 1e-5                          /* stepsize for all methods */
#define xmax 7                          /* range for calculation */
#define xmin 0
#define xstep 0.01                      /* stepsize in x  */

main()
{
   double dc, result, x;
   double f(double);                    /* function we differentiate */
   FILE *output;                        /* save data in diff.dat */
   output = fopen("diff.dat","w");   for (x=xmin; x<=xmax; x+=xstep)
   {
      fprintf(output,"%f\t", x);

      result=(f(x+h)-f(x))/h;                 /* forward difference */
      fprintf(output, "%.10f\t", result);

      result=(f(x+h/2)-f(x-h/2))/h;           /* central difference */
      fprintf(output, "%.10f\t", result);

      result=(8*(f(x+h/4.0)-f(x-h/4.0))-(f(x+h/2)-f(x-h/2.0)))/(3.0*h);
      fprintf(output, "%.10f\n", result);     /* extrapolated diff */
   }
   printf("data stored in diff.dat\n");
   fclose(output);
}
/*---------------------end of main program---------------------------*/
/* the function we want to differentiate */
double f(double x)
{         return(cos(x));        }
```

```c
/*
************************************************************************
*  dla.c:         Diffusion-limited aggregation simulation (fractals) *
*                                                                      *
*  UNIX (DEC OSF, IBM AIX): cc dla.c -lm                               *
*                                                                      *
*  comment: If your compiler complains about drand48, srand48         *
*           uncomment the define statements further down.             *
*           Plot datafile without connecting datapoints with lines.   *
************************************************************************
*/
#include <stdio.h>
#include <stdlib.h>
#include <math.h>

/* if you don't have drand48 uncomment the following two lines */
/*    #define drand48 1.0/RAND_MAX*rand
      #define srand48 srand                 */

#define max 40000                          /* number of iterations  */
#define size 401                           /* size of grid array  */
#define PI 3.1415926535897932385E0
#define seed 68111                         /* seed for number generator */

main()
{
   double angle, rad = 180.0;
   int i,j, x, y, dist, dir, step, trav;
   int grid[size][size], hit;
   int gauss_ran();                       /* gaussian random number */

   FILE *output;                          /* save data in dla.dat */
   output = fopen("dla.dat","w");
```

Fig. C.10 diff.c: Differentiation with forward-, central-, and extrapolated-difference methods; dla.c: diffusion-limited aggregation simulation.

```
for(i=0; i<size; i++)                 /* clear grid */
{
   for(j=0; j<size; j++) grid[i][j] = 0;
}
grid[200][200] = 1;                   /* one particle at the center */
srand48(seed);                        /* seed number generator */
for(i=0; i<max; i++)                  /* choose starting point  */
{
   hit   = 0;
   angle = (2*PI*drand48());          /* random angle */
   x     = (200+rad*cos(angle));      /* coordinates */
   y     = (200+rad*sin(angle));
   dist = gauss_ran();         /* random-number gaussian dist. */
   if (dist<0) step=-1;        /* move forward or backward */
   else step=1;
   trav=0;
   while((hit==0)&&(x<399)&&(x>1)&&(y<399)&&(y>1)&&(trav<abs(dist)))
   {
      if(grid[x+1][y]+grid[x-1][y]+grid[x][y+1]+grid[x][y-1]>=1)
      {
          hit = 1;               /* one neighbor is occupied */
          grid[x][y] = 1;        /* particle sticks, walk is over */
      }
      else if(drand48() < 0.5) x+=step;      /* move horizontally */
           else                y+=step;      /* move vertically */
      trav++;
   }
}
for(i=0; i<size; i++)               /*  print resulting grid  */
{
   for(j=0; j<size; j++)
      { if(grid[i][j]==1) fprintf(output,"%d\t%d\n", i, j);
      }
}
printf("data stored in dla.dat\n");
fclose(output);
}
/*-------------------------end of main program----------------------*/
```

```
/* generates random numbers with gaussian distribution using the */
/* Box-Mueller method */
int gauss_ran()
{
   double fac, rr, r1, r2;
   static int old=0;                /* have to be static so information */
   static int mem;                  /* survives between function calls */

   if (old==0)                      /* no random number left from */
   {                                /* previous function call */
      do
      {
         r1= 2.0*drand48()-1.0;       /* choose random point in */
         r2= 2.0*drand48()-1.0;       /* the unit circle */
         rr= r1*r1+r2*r2;
      }while ((rr>=1)||(rr==0));

      fac=sqrt(-2*log(rr)/rr);
      mem=5000*r1*fac;                /* save for next call */
      old=1;                          /* set flag */

      return((int)(5000*r2*fac));
   }
   else                               /* return second number */
   {                                  /* from last call */
      old=0;                          /* unset flag */
      return mem;                     /* return number from last call */
   }
}
```

Fig. C.11 dla.c (*continued*).

```c
/*
*************************************************************************
*  eqheat.c:   Solution of heat equation using with finite differences *
*                                                                       *
*  UNIX (DEC OSF, IBM AIX): cc eqheat.c                                 *
*                                                                       *
*  comment: Output data is saved in 3D grid format used by gnuplot      *
*************************************************************************
*/

#include <stdlib.h>
#include <stdio.h>

#define size 101                          /* grid size */
#define max 30000                         /* iterations */
#define thc 0.12                          /* thermal conductivity */
#define sph 0.113                         /* specific heat */
#define rho 7.8                           /* density */

main()
{
   int i,j;
   double cons, u[101][2];

   FILE *output;                          /* save data in eqheat.dat */
   output = fopen("eqheat.dat","w");

   for(i=0; i<size; i++) u[i][0]=100.;   /* t=0 all points are at 100 C */
   for(j=0; j<2; j++)                     /* except the endpoints */
   {
      u[0][j]      = 0.;
      u[size-1][j] = 0.;
   }

   cons=thc/(sph*rho);                     /* material constants */

   for(i=1; i<=max; i++)                  /* loop over max timesteps */
   {
      for(j=1; j<(size-1); j++)                  /* loop over space */
      {
         u[j][1] = u[j][0]+cons*(u[j+1][0]+u[j-1][0]-2.0*u[j][0]);
      }
      if((i%1000==0) || (i==1))          /* save every 1000 time steps */
      {
         for(j=0 ; j<size; j++)
         {
            fprintf(output, "%f\n", u[j][1]);   /* gnuplot 3D format */
         }
         fprintf(output, "\n");               /* empty line for gnuplot */
      }

      for(j=0; j<size; j++) u[j][0]=u[j][1];      /*shift new to old */
   }
   printf("data stored in eqheat.dat\n");
   fclose(output);
}
```

Fig. C.12 eqheat.c: Solution of heat equation.

```
/*
**************************************************************************
*  eqstring.c: Solution of  wave equation using time stepping           *
*                                                                        *
*  UNIX (DEC OSF, IBM AIX): cc eqstring.c -lm                            *
*                                                                        *
*  comment: Output data is saved in 3D grid format used by gnuplot      *
**************************************************************************
*/
#include <stdio.h>
#include <math.h>

#define rho 0.01                        /* density per length */
#define ten 40.0                        /* tension */
#define max 100                         /* time steps */

main()
{
   int i,k;
   double x[101][3];

   FILE *out;                           /* save data in string.dat */
   out  = fopen("eqstring.dat","w");

   for(i=0; i<81; i++)                  /* initial configuration */
   {
      x[i][0] = 0.00125*i;
   }
   for(i=81; i<101; i++)
   {
      x[i][0] = 0.1-0.005*(i-80);
   }
   for(i=1; i<100; i++)                 /* first time step  */
   {
      x[i][1] = x[i][0]+0.5*(x[i+1][0]+x[i-1][0]-2.0*x[i][0]);
   }
   for(k=1; k<max; k++)                         /* all later time steps  */
   {
      for(i=1; i<100; i++)
      {
         x[i][2] = 2.0*x[i][1]-x[i][0]+(x[i+1][1]+x[i-1][1]-2.0*x[i][1]);
      }
      for(i=0; i<101; i++)
      {
         x[i][0] = x[i][1];
         x[i][1] = x[i][2];
      }

      if((k%5) == 0)                               /* print every 5th point */
      {
         for(i=0; i<101; i++)
         {
            fprintf(out, "%f\n",x[i][2]);    /* gnuplot 3D grid format */
         }
         fprintf(out, "\n");                       /* empty line for gnuplot */
      }
   }
   printf("data stored in eqstring.dat\n");
   fclose(out);
}
```

Fig. C.13 eqstring.c: Solution of wave equation.

```c
/*
************************************************************************
*   exp-bad.c: A bad algorithm for calculating exponential            *
*                                                                      *
*  UNIX (DEC OSF, IBM AIX): cc exp-bad.c -lm                          *
*                                                                      *
*  related programs: exp-good.c                                       *
************************************************************************
*/
#include <stdio.h>
#include <math.h>
#define min 1E-10                                /* limit for accuracy */
#define max 10                                   /* maximum for x */
#define step 0.1                                 /* intervals */
main ()
{
   double sum, x, up, down;
   int i,j;
   FILE *output;                          /* save results in */
   output=fopen("exp-bad.dat", "w");      /* exp-bad.dat */

   for (x=0.0; x<=max; x+=step)           /* step through x */
   {
      sum = 1.;                           /* reset variables */
      i   = 0;
      do                                  /* sum terms until accuracy */
      {                                   /* is reached */
         i++;
         up = down = 1;                   /* reset variables */
         for(j=1; j<=i; j++)
         {
            up   *= -x;                   /* numerator */
            down *= j;                    /* denominator */
         }
      sum += up/down;
      }while( (sum == 0) || ((fabs ((up/down)/sum)) > min) );
      fprintf(output, "%f\t%e\n", x, sum);
   }
   printf("results saved in exp-bad.dat\n");
   fclose(output); }
```

```c
/*  exp-good.c: A good algorithm for calculating exponential          *
*                                                                      *
*  UNIX (DEC OSF, IBM AIX): cc exp-good.c -lm                         *
*                                                                      *
*  related programs: exp-bad.c.c                                      *
*/
#include <stdio.h>
#include <math.h>
#define min 1E-10                            /* limit for accuracy */
#define max 10.                              /* maximum for x */
#define step 0.1                             /* interval */
main ()
{
   double x, sum, element;
   int n;
   FILE *output;                             /* save results in */
   output=fopen("exp-good.dat", "w");        /* exp-good.dat */
   for (x=0.0; x<=max; x+=step)
   {
      sum = element = 1.;                    /* reset variables  */
      n   = 0;
      do                                     /* sum terms until */
      {                                      /* accuracy is reached */
         n++;
         element *= -x/n;                    /* calculate next element */
         sum += element;
      }while ((sum == 0) || (fabs(element/sum) > min));
      fprintf(output, "%f\t%e\n", x, sum);
   }
   printf("results saved in exp-good.dat\n");
   fclose(output);
}
```

Fig. C.14 exp-bad.c and exp-good.c: Bad and good algorithms for calculating exponential.

```c
/*
*******************************************************************
*  fern.c: Create fractal, fern-like pattern                      *
*                                                                 *
*  UNIX (DEC OSF, IBM AIX): cc fern.c                             *
*                                                                 *
*  comment: If your compiler complains about drand48, srand48     *
*           uncomment the define statements further down          *
*           Plot data without connecting datapoints with lines    *
*******************************************************************
*/
#include <stdio.h>
#include <stdlib.h>

/* if you don't have drand48 uncomment the following two lines */
/*    #define drand48 1.0/RAND_MAX*rand
      #define srand48 srand                 */

#define max 30000                /* number of iterations */
#define seed 68111               /* seed for number generator */

main()
{
   int i;
   double x, y, xn, yn, r;

   FILE *output;                          /* save data in fern.dat */
   output=fopen("fern.dat","w");

   srand48(seed);                         /* seed number generator */

   x     = 0.5;                           /* starting point */
   y     = 0.0;

   for(i=1; i<=max; i++)                  /* iterations */
   {
      r=drand48();
      if (r <= 0.02)                      /* case 1 */
      {
         xn = 0.5;
         yn = 0.27*y;
      }
      else if((r>0.02) && (r<=0.17))      /* case 2 */
      {
         xn = -0.139*x + 0.263*y + 0.57;
         yn =  0.246*x + 0.224*y - 0.036;
      }
      else if ((r>0.17) && (r<=0.3))      /* case 3 */
      {
         xn = 0.17*x  - 0.215*y + 0.408;
         yn = 0.222*x + 0.176*y + 0.0893;
      }
      else                                /* case 4 */
      {
         xn = 0.781*x + 0.034*y + 0.1075;
         yn = -0.032*x + 0.739*y + 0.27;
      }
      fprintf(output, "%f %f\n", x, y);
      x=xn;
      y=yn;
   }
   printf("data stored in fern.dat\n");
   fclose(output);
}
```

Fig. C.15 fern.c: Create a fractal fern.

```
/*
**********************************************************************
*  film.c: Ballistic deposition simulation (fractal)                 *
*                                                                    *
*  UNIX (DEC OSF, IBM AIX): cc film.c                                *
*                                                                    *
*  comment: If your compiler complains about drand48, srand48        *
*           uncomment the define statements further down             *
*           Plot data without connecting datapoints with lines       *
**********************************************************************
*/
#include <stdio.h>
#include <stdlib.h>

/* if you don't have drand48 uncomment the following two lines */
/*    #define drand48 1.0/RAND_MAX*rand
      #define srand48 srand                 */

#define max 30000                /* number of iterations */
#define seed 68111               /* seed for number generator */

main()
{
   int i, hit[200], r;

   FILE *output;                         /* save data in film.dat */
   output = fopen("film.dat","w");

   srand48(seed);

   for (i=0; i<200; i++) hit[i] = 0;            /* clear array */

   for(i=1; i<=max; i++)
   {
      r = (int)(199*drand48());                 /* r=0..199 */
      if((hit[r] >= hit[r-1]) && (hit[r] >= hit[r+1]))
      {
         hit[r]++;
      }
      else if(hit[r-1] > hit[r+1])
      {
         hit[r] = hit[r-1];
      }
      else
      {
         hit[r] = hit[r+1];
      }
      fprintf(output, "%d\t%d\n", r, hit[r]);
   }
   printf("data stored in film.dat\n");
   fclose(output);
}
```

Fig. C.16 film.c: Ballistic deposition simulation.

```c
/*
************************************************************************
*  fit.c:    Least-squares fit to decay spectrum                       *
*                                                                      *
*  UNIX (DEC OSF, IBM AIX): cc fit.c -lm                               *
************************************************************************
*/
#include <stdio.h>
#include <math.h>

#define data 12                         /* number of data points */

main()
{
   int i, j;
   double s, sx, sy, sxx, sxy, delta, inter, slope;
   double x[data], y[data], d[data];

   for (i=0; i<data; i++) x[i]=i*10+5;         /* input data x */

   y[0]=32; y[1]=17;  y[2]=21; y[3]=7;         /* input data y */
   y[4]=8;  y[5]=6;   y[6]=5;  y[7]=2;         /* y[9] set to 0.1 */
   y[8]=2;  y[9]=0.1; y[10]=4; y[11]=1;        /* so that log exists */

   for (i=0; i<data; i++) d[i]=1.;             /* input data delta y */
                                               /* estimate */

   for (i=0;i<data;i++) y[i]=log(y[i]);        /* log(y[i]) for */
                                               /* exponential fit */
   s=sx=sy=sxx=sxy=0;                          /* reset sums */

   for (i=0;i<data;i++)                               /* calculating sums */
   {
      s   +=         1 / (d[i]*d[i]);
      sx  +=      x[i] / (d[i]*d[i]);
      sy  +=      y[i] / (d[i]*d[i]);
      sxx += x[i]*x[i] / (d[i]*d[i]);
      sxy += x[i]*y[i] / (d[i]*d[i]);
   }

   delta = s*sxx-sx*sx;
   slope=  (s*sxy-sx*sy) / delta;                     /* calculating all */
   inter=(sxx*sy-sx*sxy) / delta;                     /* coefficients*/

   printf("intercept=%f\t +/- %f\n", inter, sqrt(sxx/delta) );
   printf("slope=%f\t +/- %f\n", slope, sqrt(s/delta) );
   printf("correlation=%f\n", -sx/sqrt(sxx*s));
}
```

Fig. C.17 fit.c: Least-squares fit to decay spectrum.

```c
/*
*************************************************************************
*  fourier.c: Discrete Fourier Transformation                           *
*                                                                       *
*  UNIX (DEC OSF, IBM AIX): cc fourier.c -lm                            *
*                                                                       *
*  comment: The program reads its input data from a file in the        *
*           same directory called input.dat. This file has to contain  *
*           only y(t) values separated by whitespaces which are real.  *
*           The output is the direct output from the algorithm which   *
*           will probably look very different than what you are used   *
*           to. The output has the form                                *
*               frequency index \t real part \t imaginary part         *
*  related programs: invfour.c                                          *
*************************************************************************
*/
#include <stdio.h>
#include <math.h>

#define max 1000                        /* max number of input data */
#define PI 3.1415926535897932385E0

main()
{
   double imag, real,input[max+1];
   int i=0,j,k;

   FILE *data;
   FILE *output;
   data=fopen("input.dat", "r");        /* read data from input.dat */
   output=fopen("fourier.dat", "w");    /* save data in fourier.dat */

   while ((fscanf(data, "%lf", &input[i]) !=EOF) && (i<max))
   {
      i++;                              /* reading input data */
   }

   for (j=0; j<i; j++)                  /* loop for frequency index */
   {
      real=imag=0.0;                    /* clear variables */
      for (k=0; k<i; k++)               /* loop for sums */
      {
         real+=input[k]*cos((2*PI*k*j)/i);
         imag+=input[k]*sin((2*PI*k*j)/i);
      }
      fprintf(output, "%d\t%f\t%f\n", j, real/i, imag/i );
   }
   printf("data stored in fourier.dat.\n");
   fclose(data);
   fclose(output);
}
```

Fig. C.18 fourier.c: Discrete Fourier transformation (DFT).

```
/*
***********************************************************************
*  gauss.c: Points and weights for Gaussian quadrature                *
***********************************************************************
*/

void gauss(int npts, int job, double a, double b, double x[], double w[])
{
/*     npts     number of points                                       */
/*     job = 0  rescaling uniformly between (a,b)                      */
/*           1  for integral (0,b) with 50% points inside (0, ab/(a+b))*/
/*           2  for integral (a,inf) with 50% inside (a,b+2a)          */
/*     x, w     output grid points and weights.                        */

      int     m, i, j;
      double  t, t1, pp, p1, p2, p3;
      double  pi = 3.1415926535897932385E0;
      double  eps = 3.e-10;                       /* limit for accuracy */

      m = (npts+1)/2;
      for(i=1; i<=m; i++)
      {
         t  = cos(pi*(i-0.25)/(npts+0.5));
         t1 = 1;
         while((fabs(t-t1))>=eps)
         {
            p1 = 1.0;
            p2 = 0.0;
            for(j=1; j<=npts; j++)
            {
               p3 = p2;
               p2 = p1;
               p1 = ((2*j-1)*t*p2-(j-1)*p3)/j;
            }
            pp = npts*(t*p1-p2)/(t*t-1);
             t1 = t;
             t  = t1 - p1/pp;
         }
         x[i-1] = -t;
         x[npts-i] = t;
         w[i-1]    = 2.0/((1-t*t)*pp*pp);
         w[npts-i] = w[i-1];
      }

      if(job==0)
      {
         for(i=0; i<npts ; i++)
         {
            x[i] = x[i]*(b-a)/2.0+(b+a)/2.0;
            w[i] = w[i]*(b-a)/2.0;
         }
      }
      if(job==1)
      {
         for(i=0; i<npts; i++)
         {
            x[i] = a*b*(1+x[i]) / (b+a-(b-a)*x[i]);
            w[i] = w[i]*2*a*b*b /((b+a-(b-a)*x[i])*(b+a-(b-a)*x[i]));
         }
      }
      if(job==2)
      {
         for(i=0; i<npts; i++)
         {
            x[i] = (b*x[i]+b+a+a) / (1-x[i]);
            w[i] =  w[i]*2*(a+b)  /((1-x[i])*(1-x[i]));
         }
      }
}
```

Fig. C.19 gauss.c: Points and weights for Gaussian quadrature.

```
/*
************************************************************************
*  int_10d.c: 10-D  Monte Carlo integration                            *
*                                                                      *
*  UNIX (DEC OSF, IBM AIX): cc int_10d.c -lm                           *
*                                                                      *
*  comment: If your compiler complains about drand48, srand48          *
*           uncomment the define statements further down               *
************************************************************************
*/
#include <stdio.h>
#include <stdlib.h>
#include <math.h>

/* if you don't have drand48 uncomment the following two lines */
/*    #define drand48 1.0/RAND_MAX*rand
      #define srand48 srand                  */

#define max 65536                        /* number of trials */
main()
{
   int i,j;
   double n=1.0,x,y=0;
   FILE *output;                         /* save data in int_10d.dat */
   output= fopen("int_10d.dat", "w");
   for (i=1; i<=max; i++)
   {
      x=0;                                      /* reset x */
      for (j=1; j<=10; j++) x+= drand48();      /* sum of 10 x values */
      y+=x*x;                            /* square and sum up */
      if (i%(int)(pow(2.0,n))==0)        /* save after 2, 4, 8, 16 ... */
      {
         n++;
         fprintf( output, "%i\t\t%f\n", i, y/i);
      }
   }
printf("data saved in int_10d.dat\n");
fclose(output);
}
```

```
/*
************************************************************************
*  integ.c:  Integration using trapezoid, Simpson and Gauss rules      *
*                                                                      *
*  UNIX (DEC OSF, IBM AIX): cc integ.c -lm                             *
*                                                                      *
*  comment: The derivation from the theoretical result for each method *
*           is saved in x y1 y2 y3 format.                             *
*           Program needs gauss.c in the same directory.               *
************************************************************************
*/

#include <stdio.h>
#include <math.h>
#include "gauss.c"       /* routine returns Legendre points, weights */

#define max_in  501                  /* max number of intervals */
#define vmin 0.0                     /* ranges of integration */
#define vmax 1.0
#define ME 2.7182818284590452354E0   /* Euler's number */

main()
{
   int i;
   float result;
   float f(float);                               /* subroutine containing */
                                                 /* function to integrate */
   float trapez  (int no, float min, float max);    /* trapezoid rule */
   float simpson (int no, float min, float max);    /* Simpson's rule */
   float gaussint(int no, float min, float max);    /* Gauss' rule */

   FILE *output;                                 /* save data in integ.dat */
   output = fopen("integ.dat","w");
```

Fig. C.20 int_10d.c: Ten-dimensional Monte Carlo integration; integ.c: integration using trapezoid, Simpson, and Gauss rules.

```
   for (i=3; i<=max_in; i+=2)           /* Simpson's rule requires */
   {                                    /* odd number of intervals */
      result = trapez(i, vmin, vmax);
      fprintf(output, "%i\t%e\t", i, fabs(result-1+1/ME));

      result = simpson(i, vmin, vmax);
      fprintf(output, "%e\t", fabs(result-1+1/ME));

      result = gaussint(i, vmin, vmax);
      fprintf(output, "%e\n", fabs(result-1+1/ME));
   }
   printf("data stored in integ.dat\n");
   fclose(output);
}
/*------------------------end of main program------------------------*/

/* the function we want to integrate */
float f (float x)
{
   return (exp(-x));
}

/* Integration using trapezoid rule */
float trapez (int no, float min, float max)
{
   int n;
   float interval, sum=0., x;

   interval = ((max-min) / (no-1));
   for (n=2; n<no; n++)                 /* sum the midpoints */
   {
      x    = interval * (n-1);
      sum += f(x)*interval;
   }
   sum += 0.5 *(f(min) + f(max)) * interval;    /* add the endpoints */

   return (sum);
}
```

```
/* Integration using Simpson's rule */
float simpson (int no, float min, float max)
{
   int n;
   float interval, sum=0., x;
   interval = ((max -min) /(no-1));

   for (n=2; n<no; n+=2)                /* loop for odd points  */
   {
       x = interval * (n-1);
       sum += 4 * f(x);
   }
   for (n=3; n<no; n+=2)                /* loop for even points  */
   {
      x = interval * (n-1);
      sum += 2 * f(x);
   }
   sum +=  f(min) + f(max);             /* add first and last value */
   sum *= interval/3.;                  /* then multilpy by interval*/

   return (sum);
}

/* Integration using Gauss' rule */
float gaussint (int no, float min, float max)
{
   int n;
   float quadra = 0.;
   double w[1000], x[1000];             /* for points and weights */
   gauss (no, 0, min, max, x, w);       /* returns Legendre */
                                        /* points and weights */
   for (n=0; n< no; n++)
   { quadra += f(x[n])*w[n];            /* calculating the integral */
   }
   return (quadra);
}
```

Fig. C.21 integ.c (*continued*).

```
/*
************************************************************************
*   invfour.c: Inverse Discrete Fourier Transformation                 *
*                                                                      *
*  UNIX (DEC OSF, IBM AIX): cc invfour.c -lm                           *
*                                                                      *
*  comment: input: fourier.data, format: \t real freq \t imag freq     *
*          created by fourier.c. Same format for output                *
************************************************************************
*/
#include <stdio.h>
#include <math.h>
#define max 10000
#define PI 3.1415926535897932385E0
main()
{
double imag, real, input [2][max];
int i=0,j,k;
FILE *data;
FILE *output;
data=fopen("fourier.dat", "r");          /* read data from fourier.dat */
output=fopen("invers.dat", "w");          /* save data in invers.dat */
while (fscanf(data, "%d %lf %lf", &j, &input[0][i], &input[1][i]) !=EOF)
{
    i++;                 /* input[0][x]:real, input[1][x]:imaginary */
}
for (j=0; j<i; j++)                         /*loop for the frequency index*/
{
   real=imag=0.0;                           /* clear variables */
   for (k=0; k<i; k++)                      /*loop for sum*/
   {
      real+=input[0][k]*cos(2*PI*k*j/i)+input[1][k]*sin(2*PI*k*j/i);
      imag+=input[1][k]*cos(2*PI*k*j/i)-input[0][k]*sin(2*PI*k*j/i);
   }
   fprintf(output, "%i\t %f\t%f\n", j, real, imag);
}
printf("data saved in invers.dat\n");
fclose(output);
fclose(data);
}
```

Fig. C.22 invfour.c: Inverse discrete Fourier transformation.

```
 /*
************************************************************************
*  ising.c: Ising model of magnetic dipole string                      *
*           Start with a uniform distribution and then heat it up      *
*                                                                      *
*  UNIX (DEC OSF, IBM AIX): cc ising.c -lm                             *
*                                                                      *
*  comment: If your compiler complains about drand48, srand48          *
*           uncomment the define statements further down.              *
*           Plot without conneting datapoints with lines               *
************************************************************************
*/
#include <stdio.h>
#include <stdlib.h>
#include <math.h>

/* if you don't have drand48 uncomment the following two lines */
/*    #define drand48 1.0/RAND_MAX*rand
      #define srand48 srand                */

#define max 100                             /* number of objects */
#define kt 100.0                            /* temperature */
#define J -1                                /* exchange energy */
#define seed 68111                          /* seed for srand48 */
main()
{
   int i, j, element, array[max];
   double olden, newen;

   double energy(int array[]);              /* energy of system */

   FILE *output1, *output2;                 /* save spin ups and */
   output1 = fopen("spin-up.dat" ,"w");     /* downs in two files */
   output2 = fopen("spin-down.dat", "w");

   srand48(seed);                           /* seed generator */

   for (i=0; i<max; i++) array[i]=1;        /* uniform start */

   for (i=0; i<=500; i++)                           /* time loop  */
   {
      olden   = energy(array);                      /* initial energy  */
      element = drand48()*max;                      /* pick one element */
      array[element] *= -1;                         /* change spin */
      newen   = energy(array);                      /* calculate new energy */

      if ( (newen>olden) && (exp((-newen + olden)/kt) <= drand48()) )
      {
         array[element] = array[element]*(-1);      /* reject change */
      }

      for (j=0; j<max; j++)                         /* save "map" of spins */
      {
         if (array[j]==1)  fprintf(output1, "%d  %d\n", i, j);
         if (array[j]==-1) fprintf(output2, "%d  %d\n", i, j);
      }
   }
   fclose (output1);
   fclose (output2);

   printf("data saved in spin-up.dat, spin-down.dat\n");
}
/*---------------------end of main program-------------------------*/

/* function returns energy of the system */
double energy (int array[])
{
   int i;
   double sum = 0.;

   for(i=0; i<(max-1); i++)                  /* loop through elements */
   {
      sum += array[i]*array[i+1];
   }
   return (J*sum);
}
```

Fig. C.23 ising.c: Ising model.

```
/*
************************************************************************
*  laplace.c:  Solution of Laplace's equation  with finite differences *
*                                                                      *
*  UNIX (DEC OSF, IBM AIX): cc laplace.c                               *
*                                                                      *
*  comment: Output data is saved in 3D grid format used by gnuplot     *
************************************************************************
*/
#include <stdio.h>

#define max 40                          /* number of grid points */

main()
{
   double x, p[max][max];
   int i, j, iter, y;

   FILE *output;                        /* save data in laplace.dat */
   output = fopen("laplace.dat","w");

   for(i=0; i<max; i++)                 /* clear the array  */
   {
      for (j=0; j<max; j++) p[i][j] = 0;
   }

   for(i=0; i<max; i++) p[i][0] = 100.0;        /* p[i][0] = 100 V */

   for(iter=0; iter<1000; iter++)               /* iterations */
   {
      for(i=1; i<(max-1); i++)                  /* x-direction */
      {
         for(j=1; j<(max-1); j++)               /* y-direction */
         {
            p[i][j] = 0.25*(p[i+1][j]+p[i-1][j]+p[i][j+1]+p[i][j-1]);
         }
      }
   }
   for (i=0; i<max ; i++)         /* write data gnuplot 3D format */
   {
      for (j=0; j<max; j++)
      {
         fprintf(output, "%f\n",p[i][j]);
      }
      fprintf(output, "\n");      /* empty line for gnuplot */
   }
   printf("data stored in laplace.dat\n");
   fclose(output);
}
```

Fig. C.24 laplace.c: Solution of Laplace's equation.

```
/*
************************************************************************
*  limit.c: Determine machine precision e                              *
*           i.e.  the smallest e for which 1 + e .ne. 1                *
*                                                                      *
*  UNIX (DEC OSF, IBM AIX): cc limit.c                                 *
*                                                                      *
*  related information: the float.h file                               *
*  comment: very crude program which produces lots of screen output   *
************************************************************************
*/

#include <stdio.h>

#define N 60

main()
{
   double eps=1.0, one;                              /* starting values */
   int i;

   for(i=0; i<N; i++)
   {
      eps  /= 2.;                                    /* divide by two */
      one=1.0+eps;
      printf("%.18f \t %.16e \n",one, eps);
   }
}
```

```
/*
************************************************************************
*  lineq.c: Solve matrix equation Ax=b using LAPACK routine sgesv      *
*                                                                      *
*  UNIX (IBM AIX): cc -c lineq.c                                       *
*                  f77 lineq.o -lblas -llapack                         *
*  UNIX (DEC OSF): rename sgesv to sgesv_                              *
*                  cc lineq.c -ldxml                                   *
*                                                                      *
*  comment: LAPACK has to be installed on your system                  *
************************************************************************
*/

#include<stdio.h>
#define size 3                                  /* dimension of matrix */

main()
{
   int i, j , c1, c2, pivot[size], ok;
   float A[size][size], b[size], AT[size*size];

   A[0][0] = 3.1; A[0][1] =  1.3; A[0][2] = -5.7;          /* matrix A */
   A[1][0] = 1.0; A[1][1] = -6.9; A[1][2] =  5.8;
   A[2][0] = 3.4; A[2][1] =  7.2; A[2][2] = -8.8;

   b[0] = -1.3;                                            /* vector b */
   b[1] = -0.1;
   b[2] =  1.8;

   for (i=0; i<size; i++)                   /* transform the matrix so */
   {                                      /* we can pass it to Fortran */
     for(j = 0 ; j < size ; j += 1) AT[j + size*i] = A[j][i];
   }

   c1 = size;            /* define variable so we can pass pointer */
   c2 = 1;               /* to these variables to the routine */
```

Fig. C.25 limit.c: Determination of machine precision; lineq.c: solution of matrix equation using LAPACK.

```
   sgesv(&c1, &c2, AT, &c1, pivot, b, &c1, &ok);    /* sgesv_ for DEC */
      /* parameters in the order as they appear in the function call: */
      /* order of matrix A, number of right-hand sides (b), matrix A, */
      /* leading dimension of A, array records pivoting, */
      /* result vector b on entry, x on exit, leading dimension of b */
      /* return value =0 for success*/

   if (!ok)
   {
      for (j=0; j<size; j++) printf("%e\n", b[j]);      /* print x */
   }
   else printf("An error occurred\n");
}
```

```
/*
************************************************************************
*  numerov.c: Numerov method to find eigenvalues and eigenfunctions of *
*             a particle in a potential well. As written this program  *
*             only finds odd eigenfunctions correctly and fails for    *
*             even eigenfunctions and if there is no eigenvalue in the *
*             specified range. The final function is not normalized.   *
*                                                                      *
*  UNIX (DEC OSF, IBM AIX): cc numerov.c -lm                           *
************************************************************************
*/
#include <stdio.h>
#include <math.h>

#define steps 1000                      /* half width of potential well */
#define V -0.001                        /* depth of potential well */
#define eps 1E-8                        /* accuracy for eigenvalues */
#define Emin -0.001                     /* set to find the third */
#define Emax -0.00085                   /* eigenfunction */

double k2(int i, double E);             /* returns potential at x */
double diff(double E);                  /* difference of derivatives */
void plot(double E);                    /* data for final plot */

main()
{
   double E, min, max;
   int i=0;                             /* counter for iterations */
   min=Emin;
   max=Emax;
```

Fig. C.26 lineq.c: LAPACK matrix program (*continued*); numerov.c: Numerov method to find eigenvalues.

```
  do
  {
     i++;
     E=(max+min)/2.0;                /* divide energy range */
     if (diff(max)*diff(E)>0) max=E;   /* the bisection algorithm */
     else min=E;
  }  while(fabs(diff(E))>eps);
  printf("eigenvalue E=%.10f\n", E);
  printf("after %d iterations\n", i);
  plot(E);
 }
/*---------------------- end of main program ----------------------*/

/* function returns difference between left and right wave packet */
double diff(double E)
{
  double one, two, three, plus, minus;
  int i;
  one=0.0;
  two=0.00001;
  for (i=1; i<=1500; i++)               /* left side first */
  {
     three=(2*two*(1-5./12.*k2(i,E))-
               (1.+1./12.*k2(i-1,E))*one)/(1.+1./12.*k2(i+1,E));
     one=two;
     two=three;
  }
  minus=two;                            /* value at matching point */
  one=0.0;                              /* reset starting conditions */
  two=0.00001;
  for (i=1; i<500; i++)                 /* now the right side */
  {
     three=(2*two*(1.-5./12.*k2(i,E))-
               (1.+1./12.*k2(i+1,E))*one)/(1.+1./12.*k2(i-1,E));
     one=two;
     two=three;
  }
  plus=two;                             /* value at matching point */
  return((minus-plus));
}
```

```
/* function returns k-vector at depending on position i */
double k2(int i, double E)
{
  if (i<500) return(E);                      /* outside the well */
  if (i>=500) return (E-(V));                /* inside the well */
}
/* write data for eigenfuntion into files left.dat, right.dat */
void plot(double E)
{
  double one, two, three;
  int i;
  FILE *right, *left;                        /* save data in files */
  right=fopen("right.dat", "w");
  left=fopen("left.dat", "w");
  one=0.0;
  two=0.00001;
  for (i=1; i<=1500; i++)                    /* left side first */
  {
     three=(2*two*(1-5./12.*k2(i,E))-
               (1+1./12.*k2(i-1,E))*one)/(1+1./12.*k2(i+1,E));
     fprintf(left, "%d\t%f\n", i-1000, three/2.8);

     one=two;
     two=three;
  }
  one=0.0;                                   /* reset starting conditions */
  two=0.00001;
  for (i=1; i<500; i++)                      /* now the right side */
  {
     three=(2*two*(1.-5./12.*k2(i,E))-
               (1+1./12.*k2(i+1,E))*one)/(1.+1./12.*k2(i-1,E));
     fprintf(right, "%d\t%f\n", 1000-i, three/2.8);
     one=two;
     two=three;
  }
  fclose(left);
  fclose(right);
  printf("data saved in left.dat and right.dat\n");
}
```

Fig. C.27 numerov.c: Numerov method (*continued*).

```c
/*
************************************************************************
*  over.c: Determine overflow and underflow limits                      *
*                                                                       *
*  UNIX (DEC OSF, IBM AIX): cc over.c                                   *
*                                                                       *
*  related information: the float.h file                                *
*  comment: very crude program which produces lots of screen output     *
************************************************************************
*/

#include <stdio.h>

#define N 1024                     /* might not be big enough to cause */
                                   /* over and underflow */
main()
{
   double under=1., over=1.;                 /* starting values */
   int i;

   for(i=0; i<N; i++)
   {
      under /= 2.;                           /* divide by two */
      over  *= 2.;                           /* multiply by two */
      printf("%d. under: %e over: %e \n",i+1,under,over);
   }
}
```

```c
/*
************************************************************************
*  pond.c: *Monte Carlo integration to determine pi (stone throwing)    *
*                                                                       *
*  UNIX (DEC OSF, IBM AIX): cc pond.c                                   *
*                                                                       *
*  comment: If your compiler complains about drand48, srand48           *
*           uncomment the define statements further down                *
************************************************************************
*/
#include <stdio.h>
#include <stdlib.h>

/* if you don't have drand48 uncomment the following two lines */
/*    #define drand48 1.0/RAND_MAX*rand
      #define srand48 srand               */

#define max 1000                          /* number of stones to be thrown */
#define seed 68111                        /* seed for number generator */

main()
{
   int i, pi=0;
   double x, y, area;
   FILE *output;                          /* save data in pond.dat */
   output = fopen("pond.dat","w");
   srand48(seed);                         /* seed the number generator */

   for (i=1; i<=max; i++)
   {
      x = drand48()*2-1;                  /* creates floats between */
      y = drand48()*2-1;                  /* 1 and -1 */
      if ((x*x + y*y)<1) pi++;            /* stone hit the pond */
      area=4*(double)pi/i;                /* calculate area */
      fprintf(output, "%i\t%f\n", i, area);
   }
   printf("data stored in pond.dat\n");
   fclose(output);
}
```

Fig. C.28 over.c: Determination of overflow and underflow limits; pond.c: Monte Carlo integration by stone throwing.

```
/*
************************************************************************
*  qmc.c:  Feynman path integral (quantum Monte Carlo) for               *
*               ground-state wave packet                                *
*                                                                        *
*  UNIX (DEC OSF, IBM AIX): cc qmc.c -lm                                 *
*                                                                        *
*  comment: If your compiler complains about drand48, srand48            *
*           uncomment the define statements further down.                *
*           This might take a couple of minutes.                         *
************************************************************************
*/
#include <stdio.h>
#include <stdlib.h>
#include <math.h>

/* if you don't have drand48 uncomment the following two lines */
/*    #define drand48 1.0/RAND_MAX*rand
      #define srand48 srand                */

#define max 250000                         /* number of trials */
#define seed 68111                         /* seed for number generator */

main()
{
   double change, newE, oldE, path [101];
   int i, j, element, prop [101];
   double energy( double array[]);         /* finds energy of path */
   FILE *output;                           /* save data in qmc.dat */
   output = fopen("qmc.dat","w");

   srand48(seed);                          /* seed number generator */

   for (j=0; j<=100; j++) path[j]=0.0;     /* initial path */
   for (j=0; j<=100; j++) prop[j]=0;       /* initial probability */

   oldE = energy(path);                    /* find energy of path */
   for (i=0; i<max; i++)
   {
      element = drand48()*101;                   /* pick one random element */
      change  = (int)((drand48()-0.5)*20)/10.0;   /* change -0.9..0.9 */
      path[element]+=change;                     /* change path */
      newE = energy(path);                       /* find the new energy  */

      if ((newE>oldE) && (exp(-newE+oldE) <= drand48()))
      {
         path[element] -= change;                /* reject */
      }
      for (j=0; j<=100; j++)                    /* add up probabilities  */
      {
         element = path[j]*10+50;
         prop[element]++;
      }
      oldE = newE;
   }
   for (i=0; i<=100; i++)
   {
      fprintf(output, "%d\t%f\n", i-50, (double) prop[i]/max);
   }
   printf("data stored in qmc.dat\n");
   fclose(output);
}
/*-----------------------end of main program-----------------------*/

/* function returns energy of the path configuration */
double energy (double array[])
{
   int i;
   double sum=0.;

   for (i=0; i<100; i++)
   {
      sum += pow(array[i+1]-array[i], 2.0) + array[i]*array[i];
   }
   return (sum);
}
```

Fig. C.29 qmc.c: Feynman path integration.

```
/*
*******************************************************************
*  random.c:    A simple random-number generator, not for serious work *
*                                                                 *
*  UNIX (DEC OSF, IBM AIX): cc random.c                           *
*                                                                 *
*  comment: plot without connecting datapoints with lines         *
*******************************************************************
*/
#include <stdio.h>

#define max 1000                      /* number of numbers generated */
#define seed 11                       /* seed for number generator */

main ()
{
   int i, old, newx, newy;

   FILE *output;                      /* save data in badrand.dat */
   output = fopen("badrand.dat","w");

   old = seed;                        /* the seed */

   for (i=0; i<max; i++)              /* generating #max numbers */
   {
      newx = (57*old+1) % 256;        /* x-coordinate */
      newy = (57*newx+1) % 256;       /* y-coordinate */
      fprintf (output, "%i\t%i\n", newx, newy);
      old  = newy;
   }
   printf("data stored in badrand.dat.\n");
   fclose(output);
}
```

Fig. C.30 random.c: A simple random-number generator not suitable for serious work.

```
/*
************************************************************************
*  rk4.c: 4th order Runge-Kutta solution for harmonic oscillator       *
*                                                                      *
*  UNIX (DEC OSF, IBM AIX): cc rk4.c                                   *
************************************************************************
*/
#include <stdio.h>
#define N 2                                /* number of equations */
#define dist 0.1                           /* stepsize */
#define MIN 0.0                            /* minimum x */
#define MAX 10.0                           /* maximum x */

main()
{
   void runge4(double x, double y[], double step);
   double f(double x, double y[], int i);

   double x, y[N];
   int j;
   FILE *output;                           /* save data in rk4.dat */
   output = fopen("rk4.dat","w");
   y[0] = 1.0;                             /* initial position  */
   y[1] = 0.0;                             /* initial velocity  */

   fprintf(output, "%f\t%f\n", x, y[0]);

   for(x = MIN; x <= MAX ; x += dist)
   {
      runge4(x, y, dist);
      fprintf(output, "%f\t%f\n", x, y[0]);   /* position vs time */
   }
   printf("data stored in rk4.dat\n");
   fclose(output);
}
/*-----------------------end of main program--------------------------*/
```

```
/* Runge-Kutta subroutine */
void runge4(double x, double y[], double step)
{
   double h=step/2.0,                            /* the midpoint */
          t1[N], t2[N], t3[N],                   /* temporary storage */
          k1[N], k2[N], k3[N],k4[N];             /* for Runge-Kutta  */
   int i;

   for (i=0; i<N; i++) t1[i] = y[i]+0.5*(k1[i]=step*f(x, y, i));
   for (i=0; i<N; i++) t2[i] = y[i]+0.5*(k2[i]=step*f(x+h, t1, i));
   for (i=0; i<N; i++) t3[i] = y[i]+    (k3[i]=step*f(x+h, t2, i));
   for (i=0; i<N; i++) k4[i] =                 step*f(x + step, t3, i);

   for (i=0; i<N; i++) y[i] += (k1[i]+2*k2[i]+2*k3[i]+k4[i])/6.0;
}
/*--------------------------------------------------------------------*/

/* definition of equations - this is the harmonic oscillator */
double  f(double x, double y[], int i)
{
   if (i == 0) return(y[1]);                /* RHS of first equation */
   if (i == 1) return(-y[0]);               /* RHS of second equation */
}
```

Fig. C.31 rk4.c: Fourth-order Runge–Kutta integration of ODE.

```
/*
ccccccccccccccccccccccccccccccccccccccccccccccccccccccccccccccccccccc
c  shms.cpp: Object Oriented Program for simple harmonic motion        c
c            creates Lissajous figures                                 c
c                                                                      c
c  UNIX (DEC OSF, IBM AIX): cpp shms.cpp                               c
cccccccccccccccccccccccccccccccccccccccccccccccccccccccccccccccccccc
*/
#include <stdio.h>
#include <stdlib.h>
#include <math.h>
/*---------------- ShmX Class Definition ----------------------------*/
class ShmX
{
// CLASS ShmX: simple harmonic motion in the x direction
   protected:
      double xpos(double partt);             /* makes  Ax sin(wt+d)  */
      double delttx;
      int stepsx;                        /* Time steps to write in file  */
   private:
      double Ampx,deltx,omegx,timex;
   public:
      ShmX(double Axini,double delxini,double omegxini,
           double totaltx,double dtx);
      ~ShmX(void);                  /* Class Constructor and Destructor  */
      void archive();                      /* send x vs t to disk file  */
};
```

```
/*-----------------ShmX Constructor and Destructor --------------------*/
ShmX::ShmX(double Axini,double delxini,double omegxini,
           double totaltx,double dtx)
{
// CONSTUCTOR ShmX: Initializes Amplitude, step sizes, and frequency
   Ampx   = Axini;
   deltx  = delxini;
   omegx  = omegxini;
   timex  = totaltx;
   delttx = dtx;
   stepsx = (int)(totaltx/delttx);
}
ShmX::~ShmX(void)
{
// DESTRUCTOR ShmX
  printf("Class ShmX destroyed\n");
}
/*--------------- ShmX Methods -----------------------------------*/
double ShmX::xpos(double partt)
{
// METHOD xpos: returns X=A sin(xt+d)
   return (Ampx*sin(omegx*partt + deltx));
}
void ShmX::archive()
{
// METHOD archive: Produces disk file with X at several time intervals
   FILE  *pf;
   int i;
   double xx,tt;
   pf = fopen("cshmx.dat","w");
   tt = 0.;
   for(i = 1 ; i <= stepsx ; i += 1)
   {
      xx = xpos(tt);
      fprintf(pf,"%f  %f\n",tt,xx);
      tt = tt + delttx;
   }
   fclose(pf);
}
```

Fig. C.32 shms.cpp: Object-oriented program for simple harmonic motion.

```
/*---------------- ShmY Class Definition ----------------------------*/
class ShmY
{
// CLASS ShmY: simple harmonic motion in Y direction
   protected:
      double ypos(double partt);                /* makes Ay sin(wt+d)  */
      double deltty;
      int stepsy;                         /* Time steps to write in file  */
   private:
     double Ampy, delty, omegy, timey;
   public:
      ShmY(double Ayini, double delyini, double omegyini,
           double totalty, double dty);
      ~ShmY(void);                  /* Class Constructor and Destructor  */
      void archive();                        /* send y vs t to disk file  */
};
/*---------------- ShmY Constructor and Destructor -------------------*/
ShmY::ShmY(double Ayini, double delyini, double omegyini,
           double totalty, double dty)
{
// CONSTRUCTOR ShmX: initializes Amplitud, ang. veloc., time, phase
   Ampy   = Ayini;
   delty  = delyini;
   omegy  = omegyini;
   timey  = totalty;
   deltty = dty;
   stepsy = (int)(totalty/deltty);
}
ShmY::~ShmY(void)
{
// DESTRUCTOR ShmY
   printf("Class ShmY destroyed\n");
}
```

```
/*--------------- ShmY Methods -------------------------------*/
double ShmY::ypos(double partt)
{
// METHOD ShmY: returns Y=A sin(yt+d)
   return Ampy*sin(omegy*partt+delty);
}
void ShmY::archive()
{
// METHOD SgmY: Produces disk file with  X at several time intervals
   FILE  *pf;
   int i;
   double yy, tt;

   pf = fopen("shmy.dat","w");
   tt = 0.;
   for(i = 1 ; i <= stepsy ; i += 1)
   {
      yy=ypos(tt);
      fprintf(pf,"%f  %f\n",tt,yy);
      tt=tt+deltty;
   }
   fclose(pf);
}
/*---------------- ShmXY Class Definition -----------------------*/
class ShmXY : public ShmX, public ShmY
{
// CLASS ShmXY: child class of ShmX, ShmY; 2D simple harmonic motion
   public:
    ShmXY(double Axini, double delxini, double omegxini, double totaltx,
          double dtx, double Ayini, double delyini, double omegyini,
          double totalty, double dty);
      ~ShmXY(void);
      void archive();
};
```

Fig. C.33 shms.cpp: (*continued*).

```
/*---------------- ShmXY Constructor and Destructor -------------------*/
ShmXY::ShmXY(double Axini,double delxini,double omegxini,double totaltx,
             double dtx, double Ayini, double delyini, double omegyini,
             double totalty, double dty):
         ShmX(Axini, delxini, omegxini, totaltx, dtx),
         ShmY(Ayini, delyini, omegyini, totalty, dty)
{
// CONSTRUCTOR ShmXY: no initialization, variables passed to ShmX, ShmY
}
ShmXY::~ShmXY(void)
{
// DESTRUCTOR ShmXY
   printf("Class ShmXY destroyed\n");
}
/*--------------- ShmXY Methods ----------------------------*/
void ShmXY::archive()
{
// METHOD ShmXY: Produces disk file with  X at several time intervals
   FILE  *pf;
   int i;
   double xx, yy, tt;

   pf = fopen("shms.dat","w");
   tt = 0.;
   for(i = 1 ; i <= stepsx ; i += 1)
   {
      yy = ypos(tt);
      xx = xpos(tt);
      fprintf(pf,"%f  %f\n",yy,xx);                /* y vs x output  */
      tt = tt + delttx;
   }
   fclose(pf);
}
```

```
/*---------------- Main Program ------------------------------------*/
void main()
{
   double Ainix, delxx, wxx, txx, Ainiy, delyy, wyy, tyy, dx, dy;
   double pii;
   pii    = M_PI;
   Ainix  = 3.0;
   delxx  = 0.0;
   wxx    = 2.0;
   txx    = 7.0;
   tyy    = 7.0;
   Ainiy  = 2.0;
   delyy  = pii;
   wyy    = 3.0;
   dx     = 0.1;
   dy     = 0.1;
   ShmXY shmot(Ainix, delxx, wxx, txx, dx, Ainiy, delyy, wyy, tyy, dy);
   printf("  \n");
   printf("  \n");
   shmot.archive();
}
```

Fig. C.34 shms.cpp: (*continued*).

```c
/*
*******************************************************************
*  sierpin.c: Creates Sierpinski gasket fractal                   *
*                                                                 *
*  UNIX (DEC OSF, IBM AIX): cc sierpin.c                          *
*                                                                 *
*  comment: If your compiler complains about drand48, srand48     *
*           uncomment the define statements further down.         *
*           Plot data without connecting datapoints with lines.   *
*******************************************************************
*/

#include <stdio.h>
#include <stdlib.h>

/* if you don't have drand48 uncomment the following two lines */
/*    #define drand48 1.0/RAND_MAX*rand
      #define srand48 srand                 */

#define max 30000                /* number of iterations */
#define seed 68111               /* seed for number generator */
#define a1 20.0                  /* vertex 1 */
#define b1 20.0
#define a2 320.0                 /* vertex 2 */
#define b2 20.0
#define a3 170.0                 /* vertex 3 */
#define b3 280.0

main()
{
   int i;
   double x, y, r;

   FILE *output;                        /* save data in sierpin.dat */
   output = fopen("sierpin.dat","w");
   x = 180.;                                     /* starting point */
   y = 150.;

   srand48(seed);                                /* seed number generator */

   for(i=1 ; i<=max ; i++)                       /*  draw the gasket  */
   {
      r  = drand48();
      if (r <= 0.3333)
      {
         x = 0.5*(x + a1);
         y = 0.5*(y + b1);
      }
      else if(r > 0.3333 && r <= 0.6666)
      {
         x = 0.5*(x + a2);
         y = 0.5*(y + b2);
      }
      else
      {
         x = 0.5*(x + a3);
         y = 0.5*(y + b3);
      }
      fprintf(output, "%f %f\n", x, y);
   }
   printf("data stored in sierpin.dat\n");
   fclose(output);
}
```

Fig. C.35 sierpin.c: Sierpiński gasket fractal.

```c
/*
**************************************************************************
*  soliton.c: Solve K-deV equation with finite difference method        *
*                                                                        *
*  UNIX (DEC OSF, IBM AIX): cc soliton.c -lm                             *
*                                                                        *
*  comment: Output data is saved in 3D grid format used by gnuplot      *
**************************************************************************
*/
#include <stdio.h>
#include <math.h>
#define ds 0.4                          /* delta x */
#define dt 0.1                          /* delta t */
#define max 2000                        /* time steps */
#define mu 0.1                          /* mu from KdeV equation */
#define eps 0.2                         /* epsilon from KdeV eq. */
main()
{
   int i, j, k;
   double a1,a2,a3, fac, time, u[131][3];
   FILE *output;                        /* save data in soliton.dat */
   output = fopen("soliton.dat","w");

   for(i=0; i<131; i++)                 /* initial wave form */
   { u[i][0] = 0.5*(1.0 - tanh(0.2*ds*i - 5.0)); }
   u[0][1]   =1.;                       /* end points */
   u[0][2]   =1.;
   u[130][1] =0.;
   u[130][2] =0.;

   fac  = mu*dt/(ds*ds*ds);
   time = dt;

   for(i=1; i<130; i++)                 /* first time step */
   {
      a1 = eps*dt*(u[i+1][0] + u[i][0] + u[i-1][0]) / (ds*6.0);
      if((i>1) && (i<129))
      {
         a2 = u[i+2][0] + 2.0*u[i-1][0] - 2.0*u[i+1][0] - u[i-2][0];
      }
      else a2 = u[i-1][0] - u[i+1][0];
         a3 = u[i+1][0]-u[i-1][0];
      u[i][1] = u[i][0] - a1*a3 - fac*a2/3.;
   }
   for(j=1; j<max; j++)                 /* all other time steps */
   {
      time+=dt;
      for(i=1; i<130; i++)
      {
         a1 = eps*dt*(u[i+1][1] + u[i][1] + u[i-1][1]) / (3.0*ds);

         if((i>1) && (i<129))
         { a2 = u[i+2][1] + 2.0*u[i-1][1] - 2.0*u[i+1][1] - u[i-2][1];}
         else a2 = u[i-1][1] - u[i+1][1];
         a3      = u[i+1][1] - u[i-1][1];
         u[i][2] = u[i][0] - a1*a3 - 2.*fac*a2/3.;
      }
      for(k=0; k<131; k++)              /* move one step ahead */
      {
         u[k][0] = u[k][1];
         u[k][1] = u[k][2];
      }

      if((j%200)==0)                    /* plot every 200th step */
      {
         for(k=0; k<131; k+=2)
         {
            fprintf(output, "%f\n", u[k][2]);   /* gnuplot 3D format */
         }
         fprintf(output, "\n");         /* empty line for gnuplot */
      }
   }
   printf("data stored in soliton.dat\n");
   fclose(output);
}
```

Fig. C.36 soliton.c: Solution of Kortewg–deVries equation.

```c
/*
************************************************************************
*  tree.c: Creates a fractal tree                                      *
*                                                                      *
*  UNIX (DEC OSF, IBM AIX): cc tree.c                                  *
*                                                                      *
*  comment: If your compiler complains about drand48, srand48          *
*           uncomment the define statements further down               *
*           Plot data without connecting datapoints with lines         *
************************************************************************
*/
#include <stdio.h>
#include <stdlib.h>

/* if you don't have drand48 uncomment the following two lines */
/*    #define drand48 1.0/RAND_MAX*rand
      #define srand48 srand                 */

#define max 30000                  /* number of iterations */
#define seed 68111                 /* seed for number generator */

main()
{
   int i;
   double x,y,r,xn,yn;
   FILE *output;                   /* save data in tree.dat */
   output=fopen("tree.dat","w");
   srand48(seed);                  /* seed number generator */

   x     = 0.5;                    /* initial position */
   y     = 0.0;
    for(i=1 ; i<=max; i++)
   {
      r = drand48();
      if (r<=0.1)
      {
         xn = 0.05*x;
         yn = 0.6*y;
      }
     else if((r>0.1) && (r<0.2))
      {
         xn  = 0.05*x;
         yn  = -0.5*y+1.0;
      }
      else if ((r>0.2) && (r<0.4))
      {
         xn  = 0.46*x-0.32*y;
         yn  = 0.39*x+0.38*y+0.6;
      }
      else if((r>0.4) && (r<0.6))
      {
         xn  = 0.47*x-0.15*y;
         yn  = 0.17*x+0.42*y+1.1;
      }
      else if ((r>0.6) && (r<0.8))
      {
         xn  = 0.43*x+0.28*y;
         yn  = -0.25*x+0.45*y+1.0;
      }
      else
      {
         xn  = 0.42*x+0.26*y;
         yn  = -0.35*x+0.31*y+0.7;
      }
      fprintf(output, "%f %f\n", xn, yn);
      x = xn;
      y = yn;
   }
   printf("data stored in tree.dat\n");
   fclose(output);
}
```

Fig. C.37 tree.c: Create a fractal tree.

```c
/*
cccccccccccccccccccccccccccccccccccccccccccccccccccccccccccccccccccccccc
c  twodsol.c:  Solves the sine-Gordon equation for a 2D soliton       c
c                                                                      c
c    Christiansen & Lomdahl, Physics 2D (1981) 482-494                 c
c    U_xx + U_yy-U_tt=j(x,y)sin (U)                                    c
c        -x_0 <x< x_0,     -y_0 <y< y_0, t >=0,                        c
c    i. c. U(x,y,0)=4 arctan (exp(3-sqrt(x^2+y^2)),                    c
c    i. c. d U(x,y,0)/dt =0,   j(x,y)=1,   x_0=y_0=7                   c
cccccccccccccccccccccccccccccccccccccccccccccccccccccccccccccccccccccccc
*/
#include<stdio.h>
#include<math.h>

#define D 1200

main()
{

double u[D][D][3];
int nint;
void initial(double u[][D][3]);
void solution(double u[][D][3], int nint);

/*    input a positive integer which is proportional to the time
      you want to see the position of the wave packet.          */
/*printf(" Enter a positive integer from 1(initial time)\n");*/
/*printf("to 1800 to get wave packet position at that time:\n");*/
scanf("%d", &nint);

/* initializes the constant values and the wave */
initial(u);

/* time dependent equation is solved */
solution(u,nint);

}
```

```c
void initial(double u[][D][3])
{
double dx,dy,dt,xx,yy,dts,time, tmp;
int i,j;
dx=14./200.;
dy=dx;
dt=dx/sqrt(2.);
dts=(dt/dx)*(dt/dx);
yy=-42.;
time=0.;
for(i=0;i<=D-1;i++){
        xx=-42.;
        for(j=0;j<=D-1;j++){
                tmp=3.-sqrt(xx*xx+yy*yy);
                u[i][j][0]=4.*atan(tmp);
                xx=xx+dx;
                }
        yy=yy+dy;
        }
}

void solution(double u[][D][3], int nint)
{
double dx,dy,dt,time,a2,zz,dts,a1, tmp;
int l,m,mm,k,j,i;
dx=14./200.;
dy=dx;
dt=dx/sqrt(2.);
time=0.;
time=time+dt;
dts=(dt/dx)*(dt/dx);
tmp=0.;
for(m=1;m<=D-2;m++){
        for(l=1;l<=D-2;l++){
                a2=u[m+1][l][0]+u[m-1][l][0]+u[m][l+1][0]+u[m][l-1][0];
                tmp=.25*a2;
                u[m][l][1]=.5*(dts*a2-dt*dt*sin(tmp));
```

Fig. C.38 twodsol.c: Solution of sine–Gordon equation for a 2-D soliton.

```
                }
        }

/* The borders in the second iteration */
for(mm=1;mm<=D-2;mm++){
        u[mm][0][1]=u[mm][1][1];
        u[mm][D-1][1]=u[mm][D-2][1];
        u[0][mm][1]=u[1][mm][1];
        u[D-1][mm][1]=u[D-2][mm][1];
        }

/* The still undefined terms */
u[0][0][1]=u[1][0][1];
u[D-1][0][1]=u[D-2][0][1];
u[0][D-1][1]=u[1][D-1][1];
u[D-1][D-1][1]=u[D-2][D-1][1];
/*Third and following iterations k=1,2... */
tmp=0.;
for(k=0;k<=nint;k++){
     for(m=1;m<=D-2;m++){
            for(l=1;l<=D-2;l++){
                 a1=u[m+1][l][1]+u[m-1][l][1]+u[m][l+1][1]+u[m][l-1][1];
                 tmp=.25*a1;
                 u[m][l][2]=-u[m][l][0]+dts*a1-dt*dt*sin(tmp);
                 u[m][0][2]=u[m][1][2];
                 u[m][D-1][2]=u[m][D-2][2];
                  }
              }
                for(mm=1;mm<=D-2;mm++){
                        u[mm][0][2]=u[mm][1][2];
                        u[mm][D-1][2]=u[mm][D-2][2];
                        u[0][mm][2]=u[1][mm][2];
                        u[D-1][mm][2]=u[D-2][mm][2];
                        }
                u[0][0][2]=u[1][0][2];
                u[D-1][0][2]=u[D-2][0][2];
                u[0][D-1][2]=u[1][D-1][2];
                u[D-1][D-1][2]=u[D-2][D-1][2];

/* New iterations are now old, reuse, recycle, recover */
    for(l=0;l<=D-1;l++){
            for(m=0;m<=D-1;m++){
                    u[l][m][0]=u[l][m][1];
                    u[l][m][1]=u[l][m][2];
                    }
            }
        if(k==nint) {
            for(i=0;i<=D-1;i=i+5){
                    for(j=0;j<=D-1;j=j+5){
                            printf("%e\n",sin(u[i][j][2]/2.)
                            );
                            }
                    printf("\n");
                    }
            }
    time=time+dt;
    }
}
```

Fig. C.39 twodsol.c (*continued*).

```
/*
cccccccccccccccccccccccccccccccccccccccccccccccccccccccccccccccccccccc
c  unim1d.cpp: Object Oriented Program for uniform motion in 1D          c
c                                                                        c
c  UNIX (DEC OSF, IBM AIX): cpp unim1d.cpp                               c
cccccccccccccccccccccccccccccccccccccccccccccccccccccccccccccccccccccc
*/
#include<stdio.h>
#include<stdlib.h>
/*---------------- Um1D Class Definition ----------------------------*/
class Um1D                                          /* class is created  */
{
   private:
     double x00,delt,vx,time;                        /* initial values  */
     int steps;                          /* Time steps to write in file  */
     double x(double tt);                          /* makes x=xo+v* dt */
   public:
     Um1D(double x0, double dt, double vx0, double tott);
     ~Um1D(void);                  /* Class Constructor and Destructor  */
     void archive();                       /* send x vs t to disk file  */
};
/*---------------- Um1D Constructor and Destructor ------------------*/
Um1D::Um1D(double x0, double dt, double vx0, double tott)
{
//CONSTRUCTOR Um1D: initializes position, veloc., time, delta t
   x00    = x0;
   delt   = dt;
   vx     = vx0;
   time   = tott;
   steps  = (int)(tott/delt);
}
/*-------------------------------*/
Um1D::~Um1D(void)
{
//DESTRUCTOR: Um1D
   printf("Class Um1D destroyed\n");
}
/*--------------- Um1D Methods -------------------------------------*/
double Um1D::x(double tt)
{
// METHOD x: returns X=Xo+dt*v
   return x00+tt*vx;
}
/*-------------------------------*/
void Um1D::archive()
{
// METHOD archive: Produces disk file with X at several time intervals
   FILE  *pf;
   int i;
   double xx, tt;

   pf = fopen("unim1d.dat","w");
   tt = 0.;

   for(i = 1 ; i <= steps ; i += 1)
   {
      xx = x(tt);                /* computes X=Xo+t*v , changing Xo  */
      fprintf(pf,"%f  %f\n", tt, xx);
      tt = tt + delt;
   }
   fclose(pf);
}
/*------------------- Main Program --------------------------------*/
main()
{
   double inix,inivx,dtim,ttotal;
   inix   = 5.;
   dtim   = 0.1;
   inivx  = 10.;
   ttotal = 4.;
                         /* class constructor. Initial values given  */
   Um1D unimotx(inix,dtim,inivx,ttotal);
                               /* To obtain disk datafile of y vs x  */
   unimotx.archive();
}
```

Fig. C.40 unim1d.cpp: Object-oriented program for 1-D uniform motion.

```
/*
cccccccccccccccccccccccccccccccccccccccccccccccccccccccccccccccccccccc
c  unimot2d.cpp   Object Oriented Program for uniform motion in 2D     c
c                                                                      c
c  UNIX (DEC OSF, IBM AIX): cpp unimot2d.cpp                           c
cccccccccccccccccccccccccccccccccccccccccccccccccccccccccccccccccccccc
*/
#include <stdio.h>
#include <stdlib.h>
/*---------------- Um1D Class Definition ----------------------------*/
class Um1D                                        /* class is created  */
{
   protected:              /* so children classes may access the data  */
      double delt;
      int steps;                        /* Time steps to write in file  */
      double x(double tt);                          /* makes x=xo+v*dt  */
   private:
      double x00, vx, time;
   public:
      Um1D(double x0, double dt, double vx0, double tott);
      ~Um1D(void);                  /* Class Constructor and Destructor  */
      void archive();                        /* send x vs t to disk file  */
};
/*---------------- Um1D Constructor and Destructor ------------------*/
Um1D::Um1D(double x0, double dt, double vx0, double tott)
{
//CONSTRUCTOR Um1D: initializes position, veloc., time, delta t
   x00    = x0;
   delt   = dt;
   vx     = vx0;
   time   = tott;
   steps  = (int)(tott/delt);
}
```

```
Um1D::~Um1D(void)
{
//DESTRUCTOR: Um1D
   printf("Class Um1D destroyed\n");
}
/*--------------- Um1D Methods ----------------------------------------*/
double Um1D::x(double tt)
{
// METHOD x: returns X=Xo+dt*v
   return (x00 + tt*vx);
}
/*-------------------------------*/
void Um1D::archive()
{
// METHOD archive: Produces disk file with X at several time intervals
   FILE  *pf;
   int i;
   double xx, tt;

   pf = fopen("unim1d.dat","w");
   tt = 0.;

   for(i = 1 ; i <= steps ; i += 1)
   {
      xx = x(tt);                    /* computes X=Xo+t*v , changing Xo  */
      fprintf(pf,"%f  %f\n", tt, xx);
      tt = tt + delt;
   }
   fclose(pf);
}
```

Fig. C.41 unim2d.cpp: Object-oriented program for 2-D uniform motion.

```
/*--------------- Um2D Class Definition ---------------------------*/
class Um2D : public Um1D
{
// CLASS Um2D: child class of UmD1, for 2 dimensional uniform motion
   protected:
      double y(double tt);
   private:
      double y00, vy;
   public:
      Um2D(double x0,double dt,double vx0,double tott,
           double y0,double vy0);   /* constructor or the Um2D class  */
      ~Um2D(void);                  /* destructor of the Um2D class  */
      void archive();      /* override Um1D definition for 2D output  */
};
/*--------------- Um2D Constructor and Destructor ------------------*/
Um2D::Um2D(double x0,double dt,double vx0,double tott,
           double y0,double vy0):
      Um1D(x0,dt,vx0,tott)
{
// CONSTRUCTOR Um2D: initializes y position and y velocity
   y00   = y0;
   vy    = vy0;
}
/*------------------------------------*/
Um2D::~Um2D(void)
{
// DESTRUCTOR Um2D
   printf("Class Um2D is destroyed\n");
}
/*--------------- Um2D Methods --------------------------------*/
double Um2D::y(double tt)
{
// METHOD ypos: returns Y=Y0+dt*v
   return (y00 + tt*vy);
}
```

```
void Um2D::archive()
{
// METHOD archive: override of Um1D.archive, for 2-dim uniform motion
   FILE  *pf;
   int i;
   double xx, yy, tt;

   pf = fopen("unimot2d.dat","w");
   tt = 0.;
   for(i = 1 ; i <= steps ; i += 1)
   {
      xx = x(tt);
      yy = y(tt);
      fprintf(pf,"%f  %f\n",yy,xx);      /* data is now x vs y  */
      tt = tt + delt;
   }
   fclose(pf);
}
/*-----------------------------------------*/
/*--------------- Main Program ---------------------------------*/
 main()
{
   double inix,iniy,inivx,inivy,dtim,ttotal;
   inix   = 5.;
   dtim   = 0.1;
   inivx  = 10.;
   ttotal = 4.;
   iniy   = 3.;
   inivy  = 8.;
                     /* class constructor. Initial values given  */
   Um2D unimotxy(inix,dtim,inivx,ttotal,iniy,inivy);
   unimotxy.archive();   /* activate desired portion of object  */
}
```

Fig. C.42 unim2d.cpp (*continued*).

```
/*
************************************************************************
*  walk.c: Random walk simulation                                      *
*                                                                      *
*  UNIX (DEC OSF, IBM AIX): cc walk.c -lm                              *
*                                                                      *
*  comment: If your compiler complains about drand48, srand48          *
*           uncomment the define statements further down               *
*           Data is saved as sqrt(steps), distance                     *
************************************************************************
*/

#include <stdio.h>
#include <stdlib.h>
#include <math.h>

/* if you don't have drand48 uncomment the following two lines */
/*    #define drand48 1.0/RAND_MAX*rand
      #define srand48 srand                */

#define SQRT2 1.4142135623730950488E0
#define max 10000
#define seed 68111

main()
{
   int i, j;
   double x, y, r[max+1];

   FILE *output;                        /* save data in walk.dat */
   output = fopen("walk.dat","w");

   srand48(seed);                       /* seed the number generator */

   for (i=0; i<=max; i++) r[i]=0.0;     /* clear array */

   for (j=1; j<=100; j++)               /* average over 100 trials */
   {
      x=y=0;                            /* starting point  */
      for (i=1; i<=max; i++)
      {
         x += (drand48()-0.5)*2*SQRT2;      /* dx and dy between */
         y += (drand48()-0.5)*2*SQRT2;      /* -sqrt(2) and sqrt(2) */
         r[i] += sqrt(x*x + y*y);           /* distance from origin */
      }
   }
   for (i=0; i<=max; i++)               /* write results into file */
   {
      fprintf(output, "%f\t%f\n", sqrt(i), r[i]/100.);
   }
   printf("data stored in walk.dat.\n");
   fclose(output);
}
```

Fig. C.43 walk.c: Random-walk simulation.

Appendix D: Listing of Fortran Programs

```
cccccccccccccccccccccccccccccccccccccccccccccccccccccccccccccccccccccccc
c      area.f: Area of a circle, sample program                        c
c                                                                      c
c      UNIX (DEC OSF, IBM AIX): f77 area.f                             c
cccccccccccccccccccccccccccccccccccccccccccccccccccccccccccccccccccccccc
       Program area
c
c     area of circle, r input from terminal
      Double Precision pi, r, A
c
c      Best value of pi for IEEE floating point
       pi = 3.14159265358979323850E0
c
c      read r from standard input (terminal)
       Write(*,*) 'Enter the radius of a circle'
       Read (*,*)  r
c
c      calculate area
       A = pi * r**2
c
c      write area onto terminal screen
       Write(*, 10) 'radius r =', r, 'area =', A
 10    Format(a10, f10.5, a10, f12.5)
       Stop 'area'
       End
```

Fig. D.1 area.f: Area of a circle, sample program.

```
cccccccccccccccccccccccccccccccccccccccccccccccccccccccccccccccc
c       bessel.f:  Spherical Bessel functions via up & down recursion  c
c                                                                      c
c       UNIX (DEC OSF, IBM AIX): f77 bessel.f                          c
c                                                                      c
c       comment: data saved as: x y1 y2                                c
cccccccccccccccccccccccccccccccccccccccccccccccccccccccccccccccc
        Program bessel
        Implicit none
c       declarations
c       order of function, x range, stepsize, start for downward recur
        Real*8 step, x, xmin, xmax, up, down, t1, t2
        Integer order, start
        xmin=0.25
        xmax=40.0
        step=0.1
        order=10
        start=50
c       open output file
        Open(6, File='bessel.dat', Status='Unknown')
c       main program
        Do 10 x=xmin, xmax, step
           t1=down(x, order, start)
           t2=up(x, order)
           Write (6, *) x, t1, t2
 10     Continue
        Close(6)
        Stop 'data saved in bessel.dat'
        End
```

```
c       calculate using downward recursion
        Function down(x, order, start)
        Implicit none
        Integer k, order, start
        Real*8 down, scale, x, j(100)
c       the arbitrary start
        j(start+1)=1
        j(start)=1
        Do 20 k = start, 2, -1
           j(k-1)=((2*k-1.0)/x)*j(k)-j(k+1)
 20     Continue
c       scale so that j(1) = sin(x)/x
        scale=(sin(x)/x)/j(1)
        down=j(order+1)*scale
        Return
        End
c
c       calculate using upward recursion
        Function up(x, order)
        Implicit none
        Integer k, order
        Real*8 up, x, one, two, thr
        one=sin(x)/x
        two=(sin(x)-x*cos(x))/(x*x)
        Do 30 k=1, (order-1)
           thr=((2*k+1.0)/x)*two-one
           one=two
           two=thr
 30     Continue
        up=thr
        Return
        End
```

Fig. D.2 bessel.f: Spherical Bessel functions via up and down recursion.

```
cccccccccccccccccccccccccccccccccccccccccccccccccccccccccccccccccccccc
c  bound.f:  bound states in momentum space of delta shell potential   c
c            uses LAPACK SGEEV                                          c
c            16 grid points, l=0, lambda varuable                       c
c  WARNING:  results are NOT stable!                                    c
cccccccccccccccccccccccccccccccccccccccccccccccccccccccccccccccccccccc
        Program bound
c
        Integer n, Size, job, info, i, j
        Real b
        Parameter (Size=100, pi=3.141592, b=10.0)
        Real lambda, scale
        Double Precision Pot
        Complex x(Size),V(Size, Size)
        REAL H(Size,Size), Work(0:Size,0:Size), k(Size), w(Size)
C
C       First we get the desired potential strength lambda
C
        Write(*,*) 'enter lambda'
        Read(*,*) lambda
        Write(*,*) 'enter scaling factor'
        Read(*,*) scale
        Write(*,*) 'number of points'
        Read(*,*) n
C
C      Set up Gauss points and weights on the interval [0,inf]
C       scale is "midpoint" of integration on infinite interval
C
        Call Gauss(n,2,0.0,scale,k,w)
C
C       Set up matrix for delta potential delta(r-b)
C
        DO i=1,n
                DO j=1,n
                        Pot=-1.0*lambda*SIN(b*k(i))*SIN(b*k(j))
                        Pot=Pot/(k(i)*k(j))
                        H(i,j)=w(j)*k(j)*k(j)*2.0*Pot/pi
                        If (i .EQ. j) Then
                                H(i,j)=H(i,j)+k(i)*k(i)
                        Endif
                End Do
        End Do
        job=1
        Call sgeev(h, Size, n, x, V,Size, Work, job, info)
        Write(*,*) info
        Do i=1,n
                Write(*,*) x(i)
        End Do
        End
```

Fig. D.3 bound.f: Bound states in momentum space.

```
cccccccccccccccccccccccccccccccccccccccccccccccccccccccccccccccccccccc
c       bugs.f: Bifurcation diagram for logistic map                 c
c                                                                    c
c       UNIX (DEC OSF, IBM AIX): f77 bugs.f                          c
c                                                                    c
c       comment: plot without conneting datapoints with lines        c
cccccccccccccccccccccccccccccccccccccccccccccccccccccccccccccccccccccc
        Program BUGS
c
c       bug population - bifurcation map of m*y*(1-y)
c
        Implicit none
c       Declarations (range, resolution for m)
        Real*8 m_min, m_max, m, step, y
        Integer x
        m_min = 1.0
        m_max = 4.0
        step  = 0.01
        Open(6, File='bugs.dat', Status='Unknown')
c       Loop for m values, arbitrary starting value for y
        Do 10 m=m_min, (m_max-step), step
           y = 0.5
c          Wait until transients die out
           Do 20 x=0, 200
              y = m*y * (1 - y)
 20        Continue
c          Record 200 points
           Do 30 x=201, 401
              y=m*y * (1 - y)
              Write (6,50) m,y
 30        Continue
 10     Continue
 50     Format (f5.3,f10.6)
        Close(6)
        Stop 'data saved in bugs.dat'
        End
```

```
cccccccccccccccccccccccccccccccccccccccccccccccccccccccccccccccccccccc
c       call.f:  Creates pseudo-random numbers using drand48 or rand c
c                                                                    c
c       UNIX (DEC OSF, IBM AIX): f77 call.f                          c
c                                                                    c
c       comment: If your compiler complains about drand48, seed48    c
c           replace drand48 with rand(seed) and remove the           c
c           call to seed48                                           c
cccccccccccccccccccccccccccccccccccccccccccccccccccccccccccccccccccccc
        Program call
        Implicit none
c       declarations; must be proper for precision
        Integer seed, i
        Real*8 drand48
c       user plants seed
        Write(*,*) 'enter seed'
        Read(*,*) seed
        Call seed48(seed)
c       generate and print 10 random numbers
        Do 100 i=1,10
           Write(*,*) 'Your random number is',  drand48()
 100    Continue
        End
```

Fig. D.4 bugs.f: Bifurcation diagram for logistic map; call.f: pseudo-random numbers using drand48 or rand.

```
cccccccccccccccccccccccccccccccccccccccccccccccccccccccccccccccccccccc
c       complex.f: Dealing with complex numbers on a computer        c
c                                                                    c
c       UNIX (DEC OSF, IBM AIX): f77 complex.f                       c
cccccccccccccccccccccccccccccccccccccccccccccccccccccccccccccccccccccc
        Program complex
c
c       complex numbers and functions
        Implicit none
        Complex*16 z, zsqrt, zlog
        Real*8 i, pi, phi, x, y, zatan, zatan2
        pi=3.1415926535897932385E0
c       write header for table
        Write (*,10) 'phi','x','y','sqrt','log','atan','atan2'
        Write (*,*) ' '
c       loop for angle
        Do 100 i=0, 2.6, 0.1
           phi    = i * pi
c          calculate carthesian representation
           x      = cos(phi)
           y      = sin(phi)
           z      = cmplx(x,y)
c          call functions
           zsqrt  = sqrt(z)
           zlog   = log(z)
           zatan  = atan(y/x)
           zatan2 = atan2(y,x)
c          write results
           Write (*,20) i,'*pi',x,y,zsqrt,'i',zlog,'i',zatan,zatan2
 100    Continue
 10     Format (a4, 2a9, a14, a18, a14, a10)
 20     Format (f3.1, a3, 3f9.4, f8.4, a1, f9.4, f8.4,a1, 2f9.4)
        Stop 'complex'
        End
```

```
cccccccccccccccccccccccccccccccccccccccccccccccccccccccccccccccccccccc
c       decay.f: Spontaneous radioactive decay simulation            c
c                                                                    c
c        comment: If compiler complains about drand48, seed48,       c
c                 replace drand48 with rand(seed), remove seed48     c
cccccccccccccccccccccccccccccccccccccccccccccccccccccccccccccccccccccc
        Program decay
        Implicit none
c       Declarations
        Real*8 r, drand48, lambda
        Integer i, j, h, nleft, nloop, start, seed
c       Set parameters (decay rate, init no atoms, seed), plant seed
        lambda = 0.01
        start = 1000
        seed = 11168
        h = 1
        nloop = start
        nleft = start
        call seed48(seed)
c       open output 'file'
        Open(6, File = 'decay.dat')
c      loop over times and over atoms
        Do 20 j=1,10000
          Do 10 i = 1, nleft
            r = drand48()
            IF (r .LE. lambda) THEN
              nloop = nloop -1
            EndIF
 10       Continue
c         atom loop ends
          nleft = nloop
          Write (6,*) h, ' ', Real(nleft)/start
          h = h + 1
          If (nleft .eq. 0) Goto 30
 20     Continue
 30     Close(6)
        Stop 'data saved in decay.dat'
        End
```

Fig. D.5 complex.f: Dealing with complex numbers; decay.f: spontaneous radioactive decay simulation.

```
cccccccccccccccccccccccccccccccccccccccccccccccccccccccccccccccccccccccc
c        diff.f:  Differentiation  using forward, central and          c
c                 extrapolated difference methods                      c
c                                                                      c
c       UNIX (DEC OSF, IBM AIX): f77 diff.f                            c
c                                                                      c
c       comment: results saved as x y1 y2 y3                           c
cccccccccccccccccccccccccccccccccccccccccccccccccccccccccccccccccccccccc
        Program diff
        Implicit None
c       Declarations
c       h stepsize for approximation, xrange and xstepsize
c
        Real*8 f, h, result(3), x, xmin, xmax, xstep
        Open(6, File='diff.dat', Status='Unknown')
        h     = 1.e-5
        xmin  = 0.0
        xmax  = 7.0
        xstep = 0.01
        Do 10 x=xmin, xmax, xstep
           result(1) = (f(x+h) - f(x))/h
           result(2) = (f(x+h/2) - f(x-h/2))/h
           result(3) = (8*(f(x+h/4)-f(x-h/4)) - (f(x+h/2)-f(x-h/2)))/(3*h)
           Write (6, 20) x, result(1), result(2), result(3)
 10     Continue
 20     Format(F5.3, TR4, F10.8, TR4, F10.8, TR4, F10.8)
        Close(6)
        Stop 'data saved in diff.dat'
        End
c
c       the function we want to integrate
        Function f(x)
           Implicit none
           Real*8 f, x
           f = cos(x)
           Return
        End
```

Fig. D.6 diff.f: Differentiation using forward-, central-, and extrapolated-difference methods.

```
cccccccccccccccccccccccccccccccccccccccccccccccccccccccccccccccccccccccc
c       eqheat.f: Solution of heat equation with finite differences    c
c                                                                      c
c       UNIX (DEC OSF, IBM AIX): f77 eqheat.f                          c
c                                                                      c
c       comment: Output data saved in 3D grid format of gnuplot        c
cccccccccccccccccccccccccccccccccccccccccccccccccccccccccccccccccccccccc
        Program heat
        Implicit None
        Double Precision cons, ro, sph, thk, u(101,2)
        Integer i, k, max
        Open(9,FILE='eqheat.dat',STATUS='UNKNOWN')
c       specific heat, thermal conductivity and density for iron
        sph=0.113
        thk=0.12
        ro=7.8
        cons = thk/(sph*ro)
c       number of iterations
        max=30000
c       At t=0 (i=1) all points are at 100 C
        Do 10 i=1,100
          u(i,1) = 100.0
 10     Continue
c       except the endpoints which are always zero
        Do 20 i=1,2
          u(1,i)   = 0.0
          u(101,i) = 0.0
 20     Continue
c
c       now start solving
c       loop over time
        Do 100 k=1,max
c         loop over space, endpoints stay fixed
          Do 30 i=2,100
            u(i,2) = u(i,1) + cons*(u(i+1,1) + u(i-1,1)-2*u(i,1))
 30       Continue
c
c         we want to know the temperatures every 1000 time steps
          If((MOD(k,1000).eq.0).or.(k.eq.1)) Then
            Do 40 i=1,101,2
              Write(9,22)u(i,2)
 40         Continue
            Write (9,22)
          EndIf
c         recycle, new values are now old.
          Do 50 i=2,100
            u(i,1) = u(i,2)
 50       Continue
 100    Continue
 22     Format (f10.6)
        Close(9)
        Stop 'data saved in eqheat.dat'
        End
```

Fig. D.7 eqheat.f: Solution of heat equation.

```
cccccccccccccccccccccccccccccccccccccccccccccccccccccccccccccccccccccccc
c       eqstring.f: Solution of  wave equation using time stepping     c
c                                                                      c
c       UNIX (DEC OSF, IBM AIX): f77 eqstring.f                        c
c                                                                      c
c       comment: Output data saved in 3D grid format used by gnuplot   c
cccccccccccccccccccccccccccccccccccccccccccccccccccccccccccccccccccccccc
        Program string
        Implicit None
        Real*8 x(101,3)
        Integer i, k, max
        max=100
        Open(9,FILE='eqstring.dat',STATUS='UNKNOWN')
c       initialize values
        Do 10 i=1,80
           x(i,1) = 0.00125*i
 10     Continue
        Do 20 i=81,101
          x(i,1) = 0.1-0.005*(i-81)
 20     Continue
c       the first step ahead in time
        Do 30 i=2,100
          x(i,2) = x(i,1)+0.5*(x(i+1,1)+x(i-1,1)-2.0*x(i,1))
 30     Continue
c       all other time steps
        Do 40 k=1,max
          Do 50 i=2,100
            x(i,3)=2.0*x(i,2)-x(i,1)+(x(i+1,2)+x(i-1,2)-2.0*x(i,2))
 50       Continue
          Do 60 i=1,101
c           The new iteration in time is now the old
            x(i,1) = x(i,2)
            x(i,2) = x(i,3)
 60       Continue
c           write plot data every 10 time steps
            If(MOD(k,10).EQ.0)then
              Do 70 i=1,101
                Write(9,11) x(i,3)
 70           Continue
              Write(9,11)
            EndIf
 40       Continue
 11     Format (e12.6)
          Close(9)
          Stop 'data saved in eqstring.dat'
          End
```

Fig. D.8 eqstring.f: Solution of wave equation for string.

```
cccccccccccccccccccccccccccccccccccccccccccccccccccccccccccccccccccccccc
c       exp-bad.f: A bad algorithm for calculating exponen              c
c                                                                       c
c       UNIX (DEC OSF, IBM AIX): f77 exp-bad.f                          c
c                                                                       c
c       related programs: exp-good.f                                    c
cccccccccccccccccccccccccccccccccccccccccccccccccccccccccccccccccccccccc
        Program expbad
c       calculating e^-x as a finite sum / bad algorithm
        Implicit none
c       declarations
c       min=accuracy limit, step in x, max in x, up numer, down denomin
        Real*8 down, min, max, step, sum, up, x
        Integer i, n
        min  = 1E-10
        max  = 10.
        step = 0.1
        Open(6, File='exp-bad.dat', Status='Unknown')
c       execution
        Do 10 x=0, max, step
           sum = 1
           Do 20 n=1, 10000
              up   = 1
              down = 1
              Do 30 i=1, n
                 up   = -up*x
                 down = down*i
 30           Continue
              sum = sum + up/down
              If ((abs((up/down)/sum) .lt. min) .AND. (sum .NE. 0)) then
                 Write (6,*) x, sum
c                 notice, no while in fortran so:
                 GoTo 10
              Endif
 20        Continue
 10     Continue
        Close(6)
        Stop 'data saved in exp-bad.dat'
        End
```

```
cccccccccccccccccccccccccccccccccccccccccccccccccccccccccccccccccccccccc
c       exp-good.f:  good algorithm for calculating exponential         c
c                                                                       c
c       UNIX (DEC OSF, IBM AIX): f77 exp-good.f                         c
c                                                                       c
c       related programs: exp-bad.f                                     c
cccccccccccccccccccccccccccccccccccccccccccccccccccccccccccccccccccccccc
        Program expgood
c
c       calculating e^-x as a finite sum / good algorithm
c
        Implicit none
c       declarations:
c       min:limit for accuracy, max in x, step in x
        Real*8 element, min, max, step, sum, x
        Integer i, j, n
        min  = 1E-10
        max  = 10.0
        step = 0.1
        Open(6, File='exp-good.dat', Status='Unknown')
c       execution
        Do 10 x=0, max, step
           sum     = 1
           element = 1
           Do 20 n=1, 10000
              element = element*(-x)/n
              sum     = sum + element
              if ((abs(element/sum) .lt. min) .AND. (sum .NE. 0)) then
                 Write (6,*) x, sum
c                notice: no while in standard fortran, therefore
                 GoTo 10
              Endif
 20        Continue
 10     Continue
        Close(6)
        Stop 'data saved in exp-good.dat'
        End
```

Fig. D.9 exp-bad.f, exp-bad.f: Bad and good algorithms for calculating exponential series.

```
cccccccccccccccccccccccccccccccccccccccccccccccccccccccccccccccccccccc
c     fit.f: Least-squares fit to decay spectrum                     c
c                                                                    c
c     UNIX (DEC OSF, IBM AIX): f77 fit.f                             c
c                                                                    c
cccccccccccccccccccccccccccccccccccccccccccccccccccccccccccccccccccccc
      Program fit
      Implicit none
c
c     declarations
      Integer i, j
      Real*8 s, sx, sy, sxx, sxy, delta, inter, slope
      Real*8 x(12), y(12), d(12)
c
c     input value y - exponential fit y > 0
      Data y /32, 17, 21, 7, 8, 6, 5, 2, 2, 0.1, 4, 1/
c
c     input values x
      Do 10 i=1, 12
         x(i)=i*10-5
 10   Continue
c
c     input value delta y - estimate
      Do 11 i=1, 12
         d(i)=1.0
 11   Continue
c
c     take logs of y values for exponential fit
      Do 20 i=1, 12
         y(i)=log(y(i))
 20   Continue
c
c    calculate all the sums
      Do 30 i=1, 12
         s   = s   +          1 / (d(i)*d(i))
         sx  = sx  +       x(i) / (d(i)*d(i))
         sy  = sy  +       y(i) / (d(i)*d(i))
         sxx = sxx + x(i)*x(i) / (d(i)*d(i))
         sxy = sxy + x(i)*y(i) / (d(i)*d(i))
 30   Continue
c
c     calculate the coefficients
      delta= s*sxx-sx*sx
      slope=  (s*sxy-sx*sy) / delta
      inter=(sxx*sy-sx*sxy) / delta
      Write(*,*) 'intercept=', inter
      Write(*,*) 'slope=', slope
      Write(*,*) 'correlation=', -sx/sqrt(sxx*s)
      Stop 'fit'
      End
```

Fig. D.10 fit.f: Least-squares fit to decay spectrum.

```
cccccccccccccccccccccccccccccccccccccccccccccccccccccccccccccccccccccccc
c       fourier.f: Calculates a discrete Fourier Transformation        c
c                                                                      c
c       UNIX (DEC OSF, IBM AIX): f77 fourier.f                         c
c                                                                      c
c       comment: Input data from file input.dat containing y(t) values c
c                separated by spaces. Output has form                  c
c                frequency index \t real part \t imaginary part        c
cccccccccccccccccccccccccccccccccccccccccccccccccccccccccccccccccccccccc
        Program fourier
        Implicit none
        Integer max
        Real*8 pi
        Parameter (max=1000,pi=3.1415926535897932385E0)
        Integer i, j, k
        Real*8 input(max), real, imag
        Open(9, File='fourier.dat', Status='UNKNOWN')
c       read data from file until end-of-file or max values
        Open(8, File='input.dat', Status='OLD')
        Do 10 i=1,max
           Read(8,*,END=20) input(i)
 10     Continue
c       loop for frequency index
 20     Do 30 j=1,i
          real=0
          imag=0
c         loop for sums
          Do 40 k=1,i
            real=real+input(k)*cos((2*pi*k*j)/i)
            imag=imag+input(k)*sin((2*pi*k*j)/i)
 40       Continue
          Write (9,*) j, real/i, imag/i
 30     Continue
        Close(8)
        Close(9)
        Stop 'data saved in fourier.dat'
        End
```

Fig. D.11 fourier.f: Discrete Fourier transformation (DFT).

```
cccccccccccccccccccccccccccccccccccccccccccccccccccccccccccccccccccccc
c     gauss.f: Points and weights for Gaussian quadrature            c
c                                                                    c
cccccccccccccccccccccccccccccccccccccccccccccccccccccccccccccccccccccc
      subroutine gauss(npts,job,a,b,x,w)
c     rescale rescales the gauss-legendre grid points and weights
c
c     npts     number of points
c     job = 0  rescalling uniformly between (a,b)
c           1  for integral (0,b) with 50% points inside (0, ab/(a+b))
c           2  for integral (a,inf) with 50% inside (a,b+2a)
c     x, w     output grid points and weights.
c
      integer npts,job,m,i,j
      real*8 x(npts),w(npts),a,b,xi
      real*8 t,t1,pp,p1,p2,p3,aj
      real*8 eps,pi,zero,two,one,half,quarter
      parameter (pi = 3.14159265358979323846264338328, eps = 3.0E-14)
      parameter (zero=0.0d0,one=1.0d0,two=2.0d0)
      parameter (half=0.5d0,quarter=0.25d0)
c
c     FIRST EXECTUABLE *************************************************
c
      m=(npts+1)/2
      do 1020 i=1,m
         t=cos(pi*(i-quarter)/(npts+half))
 1000 continue
         p1=one
         p2=zero
         aj=zero
         do 1010 j=1,npts
            p3=p2
            p2=p1
            aj=aj+one
            p1=((two*aj-one)*t*p2-(aj-one)*p3)/aj
 1010    continue
         pp=npts*(t*p1-p2)/(t*t-one)
         t1=t
         t=t1-p1/pp
c
         if(abs(t-t1).gt.eps) goto 1000
         x(i)=-t
         x(npts+1-i)=t
         w(i)=two/((one-t*t)*pp*pp)
         w(npts+1-i)=w(i)
 1020 continue
c     rescale the grid points
      if (job.eq.0) then
c     scale to (a,b) uniformly
         do 1030 i=1,npts
            x(i)=x(i)*(b-a)/two+(b+a)/two
            w(i)=w(i)*(b-a)/two
 1030    continue
      elseif (job.eq.1) then
c     scale to (0,b) with 50% points inside (0,ab/(a+b))
         do 1040 i=1,npts
            xi=x(i)
            x(i)=a*b*(one+xi)/(b+a-(b-a)*xi)
            w(i)=w(i)*two*a*b*b/((b+a-(b-a)*xi)*(b+a-(b-a)*xi))
 1040    continue
      elseif (job.eq.2) then
c     scale to (a,inf) with 50% points inside (a,b+2a)
         do 1050 i=1,npts
            xi=x(i)
            x(i)=(b*xi+b+a+a)/(one-xi)
            w(i)=w(i)*two*(a+b)/((one-xi)*(one-xi))
 1050    continue
      else
         pause 'Wrong value of job'
      endif
      return
      end
```

Fig. D.12 gauss.f: Points and weights for Gaussian quadrature.

```
cccccccccccccccccccccccccccccccccccccccccccccccccccccccccccccccccccccc
c       harmos.f: Solves time-dependent Schroedinger equation for     c
c       Gaussian wave packet in a quadratic potential                  c
c                                                                     c
c       UNIX (DEC OSF, IBM AIX): f77 harmos.f                         c
c                                                                     c
c       comment: Output data saved in 3D grid format used by gnuplot  c
c                This might take a couple of minutes.                 c
cccccccccccccccccccccccccccccccccccccccccccccccccccccccccccccccccccccc
        Program harmos
        Implicit None
        Real*8 psr(751,2), psi(751,2), v(751), p2(751)
        Real*8 pi,dx,k0,dt,x
        Complex exc,zi
        Integer max,i,n
        Open(9,FILE='harmos.dat',STATUS='UNKNOWN')
        pi   = 3.1415926535897932385E0
        zi   = cmplx(0.0,1.0)
        dx   = 0.02
c       k0 is the initial momentum given to the wave packet
        k0   = 3 * pi
        dt   = dx*dx/4.0
        max   = 750
c       initial conditions
        x   = -7.5
        Do 10 i=1,max+1
          exc       = exp(zi*k0*x)
c         real part of initial wave packet
          psr(i,1) = real(exc*exp(-0.5*(x/0.5)**2.0))
c         imaginary part of initial wave packet
          psi(i,1) = imag(exc*exp(-0.5*(x/0.5)**2.0))
c         the potential
          v(i)      = 5.0*x*x
          x         = x + dx
 10     Continue
c
c       now propagate solution through time
c
        Do 40 n=1,20000
c       the real part of the wave packet is computed here
c       and the probability P2
          Do 50 i=2,max
            psr(i,2) = psr(i,1)-dt*(psi(i+1,1)+psi(i-1,1)
     1                 -2.D0*psi(i,1))/(dx*dx)+dt*v(i)*psi(i,1)
             p2(i)  = psr(i,1)*psr(i,2)+psi(i,1)*psi(i,1)
 50       Continue
c         same thing for the imaginary part of the wave packet
          Do 60 i=2,max
            psi(i,2) = psi(i,1)+dt*(psr(i+1,2)+psr(i-1,2)
     1                 -2.D0*psr(i,2))/(dx*dx)-dt*v(i)*psr(i,2)
 60       Continue
c         every 2000 time steps we look at probability density
          If((n.eq.1).or.(MOD(n,2000).eq.0)) Then
            Do 80 i=1,max+1,10
              Write(9,11)p2(i) + 0.0015*v(i)
 80         Continue
            Write(9,11)
          EndIf
c         new iterations are now the old ones
          Do 70 i=1,max+1
            psi(i,1) = psi(i,2)
            psr(i,1) = psr(i,2)
 70       Continue
 40     Continue
 11     Format(E12.6)
        Close(9)
        Stop 'data saved in harmos.dat'
        End
```

Fig. D.13 harmos.f: Solution of time-dependent Schrödinger equation.

```
cccccccccccccccccccccccccccccccccccccccccccccccccccccccccccccccc
c       int_10d.f: Ten dimensional integration using Monte Carlo        c
c                                                                       c
c       UNIX (DEC OSF, IBM AIX): f77 int_10d.f                          c
c                                                                       c
c       comment: If your compiler complains about drand48, seed48       c
c                replace drand48 with rand(seed)  and remove the        c
c                call to seed48                                         c
cccccccccccccccccccccccccccccccccccccccccccccccccccccccccccccccc
        Program int10d
        Implicit none
c
c       Declarations
        Real*8 drand48, x, y
        Integer i, j, n
        Open(6, File='int_10d.dat', Status='Unknown')
        Call seed48(68111)
c       Outer loops determines the number of trials = accuracy
        Do 10 i=1, 65536
          x=0
c
c         Add up ten random numbers
          Do 20 j=1,10
            x=x+drand48()
 20       Continue
c         square and add up
          y=y+x*x
c         save result for 2,4,8,16 .....
          if (mod(i,2**n) .eq. 0) then
            n=n+1
            Write (6,*) i, y/i
          endif
 10     Continue
        Close(6)
        Stop 'data saved in int_10d.dat'
        End
```

```
cccccccccccccccccccccccccccccccccccccccccccccccccccccccccccccccc
c  integ.f: Integrate exp(-x) using trapezoid, Simpson and Gauss rules c
c                                                                      c
c        UNIX (DEC OSF, IBM AIX): f77 integ.f gauss.f                  c
c                                                                      c
c  comment: gauss.f contains routine to calculate Legendre points and  c
c           weights and has to be in the same directory.               c
c           The derivation from the theoretical result for each method c
c           is saved in x y1 y2 format.                                c
cccccccccccccccccccccccccccccccccccccccccccccccccccccccccccccccc
        Program integrate
        Implicit none
c       declarations
        Real*8 trapez, simpson, quad, r1, r2, r3
        Real*8 theo, vmin, vmax
        Integer i
c
c       theoretical result, integration range
        theo = 0.632120558829
        vmin=0.0
        vmax=1.0
        Open(6, File='integ.dat', Status='Unknown')
c       calculate integral using both methods for steps = 3..501
        Do 50 i=3, 501 , 2
          r1=trapez(i, vmin, vmax)
          r1=abs(r1-theo)
          r2=simpson(i,vmin, vmax)
          r2=abs(r2-theo)
          r3=quad(i,vmin, vmax)
          r3=abs(r3-theo)
          write(6,*) i, r1, r2, r3
 50     Continue
        Close(6)
        Stop 'data saved in integ.dat'
        End
```

Fig. D.14 int_10d.f: Ten-dimensional Monte Carlo integration; integ.f: integration rules.

```
c
c       the function we want to integrate
        Function f(x)
        Implicit none
        Real*8 f, x
          f=exp(-x)
        Return
        End
c
c       trapezoid rule
        Function trapez(i, min, max)
        Implicit none
        Integer i, n
        Real*8 f, interval, min, max, trapez, x
        trapez=0
        interval = ((max-min) / (i-1))
c       sum the midpoints
        Do 21 n=2, (i-1)
          x = interval * (n-1)
          trapez = trapez + f(x)*interval
 21     Continue
c       add the endpoints
        trapez = trapez+0.5*(f(min)+f(max))*interval
        Return
        End
c
c       Simpson's rule
        Function simpson(i, min, max)
        Implicit none
        Integer i, n
        Real*8 f, interval, min, max, simpson, x
        simpson=0
        interval = ((max-min) / (i-1))
c       loop for odd points
        Do 31 n=2, (i-1), 2
          x = interval * (n-1)
          simpson = simpson + 4*f(x)
 31     Continue
c       loop for even points
        Do 32 n=3, (i-1), 2
          x = interval * (n-1)
          simpson = simpson + 2*f(x)
 32     Continue
c       add the endpoints
        simpson = simpson+f(min)+f(max)
        simpson=simpson*interval/3
        Return
        End
c
c       Gauss' rule
        Function quad(i, min, max)
        Implicit none
        Real*8 w(1000), x(1000)
        Real*8 f, min, max, quad
        Integer i, job, n
        quad=0
        job=0
        call gauss(i, job, min, max, x, w)
        Do 41 n=1, i
          quad=quad+f(x(n))*w(n)
 41     Continue
        Return
        End
```

Fig. D.15 integ.f: Integration rules (*continued*).

```
cccccccccccccccccccccccccccccccccccccccccccccccccccccccccccccccc
c       ising.f: Ising model of magnetic dipole strin           c
c                We start with uniform distribution and then heat up  c
c                the system.                                     c
c       UNIX (DEC OSF, IBM AIX): f77 ising.c                     c
c                                                                c
c       comment: If your compiler complains about drand48, seed48  c
c                replace drand48 with rand(seed)  and remove the  c
c                call to seed48                                  c
c                Plot without conneting datapoints with lines    c
cccccccccccccccccccccccccccccccccccccccccccccccccccccccccccccccc
        Program Ising
        Implicit none
        Integer max
        Parameter(max=100)
        Integer element, i, spins(max), seed, t
        Real*8 drand48, energy, kt, new, j, old
c       define number temperature, exchange energy, random seed
        Parameter(kt=100, j=-1, seed=68111)
c       open files, seed generator
        Open(8, FILE='spin-up.dat', Status='Unknown')
        Open(9, FILE='spin-do.dat', Status='Unknown')
        Call seed48(seed)
c
c       First generate a uniform configuration of spins
        Do 10 i=1,max
           spins(i) = 1
 10     Continue
c
c       step through time
        Do 20 t=1, 500
c         energy of the system
          old=energy(spins, j, max)
c         pick one element
          element=drand48()*max+1
c         change spin
          spins(element)=spins(element)*(-1)
c         calculate new energy
          new=energy(spins, j, max)
c         reject change if new energy is greater and the Boltzmann factor
c         is less than another random number
          If ((new.GT.old) .AND. (exp((-new+old)/kt).LT.drand48())) Then
             spins(element)=spins(element)*(-1)
          Endif
c         save a map of spins
          Do 30 i=1,max
             If (spins(i).EQ.1) Then
                Write(8,*) t, i
             Endif
             If (spins(i).EQ.(-1)) Then
                Write(9,*) t, i
             Endif
 30       Continue
 20     Continue
        Close(8)
        Close(9)
        Stop 'data saved in spin-up.dat, spin-do.dat'
        End
c
c       function calculates energy of the system
        Function energy(array, j, max)
        Implicit none
        Integer array(max), i, max
        Real*8 energy, j
        energy=0
        Do 22 i=1,(max-1)
           energy=energy+array(i)*array(i+1)
 22     Continue
        Return
        End
```

Fig. D.16 ising.f: Ising model simulation.

```
cccccccccccccccccccccccccccccccccccccccccccccccccccccccccccccccccccccccc
c       lagrange.f: Langrange interpolation of cross table             c
c                                                                      c
c       UNIX (DEC OSF, IBM AIX): f77 lagrange.f                        c
cccccccccccccccccccccccccccccccccccccccccccccccccccccccccccccccccccccccc
        Program lagrange
        Implicit none
c
c       Declarations
        Real*8 inter, x, xin(9), yin(9)
        Integer i, end
        end=9
        Open(6, File='lagrange.dat', Status='Unknown')
c
c       Input data
        Data xin /0, 25, 50, 75, 100, 125, 150, 175, 200/
        Data yin /10.6, 16, 45, 83.5, 52.8, 19.9, 10.8, 8.25, 4.7/
c
c       Calculate f(x)
        Do 20 i=0, 1000
          x=i*0.2
          Write (6,*) x, inter(xin, yin, end, x)
 20     Continue
        Close(6)
        Stop 'data saved in lagrange.dat'
        End
```

```
c
c       Function inter
c       Evaluates the interpolation function at x
c
        Function inter(xin, yin, end, x)
        Implicit none
c       declarations
        Integer i, j, end
        Real*8 inter, lambda(10), xin(10), yin(10), x
        inter = 0
        Do 200 i=1, end
          lambda(i) = 1
          Do 300 j=1, end
            If (i .NE. j) THEN
              lambda(i) = lambda(i) * ((x - xin(j))/(xin(i) - xin(j)))
            EndIf
 300      Continue
          inter = inter + (yin(i) * lambda(i))
 200    Continue
        Return
        End
```

Fig. D.17 lagrange.f: Langrange interpolation.

```
cccccccccccccccccccccccccccccccccccccccccccccccccccccccccccccccccccccc
c       laplace.f: Solution of Laplace's equation, finite differences  c
c                                                                      c
c       UNIX (DEC OSF, IBM AIX): f77 laplace.f                         c
c                                                                      c
c       comment: Output data saved in 3D grid format used by gnuplot   c
cccccccccccccccccccccccccccccccccccccccccccccccccccccccccccccccccccccc
        Program laplace
        Implicit none
        Integer max
        Parameter(max=40)
        Real*8 x, p(max,max)
        Integer i, j, iter, y
c       open output file
        Open(8, File='laplace.dat', Status='Unknown')
c       the side with constant potential
        Do 10 i=1, max
           p(i,1)=100.0
 10     Continue
c
c       iteration algorithm
        Do 20 iter=1, 1000
          Do 30 i=2,(max-1)
            Do 40 j=2,(max-1)
              p(i,j)=0.25*(p(i+1,j)+p(i-1,j)+p(i,j+1)+p(i,j-1))
 40         Continue
 30       Continue
 20     Continue
c
c       write data gnuplot 3D format
        Do 50 i=1, max
          Do 60 j=1, max
            Write (8,22) p(i,j)
 60       Continue
        Write (8,22)
 50     Continue
 22     Format(f10.6)
        Close(8)
        Stop 'data saved in laplace.dat'
        End
```

```
cccccccccccccccccccccccccccccccccccccccccccccccccccccccccccccccccccccc
c       limit.f: determines the machine precision                      c
c                                                                      c
c       UNIX (DEC OSF, IBM AIX): f77 limit.f                           c
c                                                                      c
c       comment: very crude program, produces lots of screen output    c
cccccccccccccccccccccccccccccccccccccccccccccccccccccccccccccccccccccc
        Program limit
        Implicit none
c
c       determine the machine precision
c
        Integer I, N
        Real*8 eps, one
c       number of iterations N
        N=60
c       set initial values
        eps = 1
        one = 1
c       add eps to one and print result
        Do 15, I = 1, N
           eps = eps / 2
           one = 1 + eps
           Write (*,*) I, one, eps
 15     Continue
        Stop 'limit'
        End
```

Fig. D.18 laplace.f: Solution of Laplace's equation; limit.f: determination of precision.

```
ccccccccccccccccccccccccccccccccccccccccccccccccccccccccccccccccccccccc
c      over.f: determine overflow and underflow limits                c
c                                                                     c
c     UNIX (DEC OSF, IBM AIX): f77 over.f                             c
c                                                                     c
c     comment: very crude program, produces lots of screen output     c
ccccccccccccccccccccccccccccccccccccccccccccccccccccccccccccccccccccccc
      Program overflow
      Implicit none
c
c     determine where overflow and underflow occur
c
      Integer I, N
      Real*8 under, over
c     number of iterations N, might not be big enough
      N=1024
c     set initial values
      under = 1
      over  = 1
c     calculate underflow and overflow, print output to screen
      Do 15, I = 1, N
         under = under / 2
         over = over * 2
         Write (*,*) I, over, under
 15   Continue
      Stop 'over'
      End
```

```
ccccccccccccccccccccccccccccccccccccccccccccccccccccccccccccccccccccccc
c       pond.f: Calculate pi vi stone throwing                        c
c                                                                     c
c       UNIX (DEC OSF, IBM AIX): f77 pond.f                           c
c                                                                     c
c       comment: If your compiler complains about drand48, seed48     c
c                replace drand48 with rand(seed)  and remove          c
c                call to seed48                                       c
ccccccccccccccccccccccccccccccccccccccccccccccccccccccccccccccccccccccc
        Program pond
        Implicit none
c       declarations
c       drand48 number generator, max number of stones, needs seed
        Real*8 area, x, y, drand48
        Integer i, max, pi, seed
        max=2000
        seed=68111
c       open file, set initial value, seed generator
        Open(6, File='pond.dat', Status='Unknown')
        pi=0
        Call seed48(seed)
c       execute
        Do 10 i=1, max
           x = drand48()*2-1
           y = drand48()*2-1
           If ((x*x + y*y) .LE. 1) Then
              pi = pi+1
           Endif
           area = 4.0 * pi/Real(i)
           Write(6,*) i, area
 10     Continue
        Close(6)
        Stop 'data saved in pond.dat'
        End
```

Fig. D.19 over.f: Determine overflow and underflow limits; pond.f: calculate π by throwing stones into a pond.

```
cccccccccccccccccccccccccccccccccccccccccccccccccccccccccccccccccccccc
c       qmc.f:    Feynman path integral (quantum Monte Carlo) for      c
c            ground-state wave packet                                 c
c                                                                      c
c       UNIX (DEC OSF, IBM AIX): f77 qmc.f                             c
c                                                                      c
c       comment: If your compiler complains about drand48, srand48     c
c                 uncomment the define statements further down.        c
c                 This might take a couple of minutes.                 c
cccccccccccccccccccccccccccccccccccccccccccccccccccccccccccccccccccccc
        Program qmc
        Implicit none
        Integer i, j, max, element, prop(100)
        Real*8 change, drand48, energy, newE, oldE, out, path(100)
        max  = 250000
        Open(9, FILE='qmc.dat', Status='Unknown')
        call seed48(68111)
c       initial path and initial probability
        Do 10 j=1,100
          path(j)=0.0
          prop(j)=0
 10     Continue
c       find energy of initial path
        oldE = energy(path, 100)
        Do 20 i=1,max
c         pick one random element
          element = drand48()*100+1
c         change it by an random value -0.9..0.9
          change  = ((drand48()-0.5)*2)
          path(element)=path(element)+change
c         find the new energy
          newE=energy(path, 100)
c         reject change if new energy is greater and the Boltzmann factor
c         is less than another random number
          If ((newE.GT.oldE) .AND. (exp(-newE+oldE).LT.drand48())) Then
             path(element)=path(element)-change
          Endif
c         add up probabilities
          Do 30 j=1,100
            element=path(j)*10+50
            prop(element)=prop(element)+1
 30       Continue
          oldE = newE
 20     Continue
c       write output data to file
        Do 40 j=1,100
          out=prop(j)
          Write(9,*) j-50, out/max
 40     Continue
        Close(9)
        Stop 'data saved in qmc.dat'
        End
c
c       function calculates energy of the system
        Function energy(array, max)
        Implicit none
        Integer i, max
        Real*8 energy, array(max)
        energy=0
        Do 50 i=1,(max-1)
           energy=energy + (array(i+1)-array(i))**2 + array(i)**2
 50     Continue
        Return
        End
```

Fig. D.20 qmc.f: Feynman path integration with Metropolis algorithm.

```
cccccccccccccccccccccccccccccccccccccccccccccccccccccccccccccccccccccccc
c       random.f: A very simple random-number generator                 c
c                  - not suitable for serious work                      c
c                                                                       c
c       UNIX (DEC OSF, IBM AIX): f77 random.f                           c
c                                                                       c
c       comment: plot without connecting datapoints with lines          c
cccccccccccccccccccccccccccccccccccccccccccccccccccccccccccccccccccccccc
        Program random
        Implicit none
c       declarations
        Integer i, number, old, seed, x, y
c       set parameters (seed for generator, number of generated numbers)
        seed = 11
        number = 1000
c       open output file, seed number generator
        Open(6, FILE='random.dat', Status='Unknown')
        old = seed
c       execution
        Do 10 i = 1, number
           x = Mod((57*old+1), 256)
           y = Mod((57*x+1), 256)
           Write (6,*) x, y
           old=y
 10   Continue
c
      Close(6)
      Stop 'data saved in random.dat'
      End
```

Fig. D.21 random.f: A random-number generator not suitable for serious work.

```
cccccccccccccccccccccccccccccccccccccccccccccccccccccccccccccccccccccccc
c       rk4.f:  4th order Runge-Kutta solution for harmonic oscillator c
c                                                                      c
c       UNIX (DEC OSF, IBM AIX): f77 rk4.f                             c
cccccccccccccccccccccccccccccccccccccccccccccccccccccccccccccccccccccccc
        Program oscillator
        Implicit none
c       declarations
c       n: number of equations, min/max in x, dist:length of x-steps
c       y(1): initial position, y(2):initial velocity
        Real*8 dist, min, max, x, y(2)
        Integer n
        n=2
        min=0.0
        max=10.0
        dist=0.1
        y(1)=1.0
        y(2)=0.0
c       open file
        Open(6, File='rk4.dat', Status='Unknown')
c       do n steps of Runga-Kutta algorithm
        Do 60 x=min, max, dist
           Call rk4(x, dist, y, n)
           Write (6,*) x, y(1)
 60     Continue
c
        Close(6)
        Stop 'data saved in rk4.dat'
        End
c------------------------end of main program------------------------
c
c       fourth-order Runge-Kutta subroutine
        Subroutine rk4(x, xstep, y, n)
        Implicit none
c       declarations
        Real*8 deriv, h, x, xstep, y(5)
        Real*8 k1(5), k2(5),k3(5), k4(5), t1(5), t2(5), t3(5)
        Integer i, n
        h=xstep/2.0
        Do 10 i = 1,n
           k1(i) = xstep * deriv(x, y, i)
           t1(i) = y(i) + 0.5*k1(i)
 10     Continue
        Do 20 i = 1,n
           k2(i) = xstep * deriv(x+h, t1, i)
           t2(i) = y(i) + 0.5*k2(i)
 20     Continue
        Do 30 i = 1,n
           k3(i) = xstep * deriv(x+h, t2, i)
           t3(i) = y(i) + k3(i)
 30     Continue
        Do 40 i = 1,n
           k4(i) = xstep * deriv(x+xstep, t3, i)
           y(i) = y(i) + (k1(i) + (2.*(k2(i) + k3(i))) + k4(i))/6.0
 40     Continue
c
        Return
        End
c       function which returns the derivatives
        Function deriv(x, temp, i)
        Implicit none
c       declarations
        Real*8 deriv, x, temp(2)
        Integer i
c
        If (i .EQ. 1) deriv=temp(2)
        If (i .EQ. 2) deriv=-temp(1)
        Return
        End
```

Fig. D.22 rk4.f: Fourth-order Runge–Kutta solution for harmonic oscillator.

```
cccccccccccccccccccccccccccccccccccccccccccccccccccccccccccccccccccccccc
c                                                                      c
c       scatt.f:  scattering in momentum space from delta shell       c
c                 potential, LU decomposition with partial pivoting.   c
c                                                                      c
c       comment:  uses gauss.f, LUfactor, LUSolve (included)          c
cccccccccccccccccccccccccccccccccccccccccccccccccccccccccccccccccccccccc
c
        Program scatt.f
        Integer n,Size,i,j,Row,Column
        Double Precision b,Pot
        Parameter (Size=300,pi=3.14159265358979323846 26,b=10.0)
        Double Precision lambda,scale,ko,Temp
        Double Precision F(Size,Size),k(Size),w(Size),D(Size),r(Size)
        Double Precision V(Size),L(Size,Size),U(Size,Size),P(Size,Size)
        Integer PivotInfo(Size)
c
c       get potential strength lambda
c
        Write(*,*) 'enter lambda'
        Read(*,*) lambda
        Write(*,*) 'enter scaling factor'
        Read(*,*) scale
        Write(*,*) 'enter ko'
        Read(*,*) ko
        Write(*,*) 'enter grid size'
        Read(*,*) n
c
c       Set up Gaussian integration points and weights
c       interval [0,inf], mid-point= scale, k(N+1)=ko
c
        Call gauss(n,2,0d0,scale,k,w)
c
c       Set up D matrix
c
        Do i=1,n
                D(i)=2.0d0/pi*w(i)*k(i)*k(i)/(k(i)*k(i)-ko*ko)
        End Do
        D(n+1)=0.0
        Do j=1,n
                D(n+1)=D(n+1)+w(j)*ko*ko/(k(j)*k(j)-ko*ko)
        End Do
        D(n+1)=D(n+1)*(-2.0d0/pi)
c       Set up F matrix and V vector
        Do i=1,n+1
                Do j=1,n+1
                        Pot=-b*b*lambda*SIN(b*k(i))*SIN(b*k(j))
                        Pot=Pot/(k(i)*b*k(j)*b)
                        F(i,j)=Pot*D(j)
                        IF (i .EQ. j) Then
                                F(i,j)=F(i,j)+1.0d0
                        Endif
                End Do
                V(i)=Pot
        End Do
c
c       LU factorization.  Put LU factors of F in corresponding matrix
c       (not efficient but easy to follow).
c       Store partial pivoting info
c
        Call LUfactor(F,n+1,Size,L,U,PivotInfo)
c
c       Pivot and solve
c       Set P to identity matrix
c
```

Fig. D.23 scatt.f: Momentum-space quantum scattering.

```
        Do Row = 1,n+1
                Do Column = 1,n+1
                P(Row,Column)=0
                If (Row .EQ. Column) P(Row,Column)=1
                End Do
        End Do
C
C       Interchange rows to get true P matrix
C
        Do Row = 1,  n+1
                Do Column = 1,  n+1
                Temp=P(Row,Column)
                P(Row,Column)=P(PivotInfo(Row),Column)
                P(PivotInfo(Row),Column)=Temp
                End Do
        End Do
        Call LUSolve(V,L,U,n+1,Size,PivotInfo,r)
C
C       Output results
C
        Write(*,*) ko*ko, DATAN(-r(n+1)*ko)
        End
```

```
        Subroutine LUfactor(A,n,Size,L,U,PivotInfo)
C
C       LU factorization and partial pivoting of matrix
C       A in preparation for solving Ax=b
C
        Integer n,Column,CurrentPivotRow, CurrentRow, SwapCol, Row
        Integer ElimCol,Size
        Double Precision A(Size,Size), L(Size,Size), U(Size,Size)
        Integer PivotInfo(Size)
        Double Precision CurrentPivotValue, Swap
        Do Column = 1, n-1
           CurrentPivotRow=Column
           CurrentPivotValue=A(CurrentPivotRow,Column)
C          Determine row which provides the largest pivot
           Do CurrentRow = Column+1, n
             If( DABS(A(CurrentRow,Column)) .gt. CurrentPivotValue ) Then
                 CurrentPivotValue=DABS(A(CurrentRow,Column))
                 CurrentPivotRow=CurrentRow
             Endif
           End Do
           PivotInfo(Column)=CurrentPivotRow
C          Swap rows to get largest value at pivot position
           Do SwapCol = Column, n
                  Swap = A(Column,SwapCol)
                  A(Column,SwapCol) = A(PivotInfo(Column),SwapCol)
                  A(PivotInfo(Column),Swapcol) = Swap
                End Do
C
C       Do  Gaussian Elimination
C       Get upper triangular A and un-pivoted lower triangular L
C
        Do Row = Column+1, n
           L(Row,Column)= A(Row,Column)/A(Column,Column)
           Do ElimCol = Column+1, n
           A(Row,ElimCol)=A(Row,ElimCol)-L(Row,Column)*A(Column,ElimCol)
                  End Do
        End Do
        End Do
```

Fig. D.24 scatt.f: Momentum-space scattering, (*continued*); LUfactor.f: LU decomposition.

```
C
C       Make sure bottom right value doesn't get pivoted to zero
C
        PivotInfo(n)=n
C
C       Now pivot the L
C
        Do Row = 2, n-1
           DO Column = 1,Row-1
                   Swap = L(Row,Column)
                   L(Row,Column) = L(PivotInfo(Row),Column)
                   L(PivotInfo(Row),Column) = Swap
           End Do
        End Do
C
C       Now clean up L and U
C
        Do Column = 1, n
           Do Row =  1,Column
                   U(Row,Column) = A(Row,Column)
                   L(Row,Column) = 0
                   If (Row .EQ. Column) L(Row,Column)=1
           End Do
           Do Row = Column+1,n
              U(Row,Column)=0
           End Do
        End Do
        Return
        End
```

```
        Subroutine LUSolve(b,L,U,n,Size,PivotInfo,x)
C
C       Part of an LU decomposition, with partial pivoting
C       to solve Ax=b matrix problems
C
        Integer n,Size, Row, Column
        Double Precision b(Size), x(Size)
        Integer PivotInfo(Size)
        Double Precision L(Size,Size), U(Size,Size)
        Double Precision Temp
C              Interchange rows of b to take care of pivoting
        Do Row = 1, n
                Temp=b(Row)
                b(Row)=b(PivotInfo(Row))
                b(PivotInfo(Row))=Temp
        End Do
C
C             Solve Ly=b, where y=Ux, by forward elimination
C             Since L has ones along the diagonal y(1)=b(1)
C             Store y in b
        Do Row = 2,n
                DO Column = 1,Row-1
                        b(Row)=b(Row)-L(Row,Column)*b(Column)
                END DO
                b(Row)=b(Row)/L(Row,Row)
        End Do
C
C       Now solve Ux=y by back substitution
C
        x(n)=b(n)/U(n,n)
        Do Row = n-1,1,-1
                x(Row)=b(Row)
                Do Column = Row+1,n
                        x(Row)=x(Row)-U(Row,Column)*x(Column)
                End Do
                x(Row)=x(Row)/U(Row,Row)
        End Do
        Return
        End
```

Fig. D.25 LUfactor.f: (*continued*); LUSolve.f.

```
cccccccccccccccccccccccccccccccccccccccccccccccccccccccccccccccccccccc
c       slit.f: Solves the time-dependent Schroedinger equation for    c
c               two-dimensional Gaussian wave packet entering a slit    c
c                                                                      c
c       UNIX (DEC OSF, IBM AIX): f77 slit.f                            c
c                                                                      c
c       comment: Output data saved in 3D grid format used by gnuplot   c
c                This might take some minutes.                         c
cccccccccccccccccccccccccccccccccccccccccccccccccccccccccccccccccccccc
        Program slit
        Implicit None
        Real*8 psr(91,91,2),psi(91,91,2),v(91,91),p2(91,91)
        Real*8 a1,a2,dt,dx,k0x,k0y,x0,y0,x,y
        Integer i,j,max,n,time
        complex exc,zi
c       input a positive integer which is proportional to the time
c       you want to see the position of the wave packet.
        Write(*,*)'Enter a positive integer from 1(initial time)'
        Write(*,*)'to 800 to get wave packet position at that time'
        Read(*,*)time
        Write(*,*)'processing data for time',time
        Open(9,FILE='slit.dat',STATUS='UNKNOWN')
c       initializes the constant values and the wave packet
        zi   = cmplx(0.0D0,1.D0)
        dx   = 0.2D0
        dt   = 0.0025/(dx*dx)
c       initial momentum, position
        k0x  = 0.0D0
        k0y  = 2.5D0
        x0   = 0.0D0
        y0   = -7.D0
        max  = 90
c       initial wave packet
        y    = -9.0D0
        Do 90 j=1,max+1
          x=-9.0D0
          Do 10 i=1,max+1
            exc        = exp(zi*(k0x*x+k0y*y))
            a1 = exp(-0.5*(((x-x0))**2.0+((y-y0))**2.0))
c           real part of the initial wave packet
            psr(i,j,1) = real(a1*exc)
c           imaginay part of the initial wave packet
            psi(i,j,1) = imag(a1*exc)
            x          = x + dx
 10       Continue
          y = y + dx
 90     Continue
c
c       set up the potential slit width: 50-40=10 units
c
        Do 220 j=1,max+1
          Do 190 i=1,max+1
            If((j.eq.35).and.((i.le.40).or.(i.ge.51)))Then
              v(i,j) = 0.5
            Else
              v(i,j) = 0.0
            EndIf
 190      Continue
 220    Continue
c
c       propagate solution through time
c
        Do 40 n=1,time
c         compute real part of wave packet and probability
          Do 150 j=2,max
            Do 50 i=2,max
              a2 = v(i,j)*psi(i,j,1)+2.0D0*dt*psi(i,j,1)
              a1 = psi(i+1,j,1)+psi(i-1,j,1)+psi(i,j+1,1)+psi(i,j-1,1)
                psr(i,j,2) = psr(i,j,1)-dt*a1+2.0*a2
                If(n.eq.time) Then
                  p2(i,j) = psr(i,j,1)*psr(i,j,1)+psi(i,j,1)*psi(i,j,1)
                EndIf
 50         Continue
```

Fig. D.26 slit.f: Time-dependent Schrödinger equation for 2-D wave packet passing through a slit.

```
c           at x edges derivative is zero
            psr(1,j,2)     = psr(2,j,2)
            psr(max+1,j,2) = psr(max,j,2)
 150      Continue
c         imaginary part of wave packet is next
          Do 160 j=2,max
            Do 60 i=2,max
              a2 = v(i,j)*psr(i,j,2)+2.0*dt*psr(i,j,2)
              a1 = psr(i+1,j,2)+psr(i-1,j,2)+psr(i,j-1,2)+psr(i,j+1,2)
              psi(i,j,2) = psi(i,j,1)+dt*a1-2.0*a2
 60         Continue
c           at x edges derivative is zero
            psi(1,j,2)    = psi(2,j,2)
            psi(max+1,j,2) = psi(max,j,2)
 160      Continue
c         new iterations are now the old ones, recycle
          Do 180 j=1,max+1
            Do 70 i=1,max+1
              psi(i,j,1) = psi(i,j,2)
              psr(i,j,1) = psr(i,j,2)
 70         Continue
 180      Continue
 40     Continue
c
c       write probabilities plus potential multiplied by 0.025
c       in disk file
c
        Do 200 j=2,max,3
          Do 210 i=2,max,2
            Write(9,11)p2(i,j)+v(i,j)
 210      Continue
          Write(9,11)
 200    Continue
 11     Format (E12.6)
        Close(9)
        Stop 'data saved in slit.dat'
        End
```

Fig. D.27 slit.f (*continued*).

```
cccccccccccccccccccccccccccccccccccccccccccccccccccccccccccccccccccccc
c       soliton.f: Solves the Kortewg-deVries equation using a finite  c
c                  difference method                                   c
c                                                                      c
c       UNIX (DEC OSF, IBM AIX): f77 soliton.f                         c
c                                                                      c
c       comment: Output data saved in 3D grid format used by gnuplot   c
cccccccccccccccccccccccccccccccccccccccccccccccccccccccccccccccccccccc
        Program soliton
        Implicit None
        Real*8 ds, dt, max, mu, eps, u(131,3)
        Parameter(ds=0.4, dt=0.1, max=2000, mu=0.1, eps=0.2)
c       delta t, delta x, time steps, mu and eps from KdeV equation)
        Real*8 a1, a2, a3, fac, time
        Integer i, j, k
        Open (9,FILE='soliton.dat',STATUS='UNKNOWN')
c       Initial condition
        Do 10 i=1,131
          u(i,1) = 0.5*(1.0-tanh(0.2*ds*(i-1)-5.0))
 10     Continue
c       the endpoints
        u(1,2)   = 1.0
        u(1,3)   = 1.0
        u(131,2) = 0.0
        u(131,3) = 0.0
        fac  = mu*dt/(ds**3.0)
        time = dt
c       the first step
        Do 20 i=2,130
          a1 = eps*dt*(u(i+1,1)+u(i,1)+u(i-1,1))/(ds*6.0D0)
          If((i.gt.2).and.(i.le.129)) Then
             a2 = u(i+2,1)+2.0*u(i-1,1)-2.0*u(i+1,1)-u(i-2,1)
          Endif
          If((i.eq.2).or.(i.eq.130)) Then
            a2 = u(i-1,1)-u(i+1,1)
          Endif
          a3 = u(i+1,1)-u(i-1,1)
          u(i,2)   = u(i,1)- a1*a3-fac*a2/3.D0
 20     Continue
c       all other time steps
        Do 30 j=1,max
          Do 40 i=2,130
            a1 = eps*dt*(u(i+1,2)+u(i,2)+u(i-1,2))/(3.0D0*ds)
            If((i.gt.2).and.(i.le.129)) Then
              a2 = u(i+2,2)+2.0D0*u(i-1,2)-2.0D0*u(i+1,2)-u(i-2,2)
            Endif
            If((i.eq.2).or.(i.eq.130)) Then
              a2 = u(i-1,2)-u(i+1,2)
            Endif
            a3 = u(i+1,2)-u(i-1,2)
            u(i,3) = u(i,1)- a1*a3-2.D0*fac*a2/3.D0
            u(1,3) = 1.0D0
 40       Continue
c         new iterations are now  old, recycle, reuse
          Do 50 k=1,131
             u(k,1) = u(k,2)
             u(k,2) = u(k,3)
 50       Continue
c         every 200 time steps we want to write result to file
          If(MOD(j,200).eq.0) Then
             Do 60 k=1,131
               Write(9,22)u(k,3)
 60          Continue
            Write(9,22)
          EndIf
          time = time + dt
 30     Continue
 22     Format(f10.6)
        Close(9)
        Stop 'data saved in soliton.dat'
        End
```

Fig. D.28 soliton.f: Solution of Kortewg–deVries equation using finite-difference method.

```
cccccccccccccccccccccccccccccccccccccccccccccccccccccccccccccccccccccccc
c       spline.f: uses SLATEC's DBINT4 and DBVALU to interpolate       c
c                 a set of x-y values using cubic splines              c
c                                                                      c
c       UNIX (DEC OSF, IBM AIX): f77 spline.f -lslatec                 c
c                                                                      c
c       comment: you need the SLATEC library compiled as libslatec.a   c
cccccccccccccccccccccccccccccccccccccccccccccccccccccccccccccccccccccccc
        Program spline
        Implicit none
        Integer NDATA
        Parameter(NDATA=5)
        Real*8 BCOEF(NDATA), X(NDATA), Y(NDATA), T(NDATA+4)
        Real*8 W(5*(NDATA+2)), W2(3*4)
        Real*8 DBVALU, FBCL, FBCR, IN, VAL
        Integer I, IBCL, IBCR, IDERIV, INBV, K, N, KNTOPT
c       open output file
        Open(6, File='spline.dat', Status='Unknown')
c
c       input values
        X(1)=1.0
        X(2)=2.0
        X(3)=3.0
        X(4)=4.0
        X(5)=5.0
        Y(1)=2.0
        Y(2)=4.1
        Y(3)=3.0
        Y(4)=5.0
        Y(5)=2.0
c
c       natural splines, set second derivatives at end points to zero
c
        IBCL = 2
        IBCR = 2
        FBCL = 0.0
        FBCR = 0.0
        KNTOPT = 1
c
c       find the spline coefficients
c
        Call DBINT4 (X, Y, NDATA, IBCL, IBCR, FBCL, FBCR, KNTOPT,
     +          T, BCOEF, N, K, W)
c
c       create plot data using DBVALU
c       initialization, starting x-value, we want the spline itself
        INBV=1
        IN=1.0
        IDERIV=0
        Do 10 I=1,101
           VAL = DBVALU (T, BCOEF, N, K, IDERIV, IN, INBV, W2)
           Write (6,*) IN, VAL
           IN = IN+0.04
 10     Continue
        Close(6)
        Stop 'data saved in spline.dat'
        End
```

Fig. D.29 spline.f: Cubic spline interpolation using SLATEC's DBINT4 and DBVALU.

```fortran
cccccccccccccccccccccccccccccccccccccccccccccccccccccccccccccccccccccc
c       sqwell.f: Solution of time-dependent Schroedinger equation for c
c                 wave packet in infinite square well.                 c
c                                                                      c
c       UNIX (DEC OSF, IBM AIX): f77 sqwell.f                          c
c                                                                      c
c       comment: Output data saved in 3D grid format used by gnuplot  c
c                This might take some minutes                          c
cccccccccccccccccccccccccccccccccccccccccccccccccccccccccccccccccccccc
        Program sqwell
        Implicit None
        Real*8 psr(751,2),psi(751,2),p2(751)
        Real*8 dx,k0,dt,x,pi
        Integer i,n,max
        Complex exc,zi
        Common /values/dx,dt
        Open(9,FILE='sqwell.dat',STATUS='UNKNOWN')
        max   = 750
        pi    = 3.14159265358979323846
        zi    = CMPLX(0.0D0,1.D0)
        dx    = 0.02D0
        k0    = 17.D0*pi
        dt    = dx*dx
c       initial conditions
        x   = 0.0
        Do 10 i=1,max+1
          exc       = exp(zi*k0*x)
c         real part of the initial wave packet at t=0
          psr(i,1) = real(exc*exp(-0.5*(2.0*(x-5.0))**2.0))
c         imaginary part of the initial wave packet at t=0
          psi(i,1) = imag(exc*exp(-0.5*(2.0*(x-5.0))**2.0))
          x         = x + dx
 10     Continue
c
c       now propagate solution through time
c
        Do 40 n=1,6000
c         the real part of the wave packet and the probability
          Do 50 i=2,max
            psr(i,2) = psr(i,1) - dt*(psi(i+1,1) + psi(i-1,1)
     1                 -2.0*psi(i,1))/(2.0*dx*dx)
            p2(i)    = psr(i,1)*psr(i,2)+psi(i,1)*psi(i,1)
 50       Continue
c         the imaginary part of the wave packet
          Do 60 i=2,max
            psi(i,2) = psi(i,1) + dt*(psr(i+1,2) + psr(i-1,2)
     1                -2.D0*psr(i,2))/(2.0D0*dx*dx)
 60       Continue
c         only at certain time instants we want probability
c
          If(MOD(n,300).eq.0) THEN
            Do 80 i=1,max+1,15
              write(9,11)p2(i)
 80         Continue
            write(9,11)
          EndIf
c         new iterations are now the old ones
          Do 70 i=1,max+1
            psi(i,1) = psi(i,2)
            psr(i,1) = psr(i,2)
 70       Continue
 40     Continue
 11     Format (E12.6)
        Close(9)
        Stop 'data saved in sqwell.dat'
        End
```

Fig. D.30 sqwell.f: Solution of time-dependent Schrödinger equation for wave packet in infinite square well.

```
cccccccccccccccccccccccccccccccccccccccccccccccccccccccccccccccccccccc
c      tune.f: a matrix algebra program to be tuned for performace     c
c                                                                      c
c      UNIX (DEC OSF, IBM AIX): f77 tune.f                             c
cccccccccccccccccccccccccccccccccccccccccccccccccccccccccccccccccccccc
c
       Program  tune
       Parameter (ldim = 2050)
       Implicit Double Precision (a-h,o-z)
       Dimension ham(ldim,ldim), coef(ldim), sigma(ldim)
c
c        set up Hamiltonian and starting vector
       Do 10 i = 1,ldim
          Do 11 j = 1,ldim
             If( Abs(j-i) .gt. 10) Then
                ham(j,i) = 0.0
             Else
                ham(j,i) = 0.3**Abs(j-i)
             EndIf
   11     Continue
          ham(i,i) = i
          coef(i) = 0.0
   10  Continue
       coef(1) = 1.0
c
c        start iterating towards the solution
       err = 1.0
         iter = 0
   20    If (iter .lt.15 .and. err. gt. 1.0e-6) Then
            iter = iter + 1
c
c          compute current energy \& norm, \& normalize
        ener = 0.0
        ovlp = 0.0
        Do 21   i = 1,ldim
           ovlp = ovlp+coef(i)*coef(i)
           sigma(i) = 0.0
           Do 30   j = 1,ldim
              sigma(i) = sigma(i) + coef(j)*ham(j,i)
   30      Continue
           ener = ener + coef(i)*sigma(i)
   21   Continue
        ener = ener/ovlp
        Do 22   I = 1,ldim
           coef(i) = coef(i)/Sqrt(ovlp)
           sigma(i) = sigma(i)/Sqrt(ovlp)
   22   Continue
c       compute update and error norm
c
        err = 0.0
        Do 23 i = 1,ldim
           If (i.eq.1) GoTo 23
           step = (sigma(i) - ener*coef(i))/(ener-ham(i,i))
           coef(i) = coef(i) + step
           err = err + step**2
   23   Continue
        err = Sqrt(err)
        Write(*,'(1x,i2,7f10.5)') iter, ener, err, coef(1)
        GoTo 20
        EndIf
        Stop
        End
```

Fig. D.31 tune.f: A matrix algebra program to be tuned for performace.

```
cccccccccccccccccccccccccccccccccccccccccccccccccccccccccccccccccccccc
c     tune1.f: a matrix algebra program with basic optimization      c
c                                                                    c
c     UNIX (DEC OSF, IBM AIX): f77 tune1.f                           c
cccccccccccccccccccccccccccccccccccccccccccccccccccccccccccccccccccccc
c
      Program  tune1
      PARAMETER (ldim = 2050)
      Implicit Double Precision (a-h,o-z)
      Dimension ham(ldim,ldim),coef(ldim),sigma(ldim)
c
c       set up Hamiltonian and starting vector
c
      Do 10 i = 1,ldim
         Do 11 j = 1,ldim
             If( Abs(j-i) .gt. 10) Then
                ham(j,i) = 0.0
             Else
                ham(j,i) = 0.3**Abs(j-i)
             EndIf
 11      Continue
         ham(i,i) = i
         coef(i) = 0.0
 10   Continue
      coef(1) = 1.0
c
c      start iterating towards the solution
      err = 1.0
      iter = 0
 20   if(iter.lt.15 .and. err.gt.1.0e-6) Then
         iter = iter+1
c     compute energy, norm of current approximation,\&  normalize
c
      ener = 0.0
      ovlp = 0.0
      Do 21   i = 1,ldim
         ovlp = ovlp+coef(i)*coef(i)
         sigma(i) = 0.0
         Do 30   j = 1,ldim
            sigma(i) = sigma(i)+coef(j)*ham(j,i)
 30      Continue
        ener = ener+coef(i)*sigma(i)
 21   Continue
      ener = ener/ovlp
      fact  = 1.0/Sqrt(ovlp)
      coef(1)  =  fact*coef(1)
      err  =  0.0
      Do 22   i = 2,ldim
         t          =   fact*coef(i)
         u          =   fact*sigma(i) - ener*t
         step       =   u/(ener - ham(i,i))
         coef(i)    =   t + step
         err        =   err + step*step
 22   Continue
      err = Sqrt(err)
      Write(*,'(1x,i2,7f10.5)') iter,ener,err,coef(1)
      GoTo 20
      EndIf
      Stop
      End
```

Fig. D.32 tune1.f: A matrix algebra program with basic optimization.

```
cccccccccccccccccccccccccccccccccccccccccccccccccccccccccccccccccccccc
c     tune2.f: a matrix algebra program with vector tuning           c
c                                                                    c
c     UNIX (DEC OSF, IBM AIX): f77 tune2.f                           c
cccccccccccccccccccccccccccccccccccccccccccccccccccccccccccccccccccccc
c
      Program  tune2
      PARAMETER (ldim = 2050)
      Implicit Double Precision (a-h,o-z)
      Dimension  ham(ldim,ldim),coef(ldim),sigma(ldim),diag(ldim)
c
c       set up Hamiltonian and starting vector
c
      Do 10 i = 1,ldim
         Do 11 j = 1,ldim
             If( Abs(j-i) .gt. 10) Then
                ham(j,i) = 0.0
             Else
                ham(j,i) = 0.3**Abs(j-i)
             EndIf
 11      Continue
      ham(i,i) = i
      coef(i) = 0.0
 10   Continue
      coef(1) = 1.0
c      start iterating towards the solution
      Do 15 i = 1,ldim
         diag(i) = ham(i,i)
 15   Continue
      err = 1.0
      iter = 0
 20   If (iter.lt.15 .and. err.gt.1.0e-6) Then
         iter = iter+1
c
c     compute energy, norm of current approximation,\&  normalize
c
      ener = 0.0
      ovlp = 0.0
      Do 21   i = 1,ldim
         ovlp = ovlp+coef(i)*coef(i)
         t = 0.0
         Do 30   j = 1,ldim
            t = t + coef(j)*ham(i,j)
 30      Continue
         sigma(i) = t
         ener = ener + coef(i)*t
 21   Continue
      ener = ener/ovlp
      fact  = 1.0/Sqrt(ovlp)
      coef(1)  =  fact*coef(1)
      err  =  0.0
      Do 22   i = 2,ldim
          t         =   fact*coef(i)
          u         =   fact*sigma(i) - ener*t
          step      =   u/(ener - diag(i))
          coef(i)   =   t + step
          err       =   err + step*step
 22   Continue
      err = Sqrt(err)
      Write(*,'(1x,i2,7f10.5)') iter,ener,err,coef(1)
      GoTo 20
      EndIf
      Stop
      End
```

Fig. D.33 tune2.f: A matrix algebra program with vector tuning.

```
cccccccccccccccccccccccccccccccccccccccccccccccccccccccccccccccccccccc
c     tune3.f: a matrix algebra program with modified vector tuning  c
c                                                                    c
c     UNIX (DEC OSF, IBM AIX): f77 tune3.f                           c
cccccccccccccccccccccccccccccccccccccccccccccccccccccccccccccccccccccc
c
      Program  tune3
      PARAMETER (ldim = 2050)
      Implicit Double Precision (a-h,o-z)
      Dimension ham(ldim,ldim),coef(ldim),sigma(ldim),diag(ldim)
c
c       set up Hamiltonian and starting vector
      Do 10 i = 1,ldim
         Do 10 j = 1,ldim
             If( Abs(j-i) .gt. 10) Then
                ham(j,i) = 0.0
             Else
                ham(j,i) = 0.3**Abs(j-i)
             EndIf
 10   Continue
c
c       start iterating towards the solution
      Do 15 i = 1,ldim
         ham(i,i) = i
         coef(i) = 0.0
         diag(i) = ham(i,i)
 15   Continue
      coef(1) = 1.0
      err = 1.0
      iter = 0
 20   If(iter.lt.15 .and. err.gt.1.0e-6) Then
          iter = iter+1
c
c     compute energy, norm of current approximation,\&  normalize
c
      ener = 0.0
      ovlp = 0.0
      Do 21   i = 1,ldim
         ovlp = ovlp+coef(i)*coef(i)
         t = 0.0
         Do 30   j = 1,ldim
            t = t + coef(j)*ham(j,i)
 30      Continue
         sigma(i) = t
         ener = ener + coef(i)*t
 21   Continue
      ener = ener/ovlp
      fact  = 1.0/Sqrt(ovlp)
      coef(1)  =  fact*coef(1)
      err  =  0.0
      Do 22   i = 2,ldim
          t          =   fact*coef(i)
          u          =   fact*sigma(i) - ener*t
          step       =   u/(ener - diag(i))
          coef(i)    =   t + step
          err        =   err + step*step
 22   Continue
      err = Sqrt(err)
      Write(*,'(1x,i2,7f10.5)') iter,ener,err,coef(1)
      GoTo 20
      EndIf
      Stop
      End
```

Fig. D.34 tune3.f: A matrix algebra program with modified vector tuning.

```
cccccccccccccccccccccccccccccccccccccccccccccccccccccccccccccccccccccc
c     tune4.f: a matrix algebra program with RISC tuning              c
c                                                                     c
c     UNIX (DEC OSF, IBM AIX): f77 tune4.f                            c
cccccccccccccccccccccccccccccccccccccccccccccccccccccccccccccccccccccc
c
      Program  tune4
      PARAMETER (ldim = 2050)
      Implicit Double Precision (a-h,o-z)
      Dimension ham(ldim,ldim),coef(ldim),sigma(ldim),diag(ldim)
c        set up Hamiltonian and starting vector
      Do 10 i = 1,ldim
         Do 10 j = 1,ldim
             If( Abs(j-i) .gt. 10) Then
                ham(j,i) = 0.0
             Else
                ham(j,i) = 0.3**Abs(j-i)
             EndIf
 10   Continue
c      start iterating towards the solution
      Do 15 i = 1,ldim
         ham(i,i) = i
         coef(i)  = 0.0
         diag(i)  = ham(i,i)
 15   Continue
      coef(1) = 1.0
      err = 1.0
      iter = 0
 20   If(iter.lt.15 .and. err.gt.1.0e-6) Then
        iter = iter+1
      ener = 0.0
      ovlp1 = 0.0
      ovlp2 = 0.0
      Do 21   i = 1,ldim-1,2
         ovlp1 = ovlp1+coef(i)*coef(i)
         ovlp2 = ovlp2+coef(i+1)*coef(i+1)
         t1    = 0.0
         t2    = 0.0
         Do 30   j = 1,ldim
            t1 = t1 + coef(j)*ham(j,i)
            t2 = t2 + coef(j)*ham(j,i+1)
 30      Continue
      sigma(i)    = t1
      sigma(i+1) = t2
      ener        = ener + coef(i)*t1 + coef(i)*t2
 21   Continue
      ovlp = ovlp1 + ovlp2
      ener = ener/ovlp
      fact  = 1.0/Sqrt(ovlp)
      coef(1)  =  fact*coef(1)
      err  =  0.0
      Do 22   i = 2,ldim
         t          =  fact*coef(i)
         u          =  fact*sigma(i) - ener*t
         step       =  u/(ener - diag(i))
         coef(i)    =  t + step
         err        =  err + step*step
 22   Continue
      err = Sqrt(err)
      Write(*,'(1x,i2,7f10.5)') iter,ener,err,coef(1)
      GoTo 20
      EndIf
      Stop
      End
```

Fig. D.35 tune4.f: A matrix algebra program with RISC tuning.

```
cccccccccccccccccccccccccccccccccccccccccccccccccccccccccccccccccccccc
c     twodsol.f: Solution of sine-Gordon equation for 2-D soliton     c
c                                                                      c
c     Christiansen & Lomdahl, Physics 2D (1981) 482-494                c
c     U_xx + U_yy-U_tt=j(x,y)sin (U)                                   c
c        -x_0 <x< x_0,    -y_0 <y< y_0, t >=0,                         c
c     i. c. U(x,y,0)=4 arctan (exp(3-sqrt(x^2+y^2)),                   c
c     i. c. d U(x,y,0)/dt =0,  j(x,y)=1,  x_0=y_0=7                    c
cccccccccccccccccccccccccccccccccccccccccccccccccccccccccccccccccccccc
      PROGRAM twodsol
      Implicit NONE
      Double Precision u(201,201,3)
      Integer nint
      Open(9,File='twodsol.dat',status='new')
      write(*,*)'   Enter an integer from 1 to 100'
      write(*,*)' this number is proportional to time'
      write(*,*)'   time=0 is for the integer =1'
      read(*,*)nint
      write(*,*)'working with input =',nint
c     initializes variables and functions
      call initial(u)
c     2D soliton found for time proportional to input: nint
      call solution(u,nint)
      Stop
      End
```

```
      Subroutine initial(u)
c     initializes the constants and 2D soliton at time=0
      Implicit none
      Double Precision u(201,201,3),dx,dy,dt,xx,yy,dts,time
      Integer i,j
      Common /values/dx,dy,dt,time,dts
c     initial condition for all grid points in xy plane
      dx=14.0D0/200.0D0
      dy=dx
c     reduce number of terms this dt
      dt=dx/dsqrt(2.0D0)
      dts=(dt/dx)**2
      yy=-7.0D0
      time=0.0D0
      Do 10 i=1,201
        xx=-7.0D0
        Do 20 j=1,201
          u(i,j,1)=4.0D0*Datan(3.0D0-sqrt(xx*xx+yy*yy))
          xx=xx+dx
 20     Continue
      yy=yy+dy
 10   Continue
      Return
      End
```

Fig. D.36 twodsol.f: Solution of sine–Gordon equation for 2-D soliton.

```
      Subroutine solution(u,nint)
c     solve sine-Gordon equation for the i.c. in initial
      Implicit None
      Double Precision u(201,201,3),dx,dy,dt,time,a2,zz,dts,a1
      Integer l,m,mm,k,j,i,nint
      Common /values/dx,dy,dt,time,dts
c     these values passed by routine initial
      time=time+dt
c     2nd iteration, use dphi/dt=0 at t=0 (G(x,y,0)=0)
c     and using d U/dx=0 at -x0, x0, -y0 and y0
      Do 80 l=2,200
        Do 90 m=2,200
          a2=u(m+1,l,1)+u(m-1,l,1)+u(m,l+1,1)+u(m,l-1,1)
          u(m,l,2)=0.5*(dts*a2-dt*dt*DSIN(0.25D0*a2))
 90     Continue
 80   Continue
c     the borders in the second iteration
      Do 130 mm=2,200
        u(mm,1,2)=u(mm,2,2)
        u(mm,201,2)=u(mm,200,2)
        u(1,mm,2)=u(2,mm,2)
        u(201,mm,2)=u(200,mm,2)
 130  Continue
c     the still undefined terms
      u(1,1,2)=u(2,1,2)
      u(201,1,2)=u(200,1,2)
      u(1,201,2)=u(2,201,2)
      u(201,201,2)=u(200,201,2)
c    3rd & following iteration use input, loop goes to nint
      Do 100 k=1,nint
        Do 60 l=2,200
          Do 70 m=2,200
            a1=u(m+1,l,2)+u(m-1,l,2)+u(m,l+1,2)+u(m,l-1,2)
            u(m,l,3)=-u(m,l,1)+dts*a1-dt*dt*DSIN(0.25D0*a1)
            u(m,1,3)=u(m,2,3)
            u(m,201,3)=u(m,200,3)
 70       Continue
 60     Continue
        Do 140 mm=2,200
          u(mm,1,3)=u(mm,2,3)
          u(mm,201,3)=u(mm,200,3)
          u(1,mm,3)=u(2,mm,3)
          u(201,mm,3)=u(200,mm,3)
 140    Continue
        u(1,1,3)=u(2,1,3)
        u(201,1,3)=u(200,1,3)
        u(1,201,3)=u(2,201,3)
        u(201,201,3)=u(200,201,3)
c       new iterations are now old, reuse, recycle, recover
        Do 110 l=1,201
          Do 120 m=1,201
            u(l,m,1)=u(l,m,2)
            u(l,m,2)=u(l,m,3)
 120      Continue
 110    Continue
c       output, xxx separate spatial rows for 3D plotting,
c           need be replaced by blank lines (not even carriage return)
        If (k.eq.nint) Then
          Do 30 i=1,201,5
            Do 40 j=1,201,5
              zz=Dsin(u(i,j,3)/2.0D0)
              Write(9,*)zz
 40         Continue
            Write(9,*)'xxx'
 30       Continue
        Endif
        time=time+dt
 100  Continue
      Return
      End
```

Fig. D.37 twodsol.f: Solution of sine–Gordon equation for 2-D soliton (*continued*).

```
cccccccccccccccccccccccccccccccccccccccccccccccccccccccccccccccccccccccc
c       walk.f:        Random walk simulation                          c
c                                                                      c
c       UNIX (DEC OSF, IBM AIX): f77 walk.f                            c
c                                                                      c
c       comment: If your compiler complains about drand48, seed48      c
c          replace drand48 with rand(seed)  and remove the             c
c          call to seed48                                              c
c          Data is saved as sqrt(steps), distance                      c
cccccccccccccccccccccccccccccccccccccccccccccccccccccccccccccccccccccccc
        Program walk
        Implicit none
c       declarations
        Real*8 drand48, root2, theta, x, y, r(1:10000)
        Integer i, j, max, seed
c       set parameters (# of steps)
        max   = 10000
        seed  = 11168
        root2 = 1.4142135623730950488E0
c       open file, seed generator
        Open(6, FILE='walk.dat', Status='Unknown')
        Call seed48(seed)
c       clear array
        Do 1 j=1, max
          r(j) = 0
 1      Continue
c       average over 100 trials
        Do 10 j = 1, 100
          x = 0
          y = 0
c         take max steps
          Do 20 i = 1, max
            x = x + (drand48()-0.5)*2.0*root2
            y = y + (drand48()-0.5)*2.0*root2
            r(i) = r(i)+ Sqrt(x * x + y * y)
 20       Continue
 10     Continue
c       output data for plot of r vs sqrt(N)
        Do 30 i = 1, max
          Write (6,*) Sqrt(Real(i)), ' ', r(i)/100
 30     Continue
        Close(6)
        Stop 'data saved in walk.dat'
        End
```

Fig. D.38 walk.f: Random-walk simulation.

Appendix E: Typical Project Assignments

Please demonstrate to instructor and *then* hand in for each project a minilab report (executive summary) including the following:

Equations solved	Numerical method used
Code listing	Results (relevant graph or table)
Conclusion	Critical discussion

The prefered form is an HTML (HyperText Markup Language) document with links to codes and results. You should do *both* the implementation and the assessment for each project. Follow the instructor's requirement regarding use of the programs on the diskette. In some cases the instructor may suggest that you use the given program as a guide, while in other cases, you may be instructed to run a program and not write your own.

E.1 FIRST QUARTER (10 WEEKS)

1. §2.6, area of circle. (Just demonstrate, do not hand in.)
2. §2.13, underflow and overflow limits.

3. §2.15, machine precision.
4. §2.18, complex numbers. (C programmers may be somewhat limited.)
5. §2.23, summing series and §3.14, error behavior.
6. §3.8, spherical Bessel functions by recursion. (Optional for undergraduates.)
7. §4.8, empiral behavior of integration error.
8. §5.5, interpolation of spectrum. (As an alternative, you can use a cubic spline, as in §5.9.)
9. §5.16, fitting exponential decay, or for undergraduates, §5.17, fitting heat flow.
10. §6.4, random sequences.
11. §6.12, different random walkers. (Optional for undergraduates.)
12. §7.6, radioactive decay simulation.
13. §9.12, and §9.13, linear and nonlinear oscillations.
14. §9.14, Energy conservation with ODEs.

E.2 SECOND QUARTER (10 WEEKS)

1. §11.8, realistic pendulum with rk4. (This is really quite easy because you just need to change a few lines in your differential-equation solver.)
2. §12.8, Fourier analysis of analytic signal and then analysis of your output for a nonlinear oscillator of your choice (realistic pendulum, oscillator with large power p, or perturbed oscillator).
3. §13.5 and §13.6, nonlinear dynamical mapping and bifurcation.
4. §15.11, testing matrix calls, problems 1–3.
5. §16.5, bound states in momentum space, or §17.5, scattering in momentum space. (Optional for undergraduates.)
6. Parallel computing (PVM) tutorial from the Web.
7. §25.7, Laplace's equation. (Graduate students may want to solve one of the other PDE's.)
8. §22.6, Ising model and Metropolis algorithm.
9. §23.6, quantum path integrals. (Optional for undergraduates, maybe optional for graduate students.)

Glossary

address The numerical designation of a location in memory.

algorithm A set of rules for solving a problem independently of the software or hardware.

analog The mapping of numbers to continuous values of some physical observable, for example, speed and a car's speedometer.

architecture The overall design of a computer in terms of its major components: memory, processing, I/O, and communication.

argument A parameter passed from one program part to another or to a command.

array (matrix) A bunch of numbers stored together that can be referenced by one or more subscripts. Single-indexed arrays represent mathematical vectors; double- and higher-indexed arrays represent tensors. Each number in an array is an array element.

basic machine language Instructions telling the hardware to do basic operations such as store or add binary numbers.

bit Contraction of "binary digit"; the digits 0 or 1 used in a binary representation of numbers. Usually 8 bits are combined to form a byte, and 32 bits are combined to form a single-precision, floating-point number.

byte Eight bits of storage, the amount needed for a single character.

bus A communication channel (bunch of wires) used for transmitting information quickly among computer parts.inxxbus*

cache A small, very fast part of memory used as temporary storage between the very fast CPU registers and the fast main memory.

central processing unit (CPU) The part of a computer that accepts and acts on instructions; where calculations are done and communications controlled. Also used generically for the computer's electronics (not terminals and I/O devices).

column-major order The method used by Fortran to store matrices in which the leftmost subscript varies most rapidly and attains its maximum value before the subscript to the right is incremented.

compiler A program that translates source code from a high-level computer language to machine language or object code.

concurrent processing Same as parallel processing; simultaneous execution of several related instructions.

cycle time (clock speed) The time it takes the central processing unit (CPU) to execute the simplest instruction.

data Information stored in numerical form; plural of datum.

data dependence Two statements using or defining identical storage locations.

dependence A relation among program statements in which the results depend on the order in which the statements are executed. May prevent vectorization.

digital Representation of numbers in a discrete form (decimal, octal, or binary) but not in analog form (meter reading).

dimension of array The maximum value of each subscript of an array. The **logical dimension** is the largest value actually used by the program, the **physical dimension** is the value declared in a *Dimension* statement.

ethernet A high-speed local area network (LAN) composed of specific cable technology and communication protocols.

executable program A set of instructions that can be loaded into the computer's memory and executed.

indirect addressing The use of an array element as the subscript (index) for yet a different array. For example, $a(j(i))$.

induction variable (subscript) An integer variable which is changed by a fixed amount as an operation is performed. The index of a Do loop is an example.

floating point The representation of numbers in terms of mantissa and base raised to some power so that the decimal point floats during calculations; scientific notation.

Fortran An acronym for **for**mula **tran**slation.

instructions Orders to the hardware to do basic things such as fetch, store, and add.

instruction stack Group of instructions currently in use such as a window moving down your code as operations are performed.

kernel The inner or central part of a large program or of an operating system that does not get modified (much) when run on different computers.

linker A program that combines a number of programs to form a complete set of instructions that can be *loaded* into the computer's memory and followed by the computer.

loop A set of instructions executed repeatedly as long as some condition is met.

machine language The set of instructions understood by elementary processors.

machine precision The maximum positive number that can be added to the number stored as 1 without changing the number stored as 1.

main program A part of a that calls subprograms but cannot be called by them.

main storage The fast, electronic memory; physical memory.

massively parallel Simultaneous processing on a very large number of central processing units.

multiprocessors Computers with more than one processor.

multitasking The system by which several jobs reside in a computer's memory simultaneously. On nonparallel computers each job receives CPU time in turn.

object program (code) A program in basic machine language produced by compiling a high level language.

operating system The program that controls the computer and decides when to run applications, process I/O, and shells.

optimizer A program (or programmer) that modifies a program to make it run more quickly.

parallel (concurrent) processing Simultaneous and essentially independent processing in different central processing units. If the number of separate multiprocessors becomes very large, it is **massively parallel**.

parallelization Rewriting an existing program to run on a computer with multiple processing units.

physical memory Fast, electronic memory of a computer; main memory; physical memory stands in contrast to *virtual memory*.

pipeline (segmented) arithmetic units Assembly-line approach to central processing in which parts of the CPU simultaneously gather, store, and process data.

PostScript A standard language for sending text and graphics to printers.

program Set of actions or instructions that a machine is capable of interpreting and executing or the act of creating a program.

RAM The random access or central memory that can be reached directly.

recurrence (recursion) A statement or variable in a loop that uses the value of some variable computed in a previous iteration. May affect vectorization.

registers Very high-speed memory used by the central processing unit.

RISC Reduced Instruction Set Computer; a CPU design that increases arithmetic speed by decreasing the number of instructions the CPU must follow.

scalar A data value (number), for example, element a_4 of an array or the value of π.

scalar processing Calculations in which numbers are processed in sequence. Also, processing units (hardware) that process machine code in sequence. Different from vector and parallel processing.

section size (strip) The number of elements that can be executed with one command on vector hardware. Breaking up an array into strips is **strip mining**..

segmented arithmetic units See pipelined arithmetic units.

shell The command line interpreter; the part of the operating system with which the user interacts.

source code Program in high-level language needing compilation to run.

stride Number of array elements that are stepped through as an operation repeats.

subprogram The part of a program invoked by another program unit.

supercomputer The class of fastest and most powerful computers available.

superscalar A latter generation RISC designed for an optimal balance between compiler and machine instructions.

telnet The protocol suite for the TCP/IP internet network that permits a terminal on one host computer to seem as if directly connected to another computer on the network. Also the name of a terminal emulator program written by NCSA for using PCs on the internet.

Teraflop (tflop) 10^{12} floating-point operations per second.

vector A group of N numbers in memory arranged in one-dimensional order.

vectorization Reorganization of a program so the compiler can utilize vector hardware.

vector processing Calculations in which an entire vector of numbers is processed with one operation.

virtual memory Memory that resides on the slow, hard disk and not in the fast electronics.

visualization The production of two- and three-dimensional pictures or graphs of the numerical results of computations.

word A unit of main storage, usually 1, 2, 4, 6, or 8 bytes.

workstations A class of computers small enough in size and cost to be used by a small group or an individual in their own work location yet powerful enough for large-scale scientific and engineering applications. Typically contining a Unix operating system and good graphics.

References

[Abar 93] ABARBANEL, H. D. I., M. I. RABINOVICH, AND M. M. SUSHCHIK (1993), *Introduction to Nonlinear Dynamics for Physicists,* World Scientific, Singapore.

[A&S 64] ABRAMOWITZ, M., AND I. A. STEGUN (1964), *Handbook of Mathematical Functions,* U.S. Govt. Printing Office, Washington.

[Argy 91] ARGYRIS, J., M. HAASE, AND J. C. HEINRICH (1991), *Comput. Meth. Appl. Mech. Eng.*, **86**, 1.

[Arm 91] ARMIN, B., AND H. SHLOMO, EDS. (1991), *Fractals and Disordered Systems*, Springer-Verlag, Berlin.

[Ask 77] ASKAR, A., AND A. S. CAKMAK (1977), *J. Chem. Phys.* **68**, 2794.

[B&R 92] BEVINGTON, P. R., AND D. K. ROBINSON (1992), *Data Reduction and Error Analysis for the Physical Sciences*, McGraw-Hill, New York.

[C&P 88] CARRIER, G. F., AND C. E. PEARSON (1988), *Partial Differential Equations*, Academic Press, San Diego.

[C&L 81] CHRISTIANSEN, P. L., AND P. S. LOMDAHL (1981), *Physica* **2D,** 482.

[C&O 78] CHRISTIANSEN, P. L., AND O. H. OLSEN (1978), *Phys. Lett.* **68A,** 185; (1979), *Physica Scripta* **20**, 531.

[Cour 28] COURANT, R., K. FRIEDRICHS, AND H. LEWY (1928), *Mathematische Annalen* **100**, 32.

[B&H 95] BRIGGS, W. L., AND V. E. HENSON (1995), *The DFT, An Owner's Manual*, SIAM, Philadelphia.

[DeJ 92] DE JONG, M. L. (1992), *The Physics Teacher* **30**, 115.

[DeV 95] DEVRIES, P. L. (1995), *Am. J. Phys.* **64**, 364.

[E&P 88] EUGENE, S. H., AND M. PAUL (1988), *Nature* **335**, 405.

[Feig 79] FEIGENBAUM, M. J. (1979), *J. Stat. Physics* **21**, 669.

[Fer 90] FEREYDOON, F., AND V. TAMÁS (1990), *Comput. Phys.*, 44; (1985), *J. Phys. A* **18**, L75.

[F&W 80] FETTER, A. L., AND J. D. WALECKA (1980), *Theoretical Mechanics of Particles and Continua*, McGraw-Hill, New York.

[F&H 65] FEYNMAN, R. P., AND A. R. HIBBS (1965), *Quantum Mechanics and Path Integrals*, McGraw-Hill, New York.

[Gold 67] GOLDBERG, A., H. M. SCHEY, AND J. L. SCHWARTZ (1967), *Am. J. Phys.* **3**, 177.

[Gott 66] GOTTFRIED, K. (1966), *Quantum Mechanics*, Benjamin, New York.

[G&T 96] GOULD, H., AND J. TOBOCHNIK (1996), *An Introduction to Computer Simulation Methods*, 2nd ed., Addison-Wesley, Reading, MA.

[H&T 70] HAFTEL, M. I., AND F. TABAKIN (1970), *Nucl. Phys.* **158**, 1.

[Jack 75] JACKSON, J. D. (1975), *Classical Electrodynamics*, 2nd ed., Wiley, New York.

[Koon 86] KOONIN, S. E. (1986), *Computational Physics*, Benjamin, Menlo Park, CA.

[KdeV 95] KORTEWEG, D. J., AND G. DEVRIES (1895), *Phil. Mag.* **39**, 4.

[L 96] LANDAU, R. H. (1996), *Quantum Mechanics II, A Second Course in Quantum Theory*, 2nd ed., Wiley, New York.

[L&F 93] LANDAU, R. H., AND P. J. FINK (1993), *A Scientist's and Engineer's Guide to Workstations and Supercomputers*, Wiley, New York.

[L&L 69] LANDAU, L. D., AND E. M. LIFSHITZ (1969), *Mechanics*, Pergamon Press, 2nd ed., Oxford, UK.

[LAP 95] ANDERSON, E., Z. BAI, C. BISCHOF, J. DEMMEL, J. DONGARRA, J. DU CROZ, A. GREENBAUM, S. HAMMARLING, A. MCKENNEY, S. OSTROUCHOV, AND D. SORENSEN (1995), *Lapack Users' Guide*, 2nd ed., SIAM, Philadelphia; http://netlib.org.

[MacK 85] MACKEOWN, P. K. (1985), *Am. J. Phys.* **53**, 880.

[M&N 87] MACKEOWN, P. K., AND D. J. NEWMAN (1987), *Computational Techniques in Physics*, Adam Hilger, Bristol, UK.

[Mand 82] MANDELBROT, B. (1982), *The Fractal Geometry of Nature*, Freeman, San Francisco.

[Mann 83] MANNHEIM, P. D. (1983), *Am. J. Phys.* **51**, 328.

[M&T 88] MARION, J. B., AND S. T. THORNTON (1988), *Classical Dynamics of Particles and Systems*, 3rd ed., Harcourt Brace Jovanovich, Orlando, FL.

[M&W 65] MATHEWS, J., AND R. L. WALKER (1965), *Mathematical Methods of Physics*, Benjamin, Reading, MA.

[Metp 53] METROPOLIS, M., A. W. ROSENBLUTH, M. N. ROSENBLUTH, A. H. TELLER, AND E. TELLER (1953), *J. Chem. Phys.* **21**, 1087.

[M&L 85] MOON, F. C., AND G.-X. LI (1985),*Phys. Rev Lett.* **55**, 1439.

[NACSE] NORTHWEST ALLIANCE FOR COMPUTATIONAL SCIENCE, http://www.naces.org.

[NSF] NATION SCIENCE FOUNDATION SUPERCOMPUTER CENTERS:
Cornell Theory Center, http://www.tc.cornell.edu
Natl. Cntr. for Supercomp. Appls., http://www.ncsa.uiuc.edu
Pittsburgh Supercomputing Center, http://www.psc.edu
San Diego Supercomputing Center, http://www.sdsc.edu
Natl. Center for Atmospheric research, http://www.ucar.edu

[P&D 81] PEDERSEN, N. F., AND A. DAVIDSON (1981), *Appl. Phys. Lett.* **39**, 830.

[Penn 94] PENNA, T. J. P. (1994), *Comput. in Phys.* **9**, 341.

[P&R 95] PHATAK, S. C., AND S. S. RAO (1995), *Phys. Rev. E* **51**, 3670.

[PhT 88] *Physics Today*, December 1988.

[P&W 91] PINSON, L. J., AND R. S. WIENER (19991), *Objective-C Object-Oriented Programming Techniques*, Addison-Wesley, Reading, MA.

[P&B 89] PLISCHKE, M., AND B. BERGERSEN (1989), *Equilibrium Statistical Physics*, Prentice-Hall, Englewood Cliff, NJ.

[Potv 93] POTVIN, J. (1993), *Comput. in Phys.* **7**, 149.

[Pres 94] PRESS, W. H., B. P. FLANNERY, S. A. TEUKOLSKY, AND W. T. VETTERLING (1994), *Numerical Recipes*, Cambridge University Press, Cambridge, UK.

[Rash 90] RASBAND, S. N. (1990), *Chaotic Dynamics of Nonlinear Systems*, Wiley, New York.

[Raw 96] RAWITSCHER, G., I. KOLTRACHT, H. DAI, AND C. RIBETTI (1996), *Comput. in Phys.*, **10**, 335.

[Rhei 74] RHEINBOLD, W. C. (1974), *Methods for Solving Systems of Nonlinear Equations*, SIAM, Philadelphia.

[Russ 44] RUSSELL, J. S. (1844), *Report of the 14th Meeting of the British Association for the Advancement of Science*, John Murray, London.

[Sand 94] SANDER, E., L. M. SANDER, AND R. M. ZIFF (1994), *Comput. in Phys.* **8**, 420.

[Schk 94] SCHECK, F. (1994), *Mechanics, from Newton's Laws to Deterministic Chaos*, 2nd ed., Springer-Verlag, New York.

[Schd 87] SCHMID, E. W., G. SPITZ, AND W. LÖSCH (1987), *Theoretical Physics on the Personal Computer*, Springer-Verlag, Berlin.

[S&T 93] SINGH, P. P., AND W. J. THOMPSON (1993), *Comput. in Phys.* **7**, 388.

[Smit 91] SMITH, D. N. (1991), *Concepts of Object-Oriented Programming*, McGraw-Hill, New York.

[Stez 73] STETZ, A., J. CARROLL, N. CHIRAPATPIMOL, M. DIXIT, G. IGO, M. NASSER, D. ORTENDAHL, AND V. PEREZ-MENDEZ (1973), "Determination of the Axial Vector Form Factor in the Radiative Decay of the Pion", LBL 1707, invited paper at the Symposium of the Division of Nuclear Physics, Washington, DC, April.

[Tab 89] TABOR, M. (1989), *Chaos and Integrability in Nonlinear Dynamics*, Wiley, New York.

[Tait 90] TAIT, R. N., T. SMY, AND M. J. BRETT (1990), *Thin Solid Films* **187**, 375.

[Thom 92] THOMPSON, W. J. (1992), *Computing for Scientists and Engineers*, Wiley, New York.

[UCES] UNDERGRADUATE COMPUTATIONAL ENGINEERING AND SCIENCE, http://uces.ameslab.gov/uces/.

[Viss 91] VISSCHER, P. B. (1991), *Comput. in Phys.* **5**, 596.

[Vold 59] VOLD, M. J. (1959), *J. Collod. Sci.* **14**, 168.

[W&S 83] WITTEN, T. A., AND L. M. SANDER (1981), *Phys. Rev. Lett.* **47**, 1400; (1983), *Phys. Rev. B* **27**, 5686.

[Z&K 65] ZABUSKY, N. J., AND M. D. KRUSKAL (1965), *Phys. Rev. Lett.* **15**, 240.

Index